# Biofilms in Wastewater Treatment

*An Interdisciplinary Approach*

Edited by Stefan Wuertz, Paul L. Bishop and Peter A. Wilderer

**Published by IWA Publishing, Alliance House, 12 Caxton Street, London SW1H 0QS, UK**
Telephone: +44 (0) 20 7654 5500; Fax: +44 (0) 20 7654 5555; Email: publications@iwap.co.uk
Web: www.iwapublishing.com

First published 2003
© 2003 IWA Publishing

Printed by TJ International (Ltd), Padstow, Cornwall, UK

Apart from any fair dealing for the purposes of research or private study, or criticism or review, as permitted under the UK Copyright, Designs and Patents Act (1998), no part of this publication may be reproduced, stored or transmitted in any form or by any means, without the prior permission in writing of the publisher, or, in the case of photographic reproduction, in accordance with the terms of licences issued by the Copyright Licensing Agency in the UK, or in accordance with the terms of licenses issued by the appropriate reproduction rights organization outside the UK. Enquiries concerning reproduction outside the terms stated here should be sent to IWA Publishing at the address printed above.

The publisher makes no representation, express or implied, with regard to the accuracy of the information contained in this book and cannot accept any legal responsibility or liability for errors or omissions that may be made.

*Disclaimer*
The information provided and the opinions given in this publication are not necessarily those of IWA or of the editors, and should not be acted upon without independent consideration and professional advice. IWA and the editors will not accept responsibility for any loss or damage suffered by any person acting or refraining from acting upon any material contained in this publication.

*British Library Cataloguing in Publication Data*
A CIP catalogue record for this book is available from the British Library

*Library of Congress Cataloging-in-Publication Data*
A catalog record for this book is available from the Library of Congress

ISBN: 1 84339 007 8

# Contents

|  |  |  |
|---|---|---|
| *Preface* | | viii |
| *About the editors* | | x |
| *List of contributors* | | xi |

**PART ONE: MODELING AND SIMULATION**     1

**Modeling and simulation: Introduction**     3
*Stefan Wuertz and Christina M. Falkentoft*

**1   What do biofilm models, mechanical ducks, and artificial life have in common?** *Mathematical modeling in biofilm research*     8
*Hermann J. Eberl*

| 1.1 | Mathematics and biology: how do they go together? | 9 |
|---|---|---|
| 1.2 | What actually are mathematical models and what is mathematical modeling? | 10 |
| 1.3 | How mathematical models can be used to study complicated biofilm architectures | 16 |
| 1.4 | Beyond mathematical models: reductionism and holism, physicalism and vitalism as philosophical concepts in theoretical biology | 20 |
| 1.5 | Interdisciplinarity: why it is difficult to communicate with mathematicians | 25 |
| 1.6 | What do biofilm models, mechanical ducks, and artificial life have in common? (Conclusion) | 28 |
| 1.7 | References | 28 |

**2   Biofilm architecture: interplay of models and experiments**     32
*Slawomir W. Hermanowicz*

| 2.1 | Tools for biofilm studies | 32 |
|---|---|---|
| 2.2 | Early observations and models | 34 |
| 2.3 | Standard diffusion-reaction model | 36 |
| 2.4 | New experimental techniques | 37 |
| 2.5 | Challenges in modeling | 38 |
| 2.6 | Modeling of structural heterogeneity | 41 |
| 2.7 | Conclusions and future directions | 43 |
| 2.8 | References | 45 |

**3   Towards new mathematical models for biofilms**     49
*Volker Hösel and Volkmar Liebscher*

| 3.1 | Introduction | 49 |
|---|---|---|
| 3.2 | Strategies of mathematical modeling | 50 |
| 3.3 | Examples of mathematical models | 52 |

| | | |
|---|---|---|
| 3.4 | Established biofilm models | 54 |
| 3.5 | Models from mathematical biology | 55 |
| 3.6 | Ideas for stochastic models | 56 |
| 3.7 | Concluding remarks | 58 |
| 3.8 | References | 58 |

**4 Beyond models: requirements and chances of computational biofilms** — 60
*Hans-Joachim Bungartz and Miriam Mehl*

| | | |
|---|---|---|
| 4.1 | Introduction: simulating biofilm systems | 60 |
| 4.2 | Mathematical models | 64 |
| 4.3 | Numerical methods | 69 |
| 4.4 | Implementation | 78 |
| 4.5 | Embedding | 80 |
| 4.6 | Visualization: hot air or catalyst? | 81 |
| 4.7 | Validation | 83 |
| 4.8 | Concluding remarks | 84 |
| 4.9 | References | 85 |

**5 On the influence of fluid flow in a packed-bed biofilm reactor** — 88
*Stefan Esterl, Christoph Hartmann and Antonio Delgado*

| | | |
|---|---|---|
| 5.1 | Introduction | 88 |
| 5.2 | Some comments on specific topics | 97 |
| 5.3 | Current investigations | 101 |
| 5.4 | Influence of fluid flow and substrate concentration on biofilm architecture | 110 |
| 5.5 | Conclusions and outlook | 111 |
| 5.6 | References | 114 |

**Modeling and simulation: Conclusions** — 118
*Paul Bishop*

**PART TWO: ARCHITECTURE, POPULATION STRUCTURE AND FUNCTION** — 121

**Architecture, population structure and function: Introduction** — 123
*Stefan Wuertz*

**6 The effect of biofilm heterogeneity on metabolic processes** — 125
*Paul L. Bishop*

| | | |
|---|---|---|
| 6.1 | Introduction | 125 |
| 6.2 | Biofilm properties | 126 |
| 6.3 | Confocal laser scanning microscopy | 138 |
| 6.4 | Cell-to-cell communication | 139 |
| 6.5 | Molecular probes | 139 |
| 6.6 | Concluding remarks | 141 |
| 6.7 | References | 142 |

| | | |
|---|---|---|
| **7** | **Mass transport in heterogeneous biofilms**<br>*Zbigniew Lewandowski and Haluk Beyenal* | **147** |
| 7.1 | Introduction | 147 |
| 7.2 | Biofilm heterogeneity and biofilm models | 150 |
| 7.3 | Quantifying nutrient uptake kinetics from the nutrient concentration profiles | 153 |
| 7.4 | Quantifying mass transport mechanisms from flow velocity profiles in biofilms | 156 |
| 7.5 | Local mass transport rates in heterogeneous biofilms | 159 |
| 7.6 | The concept of biofilms composed of discrete layers | 162 |
| 7.7 | Modeling mass transport and activity in biofilms composed of discrete layers | 166 |
| 7.8 | Experimental validation of the model of biofilms composed of discrete layers | 168 |
| 7.9 | Can discretizing biofilms reflect the effect of biofilm heterogeneity? | 169 |
| 7.10 | Biofilms grown at high flow velocities | 171 |
| 7.11 | Concluding remarks | 172 |
| 7.12 | References | 175 |
| **8** | **The crucial role of extracellular polymeric substances in biofilms**<br>*Hans-Curt Flemming and Jost Wingender* | **178** |
| 8.1 | Introduction | 178 |
| 8.2 | Definition of EPS | 179 |
| 8.3 | Composition and properties | 182 |
| 8.4 | Mechanical stability mediated by EPS | 187 |
| 8.5 | Role of EPS in microbial aggregation | 188 |
| 8.6 | Function of EPS | 194 |
| 8.7 | Technical aspects of EPS | 197 |
| 8.8 | Ecological aspects | 200 |
| 8.9 | Outlook | 202 |
| 8.10 | References | 203 |
| **9** | **The importance of physicochemical properties in biofilm formation and activity**<br>*Rosario Oliveira, Joana Azeredo and Pilar Teixeira* | **211** |
| 9.1 | Introduction | 211 |
| 9.2 | How adhesion has been predicted | 212 |
| 9.3 | Surface properties relevant for adhesion | 218 |
| 9.4 | Concluding remarks | 228 |
| 9.5 | References | 228 |
| **10** | **Influence of population structure on the performance of biofilm reactors**<br>*Axel Wobus, Frank Kloep, Kerstin Röske and Isolde Röske* | **232** |
| 10.1 | Introduction | 233 |
| 10.2 | Investigation of the biotic structure of biofilms: survey of methods | 234 |
| 10.3 | Case studies | 240 |
| 10.4 | Concluding remarks | 258 |
| 10.5 | References | 259 |

| | | |
|---|---|---|
| **11** | **Detachment: an often-overlooked phenomenon in biofilm research and modeling**<br>*Eberhard Morgenroth* | **264** |
| 11.1 | Introduction | 264 |
| 11.2 | Detachment mechanisms | 265 |
| 11.3 | Influence of detachment on competition in biofilms and overall process performance | 276 |
| 11.4 | Concluding remarks | 286 |
| 11.5 | References | 288 |
| | **Architecture, population structure and function: Conclusions**<br>*Stefan Wuertz* | **291** |
| | **PART THREE: FROM FUNDAMENTALS TO PRACTICAL APPLICATIONS** | **295** |
| | **From fundamentals to practical applications: Introduction**<br>*Peter A. Wilderer* | **297** |
| **12** | **Deduction and induction in design and operation of biofilm reactors**<br>*Poul Harremoës* | **299** |
| 12.1 | Introduction | 299 |
| 12.2 | Technological development | 300 |
| 12.3 | Pragmatism versus theory-based models | 302 |
| 12.4 | Models of biofilm reactors | 306 |
| 12.5 | Model calibration and parameter estimation | 309 |
| 12.6 | Treatment plant design | 314 |
| 12.7 | Analysis of existing plant / pilot plant | 319 |
| 12.8 | Outstanding issues of engineering significance | 320 |
| 12.9 | References | 323 |
| **13** | **Effect of clay particles on biofilm composition and reactor efficiency**<br>*Luis F. Melo and Maria J. Vieira* | **325** |
| 13.1 | Introduction | 325 |
| 13.2 | Properties of clay particles | 326 |
| 13.3 | Microorganisms in soils and in aqueous solutions: interactions with clays | 329 |
| 13.4 | Microhabitats created by clay particles | 331 |
| 13.5 | Genetic exchange | 332 |
| 13.6 | Effect of particles on the toxicity of biocides | 333 |
| 13.7 | Effects of particles on biofilm physical properties | 336 |
| 13.8 | Clay particles in wastewater treatment bioreactors | 337 |
| 13.9 | Summary and future research | 338 |
| 13.10 | References | 340 |

| | Contents | vii |
|---|---|---|
| **14** | **Bioprocess engineering and microbiologists: a profit-sharing alliance** | **343** |
| | *Peter A. Wilderer and Martina Hausner* | |
| 14.1 | Postulate | 343 |
| 14.2 | Analysis of the current state of biotechnology | 345 |
| 14.3 | Chances and requirements | 350 |
| 14.4 | Concluding remarks | 354 |
| 14.5 | References | 355 |
| | Appendix: biography of microbial samples | 357 |

**From fundamentals to practical applications: Conclusions** — **374**
*Peter A. Wilderer*

**Glossary** — **377**

**Index** — **391**

# Preface

Biofilms are ubiquitous life forms consisting of microorganisms embedded in a matrix of biological origin called extracellular polymeric substances (EPS). They thrive in a wide range of environments wherever interfaces are found. In addition, biofilm-like aggregates such as activated sludge flocs can form without an obvious attachment surface.

The central theme of the book is the flow of information from experimental approaches in biofilm research to simulation and modeling of complex wastewater systems. It draws on the wealth of information already available about the function of biofilms on a technical and on a laboratory scale, covers some of the exciting basic microbiological research, and summarizes the latest efforts to predict the development of biofilms. In terms of biofilm physical structure the preferred term is 'architecture', which denotes the different components like EPS, cells and inorganic materials as well as the spatial distribution of certain microorganisms if known. The term 'structure' is used to imply microbial population structure, that is, a molecular inventory of bacteria and other microorganisms present in the biofilm.

The idea to write this book goes back to a workshop held under the auspices of the Deutsche Forschungsgemeinschaft (DFG) through its Research Center for Aerobic Biological Wastewater Treatment (SFB 411) in Garching, Germany, in November 1998. During the meeting several key questions about the architecture and function of biofilms were formulated. The participants' inability to answer these questions satisfactorily prompted further deliberations culminating in the idea to confide the results of our discussions and the major conclusions to a book. Advancements in the field since then have been incorporated and the questions have continued to stimulate scientific research.

What distinguishes the present volume from other books on biofilms is the interdisciplinary approach. Probably the greatest challenge in wastewater research lies in using the methods and the results obtained in one scientific discipline to design intelligent experiments in other disciplines, and to eventually improve the knowledge base the practitioner needs to run wastewater treatment plants.

# Preface

This book is composed of three parts, each of which begins with a short introduction to the main question followed by individual book chapters giving a complete or partial answer. Each individual book chapter attempts to answer and present different angles of looking at the problems. New methods are integrated and explained where necessary. After each section the editors provide an analysis and outlook to facilitate the reader's critical understanding.

We deliberately did not divide the book into sections like methods, basic microbiology research, wastewater engineering or up-scaling because (i) there is no shortage of good books and other publications dealing with these aspects individually, and (ii) we wanted to stress the transdisciplinary nature of biofilm research. Hence the questions posed in the different sections can only be addressed by taking into account different scientific disciplines including the science of engineering wastewater treatment plants. It is our intention that the authors will lead the way by their own examples of how biofilms should be studied.

Consequently, this book is meant to illustrate why basic research is useful and which questions still need to be answered. Nowadays there are lots of meeting grounds for scientists and engineers at international conferences and the like. One would expect there to be a continuous exchange of information. Most, however, will agree that all too often different scientific and technical languages are spoken and those who speak tend to converse in the language they know best. For this reason we provide an extensive glossary to help the reader with unfamiliar terminology, thus enabling workers from all fields interested in biofilm research to use this book as a resource. We hope that it will be useful to experienced professionals, engineers, and scientists as well as students attending advanced graduate engineering and microbiology classes who wish to broaden their horizon.

We take this opportunity to thank the contributing authors for their enthusiasm, patience, and willingness to make revisions at short notice. Their ideas and expertise were also essential to completing the glossary. We especially thank the students of Civil and Environmental Engineering at UC Davis who read the chapters and provided us with much feedback and many corrections.

| Stefan Wuertz | Paul L. Bishop | Peter A. Wilderer |
|---|---|---|
| *Davis, California* | *Cincinnati, Ohio* | *Garching, Germany* |

# About the editors

**Stefan Wuertz** is Associate Professor of Civil and Environmental Engineering at the University of California, Davis. He received a B.Sc. in Microbiology from the National University of Ireland, Galway, a Ph.D. in Environmental Sciences from the University of Massachusetts, Boston, and a Dr. habilitatus in Environmental Biotechnology from the Technical University of Munich. His principal research interests are in the areas of biological wastewater treatment, biofilm processes, bioaugmentation, application of molecular techniques to environmental engineering, public health and water reuse. He has authored or co-authored over 40 scientific publications and co-edited one book. Professor Wuertz has served as advisor to private and government organizations including the California Department of Transportation.

**Paul L. Bishop** is Associate Dean for Graduate Studies and Research and the Herman Schneider Professor of Environmental Engineering at the University of Cincinnati. He received a B.S. in Civil Engineering from Northeastern University, Boston, MA, and M.S. and Ph.D. degrees in Environmental Engineering from Purdue University. His principal research interests are in the areas of biological treatment and bioremediation, with particular emphasis on mass transport in biofilms and use of microelectrode systems. He has authored or co-authored 5 books and over 180 scientific publications. Professor Bishop is a Past President of the Association of Environmental Engineering and Science Professors in the U.S.A., and is currently Chair of the U.S.A. National Council for the International Water Association.

**Peter A. Wilderer** is Professor at the Technical University of Munich, Germany, and Director of the Institute of Water Quality Control and Waste Management. He also serves as Honorary Professor at the Advanced Wastewater Management Center of the University of Queensland, Australia. In basic research, he is mainly interested in the effects that specific reactor conditions have on the composition of microbial communities, and the resulting structure and metabolic activity of activated sludge and biofilms. He has made major contributions to the development of the sequencing batch reactor, in particular to the sequencing batch biofilm reactor technology. Professor Wilderer has authored and co-authored over 100 scientific publications, serves as editor of the journal *Water Research*, and acts as editor-in-chief of the journal *Water Science and Technology*.

# List of contributors

| | |
|---|---|
| Joana Azeredo | Centro de Engenharia Biologica – IBQF, Universidade de Minho, P-4700 Braga, Portugal |
| Haluk Beyenal | Center for Biofilm Engineering, Montana State University, EPS Building, Bozeman, MT 59717, U.S.A. |
| Paul L. Bishop | Department of Civil and Environmental Engineering, University of Cincinnati, PO Box 210071, Cincinnati, OH 45221-0071, U.S.A. |
| Hans-Joachim Bungartz | Institut für Parallele und Verteilte Systeme (IPVS), Abteilung Simulation großer Systeme, Universität Stuttgart, Universitätsstraße 38, D-70569 Stuttgart, Germany |
| Antonio Delgado | Lehrstuhl für Fluidmechanik und Prozessautomation, Technische Universität München, Weihenstephaner Steig 23, D-85350 Freising-Weihenstephan, Germany |
| Hermann Eberl | Department of Mathematics and Statistics, University of Guelph, Guelph, Ontario N1G 2W1, Canada |

List of contributors

Stefan Esterl — Lehrstuhl für Fluidmechanik und Prozessautomation, Technische Universität München, Weihenstephaner Steig 23, D-85350 Freising-Weihenstephan, Germany

Christina M. Falkentoft — Institute of Water Quality Control and Waste Management, Technical University of Munich, Am Coulombwall, D-85748 Garching, Germany

Hans-Curt Flemming — Department of Aquatic Microbiology, University of Duisburg and IWW Center for Water Research, Geibelstraße 41, D-47057 Duisburg, Germany

Poul Harremoës — Environment and Resources DTU, Technical University of Denmark, Building 118, Lyngby, DK-2800 Kgs Lyngby, Denmark

Christoph Hartmann — Lehrstuhl für Fluidmechanik und Prozessautomation, Technische Universität München, Weihenstephaner Steig 23, D-85350 Freising-Weihenstephan, Germany

Martina Hausner — Institute of Water Quality Control and Waste Management, Technical University of Munich, Am Coulombwall, D-85748 Garching, Germany

Slawomir W. Hermanowicz — Department of Civil and Environmental Engineering, University of California, Berkeley, CA 94720-1710, U.S.A.

Volker Hösel — Center for Mathematical Sciences, Chair of Applied Mathematics in Ecology and Medicine (M12), Technical University of Munich, Boltzmann-Straße 3, 85747 Garching, Germany

# List of contributors

| | |
|---|---|
| Frank Kloep | Lehrstuhl für angewandte Mikrobiologie, Technische Universität Dresden, D-01062 Dresden, Germany |
| Zbigniew Lewandowski | Department of Civil Engineering and Center for Biofilm Engineering, Montana State University, EPS Building, Room 310, Bozeman, MT 59717, U.S.A. |
| Volkmar Liebscher | Institut für Biomathematik und Biometrie, GSF-Forschungszentrum für Umwelt und Gesundheit, Postfach 1129, D-85758 Neuherberg, Germany |
| Miriam Mehl | Institut für Informatik V, Technische Universität München, Boltzmannstraße 3, D-85748 Garching, Germany |
| Luis Melo | Faculty of Engineering of the University of Porto, Chemical Engineering Department, LEPAE, P-4200-465 Porto, Portugal |
| Eberhard Morgenroth | Department of Civil and Environmental Engineering and Department of Animal Sciences, University of Illinois at Urbana-Champaign, 3219 Newmark Civil Engineering Laboratory, MC-250, 205 North Mathews Avenue, Urbana, IL 61801, U.S.A. |
| Rosario Oliveira | Centro de Engenharia Biologica – IBQF, Universidade de Minho, P-4700 Braga, Portugal |
| Isolde Röske | Lehrstuhl für angewandte Mikrobiologie, Technische Universität Dresden, D-01062 Dresden, Germany |

| | |
|---|---|
| Kerstin Röske | Department of Molecular Microbiology and Immunology, University of Missouri-Columbia, Columbia, MO, 65203, U.S.A. |
| Pilar Teixeira | Centro de Engenharia Biologica – IBQF, Universidade de Minho, P-4700 Braga, Portugal |
| Maria J. Vieira | Centro de Engenharia Biologica – IBQF, Universidade de Minho, P-4700 Braga, Portugal |
| Peter A. Wilderer | Institute of Water Quality Control and Waste Management, Technical University of Munich, Am Coulombwall, D-85748 Garching, Germany |
| Jost Wingender | Department of Aquatic Microbiology, University of Duisburg and IWW Center for Water Research, Geibelstraße 41, D-47057 Duisburg, Germany |
| Axel Wobus | Lehrstuhl für angewandte Mikrobiologie, Technische Universität Dresden, D-01062 Dresden, Germany |
| Stefan Wuertz | Department of Civil and Environmental Engineering, University of California, Davis, One Shields Avenue, Davis, CA 95616, U.S.A. |

# PART ONE

## MODELING AND SIMULATION

# Modeling and simulation: Introduction

*Stefan Wuertz and Christina M. Falkentoft*

Modeling may be defined as the identification of relevant quantities and processes and the description of these, in a mathematical or logical format (equations, rules, and so forth). Simulation is what happens once the equations that were defined via the model are converted into a set of discretized numerical algorithms and solved or approximated by a computer.

What are the purposes of modeling and simulation? Models may be used for many different purposes, such as analysis, prediction, design, framework, understanding (test hypothesis), and control. Biofilm models may be broadly classified into two categories according to their objectives (Noguera *et al.* 1999):

(1) models for *practical engineering* applications, such as design, troubleshooting, real-time operation and education;

(2) models as *research tools* for developing further understanding of specific biofilm phenomena, such as biofilm structure, population dynamics and structural heterogeneities.

At the Biofilm Workshop held in Garching in 1998, Oskar Wanner cited *'The golden rule of modeling*: models should be as simple as possible and as complex as needed'. Looking at model evolution, it is evident that all models are based on the same physical principles and that progress primarily means increased complexity. Modeling objectives should be evaluated on the desired scale of modeling—what are the relevant properties and dominant processes in a given situation? Possible simplifications should be evaluated, especially for the available database. The less simple the assumptions made, the more realistic the model, but the more information that is needed. There is little point in including a wide set of parameters in a model if these are impossible to observe or measure in practice.

© 2003 IWA Publishing. *Biofilms in wastewater treatment.* Edited by S. Wuertz, P.L. Bishop and P.A. Wilderer. ISBN: 1 84339 007 8.

Biofilm architecture has been a 'hot' topic during the past decade and continues to be so today. The original perception that biofilms are structurally homogeneous with little variation in their properties, such as porosity, pore size distribution, density and microbial populations, was mainly based on light microscopy with limited resolution and on scanning electron microscopy which involved dehydrating the biofilm before examination. Once the confocal laser scanning microscope (CLSM) was introduced, it became possible to examine non-destructively successive focal planes of living, hydrated biofilms.

At the Biofilm Workshop in Garching, the following question was discussed: *what is the most important factor determining biofilm architecture and hence transport processes within?* As could be expected, this led to a very intense debate. The session was opened by brainstorming, and almost as many suggestions for important factors were given, as researchers were present (Table 1). Parameters like microbial population, hydrodynamics/shear, substrate type, concentration and kinetics were agreed on to be highly relevant. However, no definitive answer was found, and the question is subject to an ongoing debate in the biofilm research community.

The conceptual biofilm architecture model developed by Costerton *et al.* (1994) from CLSM observations is today well known. It predicts the growth of microorganisms in a mushroom-shaped manner. The micro-colonies are situated at the top of these 'mushrooms' with a stalk of extracellular polymeric substances (EPS) and microorganisms constituting the binding link to the surface. Some of the mushrooms might fuse together, but leaving water channels open so that water from the bulk can penetrate most of the biofilm via convective flow. Water channels inside biofilms have been directly observed by tracking fluorescent particles using a combination of CLSM and classical optical microscopy. Costerton (1995) defined biofilms as 'the highest phenotypic expression of the bacterial genome' and he and his colleagues later inferred that the mushroom biofilm architecture was no accidental architecture, but directed according to growth control of the inhabiting bacteria by complex cell-to-cell communication including quorum sensing (Davies *et al.* 1998). This assumption was supported by the significant change of bacterial gene expression when cells attached to a surface. He assumed the mushroom-architecture to be a general growth mode for all biofilms; however, other researchers have subsequently questioned his assertion as illustrated in several chapters of this section.

**Table 1.** The brainstorming session at the Biofilm Workshop, 1998, produced this list of the parameters influencing biofilm architecture and hence transport processes within.

- EPS
- Detachment
- Mass transfer
- Physiology of cells
- Cell-to-cell interactions
- Internal architecture of the biofilm
- Particulate matter
- Morphology of the biofilm
- Architecture depends on the scale observed
- Microbial population
- Microbial distribution
- Grazing activity
- Hydrodynamics/shear
- Substrate type
- Substrate concentration and kinetics
- Electrostatic behavior of the environment
- Physicochemical properties of the environment
- Type of substratum
- Roughness of substratum

Some shortcomings of the biofilm investigations with CLSM are summarized below. The biofilms that have been studied were young and relatively thin, with a maximum thickness of ~200 µm, owing to the limited penetration depth of the laser light in denser and older biofilms. Many wastewater treatment biofilms are up to 2 mm thick, that is, an order of magnitude thicker than most studied films. Large porous architectures have not been reported in these thick films. They possibly began as microclusters separated by voids, but with time these colonies overlapped and grew together to form a continuous biofilm. Another aspect of criticism is that most of the investigated biofilms were monocultures or a consortium of only two or three species, whereas biofilms in wastewater treatment systems are composed of a complex mixed culture. Cell-to-cell communication through the excretion of signal molecules may promote a particular biofilm architecture, and the signal molecules may turn on metabolic systems, if one considers a single or limited number of bacterial species. However, it is unknown whether this phenomenon is of importance in mixed culture biofilms, where other species eventually fill in the channels rendering the mature biofilm more homogenous. van Loosdrecht *et al.* (1995) suggested the ratio between biofilm surface loading and shear rate to be the essential environmental factor determining the steady-state biofilm architecture. Additionally, characteristics of the individual organism, such as yield and growth rate, play a role. Based on experimental observations of biofilm formation in a biofilm airlift suspension reactor, they concluded that high loading and low shear rates lead to highly heterogeneous architectures with many pores and protuberances, whereas low loading and high shear result in thin, patchy biofilms. With an appropriate ratio, relatively smooth biofilms can be obtained.

Model complexity has increased in the past decades as a direct reflection of the advances in computational tools. In the 1970s, 1-D models of a completely homogeneous biofilm dominated. These models consisted of simple algebraic equations, which described the spatial profile of a substrate in the biofilm and which could be evaluated using a simple pocket calculator. In the 1980s these were refined to stratified, dynamic 1-D models including multisubstrate–multispecies biofilms. These models were based on a set of differential equations and demanded numerical calculations be performed by computers. A computer software program, AQUASIM, for these models was developed by Reichert (1994) and made available for biofilm research. This program has gained significant use worldwide and is under constant development according to new experimental findings, although the 1-D approach has been retained. The 1990s were the decade of multi-dimensional models, and they will be discussed in detail in this section.

Morgenroth *et al.* (2000b) concluded there to be a big gap between engineering practice and science when it comes to the application of biofilm models. With the development of more and more sophisticated models, this gap is widening, and for a useful dissemination of models and the new knowledge, it appears necessary to think for a moment about the role of modeling. The preceding years have been dominated by the inductive approach to biofilm modeling research, where the direction is from simplified to complex models. Once the complex models have been developed, and verified with experimental observations, the deductive way should be entered. Morgenroth *et al.* (2000a) took the first step in this direction comparing results from 3- and 1-D models. For a practitioner to make use of the new insights and understandings, simplified and purpose-adjusted biofilm models have to be developed. To achieve this aim, it is worthwhile to use the complex models to first identify key parameters.

The following chapters deal with different aspects of modeling and simulation. The authors describe fundamentals in terms of philosophical, mathematical and experimental considerations.

## REFERENCES

Costerton, J.W., Lewandowski, Z., de Beer, D., Caldwell, D., Korber, D. and James, G. (1994) Biofilms, the customized microniche. *J. Bacteriol.* **176**(8), 2137–2142.

Costerton, J.W. (1995) Overview of microbial biofilms. *J. Industr. Microbiol.* **15**, 137–140.

Davies, D.G., Parsek, M.R., Pearson, J.P., Iglewski, B.H., Costerton, J.W. and Greenberg, E.P. (1998) The involvement of cell-to-cell signals in the development of a bacterial biofilm. *Science* **280**, 295–298.

Morgenroth, E., Eberl, H. and van Loosdrecht, M.C.M. (2000a) Evaluating 3-D and 1-D mathematical models for mass transport in heterogenous biofilms. *Wat. Sci. Tech.* **41**(4/5), 347–356.

Morgenroth, E., van Loosdrecht, M.C.M. and Wanner, O. (2000b) Biofilm models for the practitioner. *Wat. Sci. Tech.* **41**(4/5), 509–512.

Noguera, D.R., Okabe, S. and Picioreanu, C. (1999) Biofilm modeling: present status and future directions. *Wat. Sci. Tech.* **39**(7), 273–278.

Reichert, P. (1994) AQUASIM: A tool for simulation and data analysis of aquatic systems. *Wat. Sci. Tech.* **30**(2), 21–30.

van Loosdrecht, M.C.M., Eikelboom, D., Gjaltema, A., Mulder, A., Tijhuis, L. and Heijnen, J.J. (1995) Biofilm structures. *Wat. Sci. Tech.* **32**(8), 35–43.

Wanner, O. (1995) New experimental findings and biofilm modeling concepts. *Wat. Sci. Tech.* **32**(8), 133–140.

# 1
# What do biofilm models, mechanical ducks, and artificial life have in common?
## Mathematical modeling in biofilm research

*Hermann J. Eberl*

*While I have sought to show the naturalist how a few mathematical concepts and dynamical principles may help and guide him, I have tried to show the mathematician a field for his labour.*
D'Arcy Wentworth Thompson (1917)

*She said welcome to the real world kid.*
Butch Hancock (1995)

© 2003 IWA Publishing. *Biofilms in wastewater treatment*. Edited by S. Wuertz, P.L. Bishop and P.A. Wilderer. ISBN: 1 84339 007 8.

## 1.1 MATHEMATICS AND BIOLOGY: HOW DO THEY GO TOGETHER?

Biofilm research is interdisciplinary research: biofilms occur in many different branches of science and technology, from wastewater engineering to medicine. Therefore, not surprisingly, biofilm researchers come from a broad variety of different scientific backgrounds and have a broad variety of scientific interests. Biologists may be interested in fundamental questions of bacterial behavior, environmental engineers in substrate conversion, and medical engineers in avoiding biofilm formation on artificial implants. But biofilms do not appear in mathematical systems, bacteria do not colonize abstract Banach spaces, and nobody is afraid of biofouling and biocorrosion of partial differential equations. Thus, why should mathematicians think about biofilms?

Mathematics is the language in which scientific theories are formulated. Or, rephrasing it in the words of medieval English scientist Roger Bacon: it is the door and key to science. The close relationship of mathematics on the one side and science and technology on the other one is mutual: not only are well-established mathematical methods used as tools in the study of technical or natural processes; mathematics, indeed, grows with its applications, and the applications have an important impact on mathematics' development by guiding its directions. The most famous example comes from physics. Newton had to develop differential calculus to be able to formulate his theory of gravity. But the same holds for biology, as the following examples will show.

Many mathematical results in the theory of semi-linear parabolic partial differential equations have been motivated by biological questions: for example, the spread of a favorable gene in a population. This question leads to the so-called Fisher equation, which is the canonical equation of its type and the starting point for many mathematical studies on nonlinear diffusion–reaction interactions. But not only complicated mathematical concepts like partial differential equations arise in the study of biological phenomena. Even the study of simple objects, such as iterative maps, benefited a lot from biology: In the 1970s it was found that rather simple interactions of populations or even the population dynamics of only one species (e.g. the discrete logistic growth model) can lead to virtually unpredictable behavior. The term *chaos* was coined for these effects, and studying them attracted many mathematicians. Eventually a new mathematical discipline was born.

Neural nets and genetic algorithms are two modern mathematical concepts for modeling and optimization. Their origin lies in mimicking learning processes in neurobiology and natural selection, respectively. The mathematical results obtained by studying these mathematical concepts that were originally stimulated by biology, flow back to the life sciences and often can be applied to problems in biology or bioengineering that are very different from the starting point. For example, neural nets and evolutionary algorithms are used to identify model parameters or to suggest control strategies in wastewater engineering. A more detailed survey of biological impacts in mathematics and mathematical impacts in biology can be found in *Mathematics and Biology, The Interface* on the World Wide Web at http://www.bio.vu.nl/nvtb/Interface.html. In the meantime, mathematical biology is considered to be a mathematical discipline in its own right, and research centers and research programs dedicated to the application of mathematics in the life sciences are currently being founded all over the world.

## 1.2. WHAT ACTUALLY ARE MATHEMATICAL MODELS AND WHAT IS MATHEMATICAL MODELING?

In this essay we shall try to explain how mathematics can contribute to biofilm research; in particular, how mathematics can help to explain and understand the highly heterogeneous biofilm architectures that are observed under the microscope. When we talk about the contribution of mathematics to biofilm research, we will be talking about *mathematical modeling*. (There are many other methods of applied mathematics that are very valuable for scientists and engineers of all disciplines, including biofilm research. For example, statistics and approximation theory are used to evaluate and process experimental data, optimization theory is deployed to identify model parameters or to determine best management strategies, as well as for optimal reactor design.) Mathematical models are abstract images of biological, physical, or other scientific systems. There are many different types of mathematical model, based on different mathematical concepts. In mathematical biology, the most important ones are differential and difference equations, stochastic models, and cellular automaton models. All these model concepts have different properties, and which one to choose depends on the object to be modeled and on the *a priori* information one has about the system, such as parameters. Chapter 3 of this volume contains a more detailed description and comparison of modeling concepts.

The behavior of the original system in the real world must be reflected in the behavior of its mathematical image. On the other hand, the image should not show a behavior and dependencies that cannot be found in its real world paragon. Of course, *in realiter* total equivalence of mathematical model and reality will never be observed because reality is too complex by far. Thus, mathematical models are always idealizations of reality, formulated to display some particular properties, which are of special interest for the researcher. Being idealizations of a complex process, mathematical models are always incomplete and they include empty, and for the particular purpose irrelevant, dependencies and relations. Several mathematical models might exist for one process, having different numbers of those empty relations. The modeler will always try to formulate and select a model, which has as few irrelevant relations as possible and is as complete as necessary for the scientific question he or she is interested in (Hertz 1894). A model serving very well for one purpose may be useless for another one because of being overcomplicated or incomplete. Recently, some spatial models have been suggested for biofilm growth (e.g. Wimpenny and Colasanti 1997; Picioreanu *et al.* 1998, 1999, 2001; Hermanowicz 1999, 2001; Noguerra *et al.* 1999; Eberl *et al.* 2001; Mehl 2001; Dockery and Klapper 2002). These models can be very valuable to understand local biofilm development. Owing to the complexity of multidimensional mathematics and computation, however, only small parts of a biofilm system can be described this way, but not an entire reactor. Therefore, these models are not suitable for quantitative engineering tasks, like design or control of a wastewater treatment plant, and they cannot be applied for these purposes. In return, classical zero-dimensional (*lumped*) approaches or one-dimensional models (e.g. Rittmann and McCarty 1980; Wanner and Gujer 1986; Sáez and Rittmann 1992) can deal with many more different species and substrates and they give good results on global mass conversion rates for a full reactor. However, they do not give any knowledge on the local spatial architecture of the biofilm. Thus, the extent to which simplifications and idealizations must or can be introduced depends on the particular purpose of the mathematical model. (Another very well known example can be found in the history of Physics: From the viewpoint of Einstein's relativity theory, Newton's gravity theory is wrong and only valid (though in very good approximation) in special cases of our world, now called non-relativistic. And from the viewpoint of Newton's theory, Kepler's law was wrong. It is well accepted in Physics that Einstein's theory is a better (due to being more general) model than Newton's theory, which was already a better (due to being more accurate) model than Kepler's law. But if we are only interested in a rough estimation of the moon's motion around the earth, the latter one might be sufficient in accuracy for our purpose. We will prefer it because it is much easier to evaluate than its successors, though being proven wrong from an overall point of view. However, we would never expect to learn from Kepler's Law anything about the future development of the universe. On the other hand, we prefer Newton's theory of gravity for non-relativistic calculations, because it has fewer, and for this purpose irrelevant, relations than Einstein's theory. Indeed, the non-relativistic view holds for most observations of our intuitive world, including biofilms.) But there are further important constraints for the formulation of a mathematical model. It is useful only if we are able to derive the desired information from it, i.e. if we can solve it quantitatively or discuss the qualitative behavior of model solutions. That is, if our computational and mathematical capabilities are sufficient to work with the model. Typically, a mathematical model will consist of a set of equations with some model parameters, input data, and probably initial and boundary conditions, which are connecting the spatial domain of the model with the exterior world. If the purpose of modeling is a quantitative solution, quantitatively knowing this information *a priori* is essential. Thus, one must be aware that the model formulation should not contain parameters being too difficult or even impossible to be determined. If the purpose of modeling is a qualitative analysis, the actual values of the model parameters and boundary conditions might not be so important, but the model must have a mathematical structure that allows rigorous analytical treatment. Another restriction can come from economics and technology, if the modeling work must be done with given finite resources in a given finite time.

Other restrictions are conceptual. Because mathematical models always formulate a sort of hypothesis, they must be subject to tests for falsification. One method of model falsification is direct comparison with measurement data. Those, however, often are not available with the required accuracy. Thus, often falsification tests can only be done qualitatively, by comparing model behavior with the behavior of the real world's paragon, expected by experience. An evolutionary approach can be comparison with other theories and models. Thus, competition between models takes place. Falsification never stops until a model is definitely proven wrong. This means the model must undergo new falsifications whenever new scientific knowledge of relevance for the range of model validity is gained or whenever new, stronger methods for falsification become available. Accepting a hypothesis or a model finally and forever and stopping falsification tests is leaving the scientific game (Popper 1982). Testing models for falsification is very closely linked to model extrapolation. Because if a model, which has been formulated and falsified for one approach, is applied to another one not yet covered by previous tests, it must be tested for falsification again.

The idea of using idealized mathematical models to study scientific systems and processes dates back at least to Galileo's insight that the book of nature is written in the language of mathematics and his idea to formulate abstract laws for ideal objects (von Glaserfeld 1996). Whereas physics and mathematics have been closely linked since Galileo's days, the systematic application of mathematical modeling methods in biology is much newer and started only during the 1920s. Although there were contributions to a mathematical biology before that time, they can only be considered to be isolated researches in biology with mathematical techniques (Israel 1994). This may be so because biological systems appear more complex and irregular than many physical systems and, therefore, they seem to be more difficult to idealize. Another reason may be the strong influence of theology and the idea of Creation. A third possible factor is the biologists' feeling that every species is unique. Therefore, they often prefer to emphasize the diversity of species rather than to achieve generalizations, which is one of the base concepts of mathematics. Although theoretical, mathematical, or computational physics have been considered scientific disciplines in their own right for a very long time, a comparable theoretical, mathematical, or computational biology only evolved in recent decades. Surveying this historical development, it seems quite natural that in particular the physical processes of biological systems are under consideration for classical mathematical modeling. In biofilm research, examples are nutrient consumption and mass conversion which can be described as diffusion–reaction equations on different length scales (e.g. Wanner and Gujer 1986; Wood and Whitaker 1998), hydrodynamics and mass transfer in the bulk liquid (Picioreanu et al. 2000; Eberl et al. 2000b), and detachment due to shear forces (Picioreanu et al. 2001). These are known as classic problems in applied and computational mathematics: fluid dynamics, (reactive) mass transfer, and fluid–structure interactions.

A famous and early historical example for models of a biological system from the times before systematic theoretical biology evolved is Vaucanson's mechanical duck from 1735, an automaton that moved like a duck, looked like a duck, and seemed to digest like a duck (e.g. Hillier 1976; see figure 1.1). Of course, this mechanical duck is not what we understand as a mathematical model. But the principle is the same: underlying is the thought (or hypothesis) that the motion of a duck can be described by mechanical laws and the mathematical theory of mechanics already started with Newton some decades earlier; speaking in mathematical terms, kinematics is a part of differential geometry and therefore a mathematical discipline. Therefore, the mechanical model can be seen as an analog realization of this idea like computer simulations are modern day digital realizations. Nobody will claim that Vaucanson's duck really describes the nature of a duck; e.g. already the *question* '*Why does a duck move like this?*' is excluded and also '*How can ducks reproduce themselves?*'. It is quite similar with biofilm models: the diffusion–reaction model for substrate concentrations is meant to describe mass transfer and conversion processes in biofilms. If mass transfer is not the only one thing about biofilms, this model should not be expected to tell the whole truth about biofilms. The underlying idea is: idealizing and reducing the complex system into smaller and easier to handle modules, which are of particular interest (motion of a duck, mass transfer in biofilms) and then studying them.

**Figure 1.1.** In the 18$^{th}$ century, de Vaucanson created a mechanical automaton that walked like a duck, moved like a duck, and seemed to eat and digest like a duck. This is an analog realization of a mechanical model of a living system. Note that the original automaton was destroyed long ago. This picture shows a 19$^{th}$ century painting of a mechanical duck.

With the advent and general availability of powerful computers, mathematical modeling received a new boost, and it got the status of a key technique in science and engineering under the shiny label of simulation. This means, laboratory experiments are replaced with the aid of a computer by quantitatively solving established and accepted model equations. The advantages of computer experiments are well known. They are cheaper than laboratory experiments and often can be done and repeated much faster; the conditions under which the processes are studied can be well defined and are not subject to disturbing external influences; interesting processes easily can be isolated; the system behavior can be investigated in extreme situations, which are often very difficult to generate in a laboratory reactor. Having already an adequate model description in the beginning is a necessary prerequisite for computer simulation, not its goal. To simulate fluid motion in a biofilm reactor, for example, the well-established incompressible Navier–Stokes equations (the basic equations of hydrodynamics) must be solved. This is achieved by an application of methods of computational fluid dynamics (CFD). No new mathematical model must be formulated at all, and therefore we cannot talk about modeling in this context, but of simulation. Because closed analytical and exact solutions do not exist for many problems of relevance, typically numerical methods must be applied to obtain an approximation of the desired model solution. The more complicated the mathematics of a model is and the bigger the amount of data to be mastered, the more important are computational aspects. Accurate and fast algorithms must be selected to obtain an approximate model solution with available computing resources, and an efficient implementation of these algorithms is necessary. Even if a lower dimensional model shows every important qualitative property, often three-dimensional models are needed for a realistic quantitative description. The numerical effort increases tremendously when going to a higher dimensional spatial resolution. Hence, the more dimensions we consider in a computer realization, the more important are these issues. Spatial biofilm models made up from several multidimensional partial differential equations require an enormous amount of computational work and, therefore, are naturally an application for high performance computing on powerful parallel computers (Eberl *et al.* 2000a). The computer will yield all numerical results in long columns of data that are too voluminous to be surveyed by eye. They must be processed for further interpretation, validation, and usage. In most cases, graphical representations are very helpful and necessary, at least for qualitative purposes. In three-dimensional modeling, this is a complicated problem of its own, again involving extensive mathematics and requiring huge computational resources. These three tasks together – algorithms, implementation, and visualization – are the kernel of scientific computing, a discipline at the interface between mathematics and computer science, whereas mathematical modeling is a discipline at the interface between mathematics and science. The computational issues are not considered in this essay. Chapter 4 deals with them in more detail in the framework of scientific computing.

Quantitative computer simulation is not the only reason why mathematical models are developed. A typical analytical application of mathematical models is the investigation of qualitative properties of a biological or physical system by a qualitative analysis of its mathematical representation. This way, the mechanisms governing the dynamical behavior of a system and its sensitivity to external factors can be analyzed. A recent example for qualitative mathematical analysis is the study of microbial growth in plug flow reactors with wall attachment by Ballyk and Smith (1999), Jones and Smith (2000), and Smith and Zhao (2000). Therefore, mathematical modeling should be viewed separately from simulation.

Two contrasting, but complementary, skills are needed for mathematical modeling, namely the ability to formulate a given problem in appropriate meaningful mathematical terms, and sufficient knowledge to obtain useful information from that mathematical model. The skill in model formulation lies in finding a mathematically complete description simple enough to give all the information required with sufficient accuracy. Modeling is an iterative process and no advanced mathematical model is obtained immediately, but step by step, trying different variations (Tayler 1986). Besides mathematical issues, like stability, existence and uniqueness of model solution, also the scientific meaning must be evaluated during the modeling process. In times of highly specialized scientific education, hardly anyone has sufficient capabilities in both areas involved: good and sufficient mathematics and the particular scientific or engineering application. Therefore, interdisciplinary collaboration between researchers from different fields is necessary in mathematical modeling.

## 1.3 HOW MATHEMATICAL MODELS CAN BE USED TO STUDY COMPLICATED BIOFILM ARCHITECTURES

In recent years, strong emphasis was put on the geometrical and population dynamical heterogeneity of biofilms as revealed by modern microscopy techniques. What causes this heterogeneity is not yet fully understood, and the question about the most important factors determining biofilm architecture is the guiding question of the first part of this volume. How can mathematical modeling contribute to this question? According to what was said above, it was not surprising that attempts were soon made to formulate mathematical descriptions for this phenomenon. Despite the growing number of multidimensional biofilm model studies, it often appears to be unclear to experimentally oriented researchers, how they can gain from these theoretical studies. In fact, mathematical models are used in many different ways to study natural and technical processes:

*Mathematical models as measurement tools (inverse problems)*. If the quantities of interest cannot be observed directly in an experiment, they often can be deduced indirectly from measured data using a mathematical model. An example comes from mass transfer in biofilms. It is commonly described by Fick's law, which says that the mass flux is proportional to the concentration gradient. The proportionality factor (diffusion coefficient) is a biofilm parameter. It is not directly observable. Instead, concentration values or other signals are measured and the diffusion coefficient itself is determined by data fitting such that the measured data satisfy the underlying diffusion model. This is a rather universal approach, independent of the way and how the input signals/concentrations are measured (cf. Bryers and Drummond (1998), or many other publications on experimental determination of diffusion coefficients). Typically, these inverse mathematical problems do not have a unique exact solution, because of measurement errors and model inaccuracies. Therefore, what one actually is looking for is a best approximate solution (i.e. diffusion coefficient in our example). Bryers and Drummond (1998) used the fluorescence recovery after photobleaching (FRAP) technique to identify local diffusion coefficients inside a biofilm. The FRAP recovery traces in this approach are fitted by using a standard two-dimensional mass-transfer model. Thus, they were able to identify regions of different diffusivity in a biofilm. Hereby they showed that the local mass transfer rate, and hence substrate supply, inside a biofilm system can vary drastically. This shows that a bacterial biofilm should not be considered as uniform gel-like matrix. Another example is evaluation of confocal laser scanning microscope (CLSM) data (e.g. Wuertz *et al.* 2001): directly measured, that is counted, are labeled voxels, i.e. rectangular geometrical cells. To evaluate these raw data, geometrical and topological concepts are deployed which allow quantification and insight into spatial relationships. Note that in this example, as opposed to the first one, the model is not a physical one, but a purely geometrical concept.

*Reactor design and optimal control.* In many branches of wastewater engineering, mathematical models and computer simulations are used for design and control. For example, the general configuration of a wastewater treatment plant can be described by a mathematical model. The actual dimensions of the reactors are parameters in this description. Performing computational simulations or qualitative mathematical analysis, it can be found how these parameters must be chosen for the treatment plant to show the desired performance. In adaptive control, on the other hand, an online control strategy is suggested based on measurements of the actual state of the system and theoretical model prediction for the future. Every so often this procedure is repeated after a certain time-step. A rough estimation of computational requirements (see Mehl (2001), or Chapter 4 of this volume) shows that a simulation of an entire biofilm reactor based on present multidimensional biofilm models is not feasible, even with current high-performance computing systems. Therefore, it cannot be expected in the near future to use detailed models exploiting the spatial biofilm architecture for design and operative control of wastewater treatment plants as described above. However, one may hope that some day multidimensional simulation tools can be used to design small laboratory systems for the experimental study of biofilm heterogeneity, like flow channels or Roto Torque reactors.

*Analysis of qualitative system behavior.* The application of multidimensional mathematical models to study the qualitative behavior of heterogeneous biofilm architectures is divided into two aspects: (i) mass transfer in an irregular architecture; and (ii) formation of irregular biofilm architectures. For the first approach, the biofilm architecture is assumed to be known and provided as input data. Mass transfer processes are much faster than the formation of the biofilm and quickly relax to an quasi-equilibrium. Therefore, the biofilm architecture is assumed to be non-changing within a small time-window and a steady state analysis is considered to be sufficient (Picioreanu *et al.* 1999). This assumption of a frozen steady-state is a standard technique in the analysis of systems with different time-scales. Picioreanu *et al.* (2000) and Eberl *et al.* (2000b) performed extensive two- and three-dimensional studies on the dependence of mass transfer and conversion in biofilms on hydrodynamics in the bulk liquid and the shape of the biofilm. In these studies it was shown that mass transfer and conversion in biofilms, as expressed by an average Sherwood number, is strongly affected by surface roughness and hydrodynamic conditions. For an *a priori* specified biofilm architecture, the mass transfer increases with the flow rate in the system, and for a specified hydrodynamic load, mass transfer decreases for coarser biofilm architectures. A similar but more simplified study was presented in Rittmann *et al.* (1999), who used simpler biofilm architectures and did not take the actual flow field into account but considered different hydrodynamic situations by varying the thickness of the concentration boundary layer. Basically all attempts to model the formation of heterogeneous biofilms up to now are based on hypotheses of biofilm formation by Wimpenny and Colasanti (1997) and van Loosdrecht *et al.* (1997). They claim that environmental conditions such as availability of substrates and hydrodynamics are key factors determining the actual shape of a biofilm. Many different mathematical concepts to describe biofilm formation have been suggested following these ideas. Wimpenny and Colasanti (1997), Picioreanu *et al.* (1998, 2001), Noguera *et al.* (1999), Hermanowicz (1999, 2001), and Mehl (2001) developed model local rules or cellular automata for the spatial spreading of biomass, in most cases in conjunction with a diffusion–reaction model for dissolved substrates, such as oxygen or nutrients. Kreft *et al.* (1998, 2001) suggested an individual- based modeling approach, in which each single (biological) cell is tracked. Eberl *et al.* (2001) and Dockery and Klapper (2002) suggested fully continuous, deterministic models based on partial differential equations. Even if they all are based on the same assumptions concerning the interaction of biomass and dissolved substrates, these models can be distinguished in many ways, according to their underlying concepts and the simplifying assumptions made: they do or do not include random effects; they are continuous or discrete; they make different simplifying assumptions about the transport and conversion of dissolved substrates and their boundary conditions; they take biomass density into account either as a constant throughout the biofilm, or as a dependent variable which is to be determined as part of the model solution. The one thing these models have in common: they all are able to predict spatially heterogeneous biofilm architectures as observed in the microscope, including mushroom-shaped colonies (see Figure 1.2). What does this show? It shows that simple concepts such as nutrient supply already are sufficient to make biofilms grow in highly irregular shapes and that there is no conceptual need for other, non-mechanistic principles, such as bacterial self-will, to keep channels and pores open in a biofilm. Note that this does not mean there are no further factors influencing the architecture of biofilms. With this multitude of different model studies, the '*environmental conditions*' hypothesis by Wimpenny and Colasanti (1997) and van Loosdrecht *et al.* (1997) survived an important round of falsification. As an example for biofilm formation processes that do not depend solely on environmental conditions, a diffusion–reaction model for quorum sensing in *Pseudomonas aeruginosa* biofilms was recently suggested by Dockery and Keener (2001). Because this study was one-dimensional, the full potential of their model for the heterogeneity of biofilms is still to be investigated.

*Multidimensional models as calibration tools for one-dimensional models.* Another, indirect application of multidimensional models was introduced by Morgenroth *et al.* (2000). They used the simulation results of a three-dimensional mass transfer study to calibrate a modified and extended one-dimensional model. This model was applied to qualitatively study mass transfer, the development of biofilm morphologies, and population dynamics.

Qualitative model studies as they are described here allow one to investigate a restricted number of effects in a biofilm system and predict the behavior of the system in dependence of these processes and the governing parameters. The big advantage of theoretical studies, compared with laboratory experiments, is that parameters can easily be changed and experiments can easily be repeated with only slightly modified parameter values, but under exactly the same conditions as before. Furthermore, extreme situations often can be realized easily in a model study, while they are very difficult to be studied in an experiment. Therefore, mathematical models also can be a valuable help for experimental research groups in order to plan and design experiments. Mathematical modeling *per se* will not be able to find the most significant factors determining biofilm architecture if it is not based on a thorough hypothesis about the underlying biological processes. This requires the knowledge input from experimental researchers. Thus, theoretical and experimental researchers can mutually benefit from each other. This appears to be the most promising way to study biofilm heterogeneity. So far, only few of the processes listed in the introductory words for Part 1 of this volume are studied and understood in a way that allows for a mathematical description suitable for multidimensional analysis. Among them, the interaction of mass transfer and substrate concentration and kinetics are the by far most studied ones, as explained above.

About the remaining points, mathematical modeling does not tell us anything yet. They provide a field to be explored for mathematical modeling.

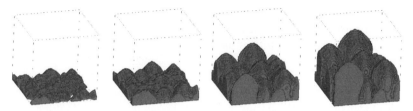

**Figure 1.2.** Formation of a spatially heterogeneous biofilm architecture. The dynamic simulation is based on the density-dependent diffusion reaction model for biomass and nutrients, suggested by Eberl *et al*. (2001). First a wavy biofilm layer develops. Later on, nutrients get limited close to the substratum and bigger biomass colonies begin to dominate over smaller ones. Thus, the typical mushroom shaped biofilm architecture develops.

## 1.4 BEYOND MATHEMATICAL MODELS: REDUCTIONISM AND HOLISM, PHYSICALISM AND VITALISM AS PHILOSOPHICAL CONCEPTS IN THEORETICAL BIOLOGY

We have seen that mathematical models only describe particular properties of a biological system like a biofilm. Thus, are we able to understand the entire biological system if we understand more and more, and finally all its single properties? And hence, is the complexity of a biofilm a mere complication? Or do new qualities emerge at each level of complexity, which are meaningless or even absent at lower levels? The first view is known as reductionism, the second one as holism. (In medicine, the term holism is often used in a different context and with a different meaning. This is not discussed here; see van den Hooff (1995).)

The underlying idea of reductionism is: reducing the complex system into smaller and easier-to-handle modules (e.g. the motion of a duck, mass transfer in biofilms) and describing them as an isolated process. Finally, when we are able to describe more and more of those idealized details, the whole system can be understood as a kind of superposition of those details. Davies (1989) claims that reductionism *'is a legacy of the age of linear physics, which began in earnest with Newton and endured until well into this century. In a linear system, such as electromagnetism, reductionism is a powerful methodology.'*

Indeed, the superposition principle is a basic technique to derive general solutions of linear systems by combining some particular solutions: The effect of the combined action of two different causes is the combination of the effects of each cause individually. The failure of superposition is a striking difference between linearity and nonlinearity (Nicolis 1995). Most biological processes and systems are described by nonlinear mathematical models, as is apparent by glancing through the voluminous textbook of Murray (1989) on *Mathematical Biology*. Consequently, Davies (1989) concludes about the applicability of a reductionist approach to life science, *'any attempts to explain their* [that is, living organisms] *qualities as wholes ... is doomed to failure'*. Instead, global issues are of key importance in nonlinear models: initial and boundary conditions, constraints, sensitivity and stability of model equations with respect to parameters and so on. These aspects connect different parts of the model or are very strongly dependent on their interaction. The behaviors of single processes are still *'necessary ingredients in providing a full explanation, but they are not sufficient; the key concepts are left out'* (Davies 1989). This global view is the mathematical concept behind holism.

From many examples in nonlinear dynamics, it is known that those global issues can introduce new effects into a model and generate a broad variety of solutions with different qualitative properties for the same mathematical model. This multiplicity of possible behaviors cannot be seen from a single ingredient, e.g. the reaction term in a reaction–diffusion process. It is the interaction of processes that matters. Depending on the actual numerical values of model parameters and boundary conditions, they might lead to self-organized systems in many biological areas, like pattern formation, chemotaxis, and population dynamics (Murray 1989). Without prescribing boundary conditions on a finite computational domain, diffusion equations may have an infinite number of mathematical solutions. A physically and biologically meaningful boundary condition specifies the corresponding physical and biological meaningful member of this set of solutions. In nonlinear diffusion–reaction equations of pattern formation, different boundary conditions can lead to totally different patterns. For example, Efendiev *et al.* (2002) show for the density-dependent diffusion–reaction model for biofilm formation introduced by Eberl *et al.* (2001) that the type of boundary condition imposed decides whether the model has a non-uniform bounded solution for all time or whether the biomass eventually will reach its maximum density everywhere (in the absence of detachment).

Hence, boundary conditions are an issue of modeling and not only an unimportant technical requirement stemming from a mathematics deficiency. Being a part of model formulation, boundary conditions also can be used to simplify the calculation. This can be done by choosing them in a way that leads to faster convergence of computer algorithms or by reducing the computational domain. An example is to prescribe explicitly the shape of an artificial concentration boundary layer around a biofilm (e.g. Noguera *et al.* 1999); thus, the time for computing the outer region can be saved. This calculatory simplification, however, often is at the expense of putting too much external force on the model. The nutrient concentration boundary layer of our example *in realiter* develops according to mass transfer and conversion and not *vice versa*. Hence, it should be derived from the model solution, and should not be its cause. Therefore, one must be aware that simulation results are always influenced by specified boundary conditions for the underlying model and their simplifications.

An example for the different behavior a nonlinear model can show in dependence of model parameters can be found in Pritchett and Dockery (2001). For their one-dimensional single-species biofilm model, they showed that a positive steady state exists if the model parameters inactivation rate is zero and if the model parameters cell death rate, specific growth rate, and bulk substrate concentration satisfy a certain condition (which basically says that growth is faster than decay) even in the absence of biomass detachment. If inactivation is considered in this model, the only steady state is the trivial one that is there is no biomass left in the system. By mathematical analysis, they could show that this steady state is unstable under some further conditions on the model parameters (which basically say that growth is faster than decay and inactivation). That is, if a little bit of biomass is added to the system, a biofilm will develop. The bulk concentration value in this model is specified as a boundary condition. Thus, this example for parameter dependent model behavior could also serve as an example for the dependence on boundary conditions, quite in the sense of holistic process interaction. Another example for qualitative analysis of parameter dependencies in nonlinear models can be found in Ballyk and Smith (1999), Jones and Smith (2000), and Smith and Zhao (2000). They determine criteria for persistence, wash out, and wall attachment of bacteria in plug flow reactors.

Inspired by mathematical physics, the reductionist approach was prevalent in the beginning of mathematical biology in the 1920s (Israel 1994). (However, the idea of studying the whole biological system by first separating its parts, analyzing them and then putting pieces together again is much older and also its shortcoming was already known, e.g. in the studies of J. W. von Goethe on *Morphology* in 1817 (von Goethe 1989). But this should not be seen immediately in the context of a theoretical biology, but rather as an example of how physical thinking influenced biological thinking for a long time, since mathematical physics was already established and accepted then.) Both views, reductionism and holism, are lines in the philosophical hypothesis of physicalism, which claims all processes in nature can be described by physical laws. Physicalism seems to be the dominating direction in biology. von Weizsäcker (1995) claims that this will be so for a long time and that the fast progress biology made in recent decades is due to the application of physical and chemical methods and concepts. A competing opinion is vitalism, i.e. the persuasion that there are additional non-physical, but vital, forces driving biological systems. It was introduced after '*biologists despaired of ever explaining biological organization in terms of ordinary physical forces*' (Davies 1989). But also in the non-animated world many examples can be found showing organized behavior and self-organization, and this shows that there can not be conceptually a need for vital forces to create those phenomena. (An often-cited example is thermal convection, the so called Benard instability: a fluid between two parallel plates initially at rest starts moving and organizes in a cyclic pattern, when the lower plate is heated above a certain value and the upper plate remains at the same temperature (e.g. Tritton 1977; Nicolis 1995).)

Vitalism was a response to reductionism (Davies 1989), whereas holism can be considered its extension and successor. (The idea of vital forces is also much older, of course, and appears already in ancient philosophy. But again we see it in the context of modern science.) Both approaches have been introduced to overcome the drawbacks and shortcomings of reductionism and explain qualities that cannot be explained by looking only at isolated single parts of the system. However, even if physical theories are found to be superior, Gierer (1996) reminds that biology should not be reduced to being a part of physics. The animated world is more expanded than the non-animated world, and the way experiences are made and cognition is gained is different in biology and in physics. In conflicts between mechanistic and non-mechanistic biology, in most cases mechanists have turned out to be right throughout history, and a failing of physical laws in biology has never been shown (Fischer 1993, and the references given therein). However, non-mechanistic thinking often provided the more creative contributions to the understanding of life because it could avoid excluding problems still too difficult for a physical approach (Gierer 1996). Embedded in a physical theory, these non-mechanistic parts can be considered as black boxes, which are still to be elucidated.

Recently, various discrete or semi-discrete models for biofilm growth have been suggested by several authors (Wimpenny and Colasanti 1997; Picioreanu *et al*. 1998; Hermanowicz 1999; Noguerra *et al*. 1999; Mehl 2001). The spatial biofilm development according to those models is governed by different sets of local rules, similar to and derived from the mathematical theory of cellular automata. The underlying idea is a partitioning of the computational domain into discrete grid cells. Each grid cell can be either occupied by biomass or not (*i.e.* it is occupied by bulk liquid). The biomass occupied grid cells form the biofilm. Biomass grows due to nutrient consumption, and will be shifted from one grid cell to another according to a local rule. Say, if a certain value is reached, then a certain (or randomly chosen) amount will be given to an appropriate neighbor grid cell, according to the state of the system in the neighborhood. Thus a biofilm architecture develops. There is much freedom for the modeler in the formulation of the exact rules according to which the model works and, consequently, there is a multitude of sets of local rules describing a multitude of possible biofilm architectures.

John von Neumann worked out the fundamental ideas behind cellular automata in the 1940s. Mathematicians and theoretical biologists adapted them to describe what could not be formulated by physical laws: the nature of life. Finally, this lead to a theory of artificial life based on cellular automata (for a popular review, see Emmeche (1994)). That is, a model of abstract and non-material life *per se*, as opposed to a model of a living organism. Cellular automata come from the idea that the complex behavior of a total system can be derived from sometimes very simple local rules of interaction for the behavior of its elements. These elements are called cells, though they are not biological cells, but lattice cells. The rules for the behavior often are chosen on a heuristic basis, biased by human behavior. The most famous cellular automaton model is John Conway's Game of Life. Active (that is, living) cells in this automaton can die because of overcrowding or isolation. New cells will be born (that is, inactive cells are activated) if the neighborhood provides appropriate conditions. It can be found in virtually every book dealing with this class of models. Among the bibliography of this essay, examples are Emmeche (1994), Gerhardt and Schuster (1995), Chopard and Droz (1998).

It is clear that there is no need to describe non-material life (that is, artificial life) by mechanistic laws. However, in the meantime cellular automata and local rules are also widely used for many '*real*' biological, chemical, or physical processes, like spatial biomass spreading during biofilm growth as mentioned above (other examples can be found in Gerhardt and Schuster (1995)). The heuristic formulation of the local rules often makes this model class very attractive for non-mathematicians because the system appears to be intuitively easier to understand than a set of partial differential equations or stochastic models. Besides the conceptual simplicity, which also leads to easy and fast implementation, cellular automatons and local rules are found to be capable of describing two principles often considered to be main characteristics of life:

self-organization and self-reproduction which are typical for many nonlinear systems described by continuous models. This is another reason why they are so popular in theoretical biology. In biofilm modeling based on mass transfer and reaction interactions, discrete local rules were induced as vitalistic black boxes for a continuum process not yet fully understood. (A hypothesis for biofilm growth based on physical principles has been formulated by Wimpenny and Colasanti (1997) and van Loosdrecht *et al.* (1997). They state that the structure of a biofilm is governed by environmental conditions like hydrodynamics of the bulk liquid and nutrient availability.) Therefore, from the viewpoint of physicalism, they should be seen as transition states in biofilm architecture modeling but not as a final theory, although they show reliable results on first sight. It is subject to be changed or updated as understanding of biological processes can be better quantified.

Finally, it should be mentioned that local rules or cellular automaton-based approaches are used in biofilm modeling in other contexts as well: an individual based model concept, instead of continuum modeling, is introduced by Kreft *et al.* (1998) on a smaller length scale. Here, the discrete objects are representations of biological cells, however, they are no longer ordered on a regular lattice. In contrast to the approaches mentioned earlier, the discrete model is used to describe a discrete process and has a much smaller vitalistic component. Another approach in biofilm literature is using cellular automata as mere discretization techniques for partial differential equations describing the physics of the system; that is, they are numerical approximations of *known* continuous models. The book by Choppard and Droz (1998) gives a general description of this approach. Picioreanu *et al.* (1999, 2000), and Eberl *et al.* (2000a) use Lattice–Boltzmann schemes for the Navier–Stokes equations in the bulk liquid of a biofilm system. Picioreanu *et al* (1999, 2000) use this approach to solve the mass transfer equations as well. The Lattice–Boltzmann method is a floating point extension of the lattice-gas cellular automaton (see the review article by Chen and Doolen (1998), and the references cited therein for an overview).

## 1.5 INTERDISCIPLINARITY: WHY IT IS DIFFICULT TO COMMUNICATE WITH MATHEMATICIANS

Mathematical modeling of biofilm processes requires both good knowledge of biology (and biotechnological needs) and good knowledge of mathematics. Owing to the strong specialization in the sciences, it is not easy for one researcher to have sufficient skills in all the participating disciplines. Therefore close collaboration between biologists, engineers and mathematicians is necessary, and the whole enterprise is an interdisciplinary task. However, whenever mathematics and mathematicians enter the ball game, one of the major problems of interdisciplinary research becomes obvious: the exchange of information, or: how to understand them and how to be understood.

The carrier of communication and exchange of information is language. Therefore, ideally a common language should be found to enable communication between two disciplines, as long as it is not possible to communicate information directly, that is without an underlying language and inevitable misunderstandings. (Attempts towards this goal have been made for a long time. For example, in the 19$^{th}$ century Frege suggested a language with the purpose of avoiding mistakes that arise during communication (Radbruch 1989). His idea was a formal language based on a similar concept as arithmetic, however not only for arithmetic but the whole of mathematics and all scientific disciplines. Frege's idea, like virtually all similar efforts, was not very successful and only in mathematics was it adopted after many decades. Eventually an entire branch of mathematics based on similar concepts was established: *mathematical logic*.) Instead, the scientific disciplines develop their own specialist languages, which make it very difficult for people without formal training in this discipline to follow.

Although many or most specialist languages are to a certain extent merely vocabulary extensions of the underlying cultural language (like English, French, or German), the language of mathematics is very different. (We also count the nomenclature system of biology among these vocabulary extensions, despite the fact that it is a very compact and efficient way of compressing information, and although historical disputes (Porter 1986) show that developing a nomenclature system is not a trivial task at all. However, the terms formed in this biological language are used in the carrier language in the same way as regular expressions, that is, in the same grammatical framework.) The nature of mathematics is gaining new information by logical deduction from existing verified knowledge, but our cultural languages lack a natural way of expressing logical deduction (Radbruch 1989). Hence, for a mathematical language, this ability had to be invented. In a translation from the mathematical language into a cultural language, this cannot be communicated one-to-one but must be paraphrased, which opens the way for new misunderstandings and mistakes. Another example is uniqueness, which is one of the driving principles of mathematics. Thus, it must be possible to express it in the language of mathematics. Colloquial languages do not provide this possibility (von Weizsäcker 1995). Therefore, along with vocabulary and terminology, the other integral part of language is special, grammar and syntax. In the mathematical language, they are based on logic, and if they are spoiled, the sentence becomes worthless or, which is even worse, wrong. On the other hand colloquial languages are rather robust in this respect and may remain understandable even with grammatical mistakes. As a matter of fact, when we are speaking in a foreign language we typically depend on this flexibility of colloquial cultural languages. The strict adherence of mathematics and the mathematical language to logic necessarily leads to sentences with litotes and other strange syntactic constructions appearing to be complicated and dispensable for non-mathematicians but which are not only necessary but clear and often even beautiful to mathematicians.

Misunderstandings arising from the use of one word in different contexts (e.g. 'cells' or 'functions' in mathematics and biology) can be detected and avoided, for example using 'grid cells' if the mathematical object is addressed in a cellular automaton context. But it is much more difficult to overcome the syntactic and grammatical problem described above and to avoid inadmissible simplifications or generalizations when making mathematical statements in a non-mathematical colloquial language. A dilemma of mathematics (at least of somewhat complicated mathematics) is, either it is correct or it is easily understandable. Both communicating parties must understand and accept this and try to find a compromise. This is very difficult because mathematics is inherent in its language, and learning the language of mathematics means becoming a mathematician. Therefore it is very complicated to explain to non-mathematicians even the nature of mathematics and what mathematicians are doing. Barrow (1996) considers this is a major contrast to other academic disciplines; a historian and an engineer will be able to explain in only a few sentences what they do, so that a layman has at least a (sometimes very) rough idea about it.

In the research areas, in which collaboration between mathematics and science or engineering has a long history, the problem of communication appears to be much easier, since common languages are developed which are part of the scientific education in these fields. Indeed, these research areas often form disciplines on the interface between both mother disciplines, like computational fluid dynamics or mathematical physics. An isolated task like biofilm modeling never can achieve a common interdisciplinary language. Therefore, for the time being collaborators from different disciplines must live with this dilemma, being aware, that it will probably not be possible to get more than an intuitive idea of what the other wants to say.

## 1.6 WHAT DO BIOFILM MODELS, MECHANICAL DUCKS, AND ARTIFICIAL LIFE HAVE IN COMMON? (CONCLUSION)

It is essential to understand that there is no general biofilm model covering all biofilm processes. Instead, every mathematical model focuses on some particular properties, like Vaucanson's mechanical duck only described the external and superficial moving of an animal without considering any vital aspects, and like cellular automata in artificial life describe self-organization and self-reproduction, without taking physical and material properties into account. In the mathematical modeling of spatial heterogeneous biofilm architectures, up to now most models are based on the hypothesis that environmental conditions like nutrient supply and hydrodynamics are the prevailing factors. Different mathematical models based on this assumption lead to the conclusion that indeed mass transfer processes might be sufficient to determine a highly irregular biofilm architecture. Other possible factors influencing biofilm architecture like cell-to-cell signaling mechanisms are so far left out in modeling. Therefore, no statement can be made about their significance yet.

Mathematical modeling requires good knowledge of mathematics and the mathematical and computational skills to use the model. It requires good understanding of the biological processes. None of these issues is trivial and, therefore, mathematical modeling in general can only be accomplished by interdisciplinary collaboration.

## 1.7   REFERENCES

Ballyk, M. and Smith, H. (1999) A model of microbial growth in a plug flow reactor with wall attachment. *Math. Biosciences* **158**, 95–126.

Barrow, J.D. (1996) *Warum die Welt mathematisch ist*. DTV, Munich. German translation of the Italian original *Perche il mondo e matematico?* Laterza & figli, 1992.

Bungartz, H.-J. and Mehl, M. (2003) Beyond models: requirements and chances of computational biofilms. (This volume, chapter 4.)

Bryers, J.D. and Drummond, F. (1998) Local macromolecule diffusion coefficients in structurally non-uniform bacterial biofilms using fluorescence recovery after photobleaching. *Biotechnol. Bioengng* **60**, 462–473.

Chen, S. and Doolen, G.D. (1998) Lattice Boltzmann method for fluid flow. *Annu. Rev. Fluid Mech.* **30**, 329–364.

Chopard, B. and Droz, M. (1998) *Cellular Automata Modelling of Physical Systems*. Cambridge University Press.

Davies, P.C.W. (1989) The Physics of Complex Organisation. In *Theoretical Biology. Epigenetic and Evolutionary Order from Complex Systems* (ed. B. Goodwin and P. Saunders), pp. 101–111. Edinburgh University Press, Edinburgh.

Dockery, J.D. and Keener, J.P. (2001) A mathematical model for quorum sensing in *Pseudomonas aeruginosa*. *Bull. Math. Biol.* **63**, 95–116.

Dockery, J. and Klapper, I. (2002) Finger formation in biofilm layers. *SIAM J. Appl. Math.* **62**(3), 853–869.

Eberl, H., Picioreanu, C. and van Loosdrecht, M.C.M. (2000a) Modeling geometrical heterogeneity in biofilms. In *High Performance Computing Systems and Applications* (ed. A. Pollard, D.J.K. Mewhort and D.F. Weaver), pp. 497–512. Kluwer Academic Publishers, Boston.

Eberl, H., Picioreanu, C. and van Loosdrecht, M.C.M. (2000b) A three-dimensional numerical study on the correlation of spatial structure, hydrodynamic conditions, and mass transfer and conversion in biofilms. *Chem. Engng Sci.* **55**, 6209–6222.

Eberl, H.J., Parker, D.F. and van Loosdrecht, M.C.M. (2001) A new deterministic spatio-temporal continuum model for biofilm development. *J. Theor. Med.* **3**, 161–175.

Efendiev, M.A., Eberl, H.J. and Zelik, S.V. (2002) Existence and longtime behavior of solutions of a nonlinear reaction–diffusion system arising in the modeling of biofilms. In *Nonlinear Systems and Related Topics*, RIMS Tokyo, vol. 1258, pp. 49–71.

Emmeche, C. (1994) *Das lebende Spiel. Wie die Natur Formen erzeugt.* Rororo Science, Reinbeck. German translation of the Danish original *Det levende Spil: Biologisk form og kunstigt liv*. Munskgaard Kopenhagen, 1991.

Fischer, E.P. (1993) Was ist Leben? – Mehr als vierzig Jahre später. Introductory essay to Erwin Schrödinger's book *What is life?* published with the German edition of *Erwin Schrödinger, Was ist Leben*. Serie Piper, Munich and Zürich.

Gerhardt. M. and Schuster, H. (1995) *Das digitale Universum. Zelluläre Automaten als Modelle der Natur.* Vieweg. (In German.)

Gierer, A. (1996) Lebensvorgänge und mechanistisches Denken. Beziehungen und Beziehungskrisen vom siebzehnten Jahrhundert bis heute. In *Die Natur is unser Modell von ihr* (ed. V. Braitenberg and I. Hosp*)*, pp. 51–65. Rororo Science, Reinbeck. (In German.)

Hancock, B. (1995) From the album *Eats Away The Night*. Sugar Hill Records, No. 1048.

Hermanowicz, S.W. (1999) Two-dimensional simulations of biofilm development: Effects of external environmental conditions. *Wat. Sci. Technol* **39**(7), 107–114.

Hermanowicz, S.W. (2001) A simple 2D biofilm model yields a variety of morphological features, *Math. Biosciences* **169**, 1–14.

Hertz, H. (1963). *Die Prinzipien der Mechanik in neuem Zusammenhange dargstellt.* Wissenschaftliche Buchgesellschaft, Darmstadt, photo-mechanical reprint of the Edition Leipzig, 1894. (In German.)

Hillier M. (1976) *Automata and Mechanical Toys. An illustrated History.* Jupiter.

Hösel, V. and Liebscher, V. (2003). Towards new mathematical models for biofilms. (This volume, chapter 4.)

Israel, G. (1994) Mathematical biology. In *Companion Encyclopedia of the History and Philosophy of the Mathematical Sciences* (ed. I. Grattan–Guinness), vol. 2, pp. 1275–1280. Routledge, London and New York.

Jones, D.A. and Smith, H. (2000) Microbial competition for nutrient and wall sites in plug flow. *SIAM J. Appl. Math.* **60**, 1576–1600.

Kreft, J.-U., Booth, G. and Wimpenny, J.W.T. (1998) BacSim, a simulator for individual-based modeling of bacterial colony growth. *Microbiology* **144**, 3275–3287.

Kreft, J.-U., Picioreanu, C., Wimpenny, J.W.T. and van Loosdrecht, M.C.M. (2001) Individual-based modeling of biofilms. *Microbiology* **147**, 2897–2912.

*Mathematics & Biology, the Interface. Challenges and Opportunities.* To be found on the World Wide Web under URL http://www.bio.vu.nl/nvtb/Interface.html.

Mehl, M. (2001) Ein interdisziplinärer Ansatz zur dreidimensionalen numerischen Simulation von Strömung, Stofftransport und Wachstum in Biofilmen auf der Mikroskala. Ph.D. thesis, Technical Univ. of Munich. (In German.)

Morgenroth, E., Eberl, H. and van Loosdrecht, M. (2000) Evaluating 3D and 1D mathematical models for mass transfer in heterogeneous biofilms. *Wat. Sci. Tech.* **41**(4/5), 347–356.

Murray, J.D. (1989) *Mathematical Biology*. Springer, Heidelberg.

Nicolis, G. (1995) *Introduction to Nonlinear Science.* Cambridge University Press.

Noguera, D.R., Pizarro, G., Stahl, D.A. and Rittmann, B.E. (1999) Simulation of multispecies biofilm development in three dimensions. *Wat. Sci. Tech.* **39**(7), 123–130.

Picioreanu, C., van Loosdrecht, M.C.M. and Heijnen, J.J. (1998) A new combined differential-discrete cellular automaton approach for biofilm modeling: application for growth in gel beads. *Biotechnol. Bioengng* **57**, 718–731.

Picioreanu, C., van Loosdrecht, M.C.M. and Heijnen, J.J. (1999) Discrete-differential modeling of biofilm structure. *Wat. Sci. Tech.* **39**(7), 115–122.

Picioreanu, C., van Loosdrecht, M.C.M. and Heijnen, J.J. (2000) A theoretical study on the effect of surface roughness on mass transport and transformation in biofilms. *Biotechnol. Bioengng* **68**, 355–369.

Picioreanu, C., van Loosdrecht, M.C.M. and Heijnen, J.J. (2001) Two-dimensional model of biofilm detachment caused by stress from liquid flow. *Biotechnol. Bioengng* **72**, 205–218.

Popper, K. (1982) *Logik der Forschung*, 7th corrected edition. J.C.B. Mohr (Paul Siebeck), Tübingen.

Porter, C.M. (1986) *The Eagle's Nest. Natural History and American Ideas, 1812–1842.* University of Alabama Press.

Pritchett, L.A. and Dockery, J.D. (2001) Steady state solutions of a one-dimensional biofilm model. *Math. Comput. Modeling* **33**, 255–263.

Radbruch, K. (1989). *Mathematik in den Geisteswissenschaften*. Vandenhoek & Rupprecht, Göttingen. (In German.)

Rittmann B.E. and McCarty, P.L. (1980) Model for steady-state-biofilm-kinetics. *Biotechnol. Bioengng* **22**, 2343–2357.

Rittmann, B.E., Pettis, M., Reeves, H.W. and Stahl, D.A. (1999) How biofilm clusters affect substrate flux and ecological selection. *Wat. Sci. Tech.* **39**(7), 99–105.

Sáez, P.B. and Rittmann, B.E. (1992) Accurate pseudoanalytical solution for steady-state biofilms. *Biotechnol. Bioengng* **39**, 790–793.

Smith, H. and Zhao, X.-Q. (2000) Microbial growth in a plug flow reactor with wall adherence and cell motility. *J. Math. Analysis Applic.* **241**, 134–155.

Tayler, A.B. (1986). *Mathematical Models in Applied Mechanics*. Clarendon Press, Oxford.

Thompson, D.'A. W. (1917) *On Growth and Form*, abridged edition 1961 (ed. J.T. Bonner). Cambridge University Press.

Tritton, D.J. (1977) *Physical fluid Dynamics.* Van Nostrand Reinhold Company, Wokingham, UK.

van den Hooff, A. (1995) Holisme als Kitsch. In *De Schok der Biologie*, pp. 143–144, Uitgeverij SUN, Nijmegen. (In Dutch.)

von Glaserfeld, E. (1996) Natur als Black-Box. In *Die Natur ist unser Modell von ihr*, (ed. V. Braitenberg & I. Hosp*)*, pp. 15–26. Rororo Science, Reinbeck. (In German.)

von Goethe, J.W. (1989) Zur Morphologie. Die Absicht eingeleitet. In *Sämtliche Werke*, vol. 12, pp. 12–17. Carl Hanser Verlag, Munich. Originally *in Zur Naturwissenschaft überhaupt, besonders zur Morphologie*, 1817. (In German.)

van Loosdrecht, M.C.M., Picioreanu, C. and Heijnen, J.J. (1997) A more unifying hypothesis for biofilm structures. *FEMS Microbiol. Ecol.* **24**, 181–183.

von Weizsäcker, C.F. (1995) *Die Einheit der Natur*, part I (*Wissenschaft, Sprache und Methode*) and part III (*Der Sinn der Kybernetik*). DTV Wissenschaft, Munich. (In German.)

Wanner, O. and Gujer, W. (1986) A multi-species biofilm model. *Biotechnol. Bioengng* **28**, 314–328.

Wimpenny, J.W.T and Colasanti, R. (1997) A unifying hypothesis for the structure of microbial biofilms based on cellular automaton models. *FEMS Microbiol. Ecol.* **22**, 1–16.

Wood, B.D. and Whitaker, S. (1998) Diffusion and reaction in biofilms. *Chem. Engng Sci.* **53**, 397–425.

Wuertz, S., Hendrickx, L., Kuehn, M., Rodenacker, K. and Hausner, M. (2001) In situ quantification of gene transfer in biofilms. *Methods Enzymol.* **336**, 129–143.

## Acknowledgements

For the work on this chapter the author was partly supported by the European Commission under contract ERBFMRX-CI 97-0114 (the TMR Network *Biological Removal of Nitrogen, from Biofilms to Bioreactors*).

# 2
# Biofilm architecture: interplay of models and experiments

*Slawomir W. Hermanowicz*

## 2.1 TOOLS FOR BIOFILM STUDIES

Two methods can be used to examine biofilm structure: models and experiments. They are not mutually exclusive; rather they form complementary tools that have been used in a synergistic fashion to expand the knowledge about biofilms, their developments and controlling factors. This chapter explores connections between various biofilm models and features observed in real biofilms. These connections will be illustrated by selected examples of specific models and related observations but without a goal of an all-encompassing review. From an overall perspective, the interactions between experiments and modeling, shown schematically in Figure 2.1, stimulated the flow of ideas leading to an ever more refined conceptual understanding of biofilms. Although the emphasis shifted through time between modeling and experiment, as indicated by the grand arrow, both components remained intimately linked and balanced.

© 2003 IWA Publishing. *Biofilms in wastewater treatment.* Edited by S. Wuertz, P.L. Bishop and P.A. Wilderer. ISBN: 1 84339 007 8.

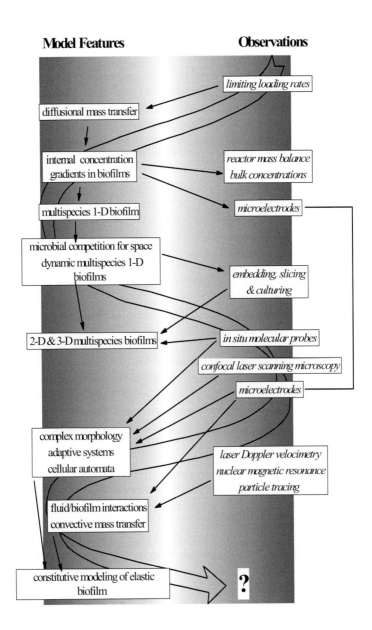

**Figure 2.1.** Interactions of biofilm models and experimental observations.

## 2.2 EARLY OBSERVATIONS AND MODELS

Although the presence of biofilms in natural and engineered microbial systems was recognized long ago, the nature of biofilms, their structure and effects are still not fully understood. One of the first observations that prompted an interest in biofilm structure was the existence of a limiting removal rate (loading rate) in biofilm reactors (see, for example, Heukelekian 1945; Schulze 1957, 1960). Several researchers have addressed the question of whether these limitations were caused by the saturation of microbial kinetics (similar to the Michaelis–Menten equation) or by mass transfer limitations through models and experimental observations.

For example, Schulze (1957, 1960) cultivated biofilms of different thickness and reported that oxygen did not penetrate through those that were thicker than 2 mm. These and similar observation lacked, however, a theoretical underpinning required to validate the limitation mechanisms. Such theory, in the form of biofilm models, was proposed in the 1960s (Atkinson et al. 1963; Atkinson and Swilley 1963; Swilley et al. 1964). The models assumed that the intrinsic reaction rate in the biofilm is of first order (as a simplification of the Michaelis–Menten or Monod kinetics). In the earliest papers from this group (Atkinson and Swilley 1963; Atkinson et al. 1963), two competing models of biofilm structure and associated hydrodynamics were considered (Figure 2.2). In the 'pseudo-homogeneous' model, liquid was supposed to flow through a 'matrix of microorganisms' whereas in the 'heterogeneous' model the microorganisms formed a dense 'surface' exposed to the flowing liquid. Remarkably, the morphology of the postulated 'pseudo-homogeneous' model is similar to real biofilms observed some 30 years later (see, for example, Stewart et al. 1995) and modeled using cellular automata (Wimpenny and Colasanti 1997; Hermanowicz 1997) as seen in Plate 2.1.

First experiments were focused on determining which of the two competing models better represented the reality. However, when the first models were created, experimental tools to verify biofilm morphology were limited. Thus, the validity of the models shown in Figure 2.2 was investigated by comparing predicted removal (uptake rates) with those calculated from experimental mass balances. For example, Atkinson et al. (1967) evaluated experimentally the effects of external hydrodynamics on glucose removal in an inclined-plate biofilm reactor. They varied the flow rate and the inclination angle, and attributed the limits of the reactor performance to the liquid phase diffusion in the external film. They also concluded that the 'heterogeneous' model (with a liquid film overlaying the impervious biofilm) provided a better description of their biofilm than the 'pseudo-homogeneous' model. Since then, the 'heterogeneous' biofilm idealization has dominated modeling efforts.

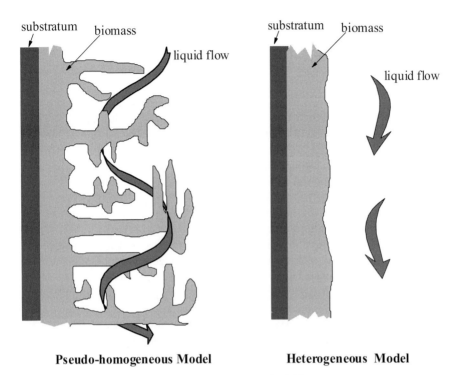

**Figure 2.2.** Two biofilm models (after Atkinson *et al.* 1967).

Diffusional mass transport was also postulated by Gulevich *et al.* (1968), who examined glucose removal in a batch reactor by a biofilm grown on a rotating disk. The effects of varying rotational speed on the removal rates were compared with the predictions of a simple model that accounted for external diffusional mass transfer. Based on the results, Gulevich and co-workers stated that 'the overall transformation' of a substrate in a rotating-disk biofilm reactor was 'diffusion dependent' and that the 'diffusional mechanism must be considered' in the design of treatment facilities.

## 2.3 STANDARD DIFFUSION-REACTION MODEL

### 2.3.1 Mathematical formulation

These early models focused on external mass transfer and described biofilm simply as a sink of substrates (reactants). They assumed either a fixed substrate concentration $C$ at the biofilm–liquid interface (typically $C = 0$) or prescribed a value of the substrate flux through the interface. Incorporation of mass transfer and reaction inside a biofilm was a natural extension of such models. Internal mass transfer was uniformly assumed to be done by molecular diffusion through a biofilm matrix. The flux of substrates and products was to follow Fick's law, with an effective diffusion coefficient $D_{eff}$ accounting for any differences between clean water and the biofilm. The values of $D_{eff}$ were estimated by using a range of experimental setups (see a review by Stewart (1998)), but primarily by measurement of fluxes entering or leaving a biofilm. The estimates of $D_{eff}$ were then calculated from assumed biofilm models.

Intrinsic microbial reaction rates $r$ inside the biofilm were variously assumed as Michaelis–Menten (e.g., Atkinson and Davies 1974), zero order (e.g., Harremoes 1976), or inhibitory Haldane (Moo-Young and Kobayashi 1972). These assumptions, combined with the diffusional mass transfer, led to a model that, with slight variations, became a standard for almost a quarter of the century (e.g., Atkinson and Davies 1974; Harremoes 1976; Williamson and McCarty 1976; Rittmann and McCarty 1980). The model can be generally described by the following equation:

$$\frac{\partial C}{\partial t} - D_{eff} \nabla^2 C + r = 0 \tag{2.1}$$

Enormous effort has been spent in producing exact analytical, approximate and numerical solutions to Equation 2.1 under different assumptions (e.g., steady state), for various boundary conditions, and for different geometries (e.g., Aris 1957; Knudsen *et al.* 1966; Do and Greenfield 1981). Invariably, the solutions depicted substrate concentration profiles that decreased from the interface inwards in the biofilm.

Despite the models generating a wealth of detailed descriptions of substrate (or product) profiles *inside* a biofilm, their verification was still based on matching the quantities external to the biofilm. Typically in the experiments, interface substrate concentrations resulting from a solution of Equation 2.1 were compared with measured bulk liquid concentrations. Alternatively, the solution of Equation 2.1 was integrated to calculate the interfacial flux from the bulk liquid to the biofilm, and the flux was matched with a corresponding component of an empirical mass balance. Given all model idealizations, computational approximations, parameter adjustment and inevitable experimental errors, a reasonable agreement between models and observations was often reported, leading to claims about the validity of specific model details.

### 2.3.2 Standard model validation

Rigorous validation of each model could be accomplished by comparing model predictions with *internal* concentration profiles experimentally measured in a biofilm. Although such measurements were attempted with oxygen microelectrodes as early as the 1960s (Whalen *et al.* 1969), their wider applications occurred only in the 1980s (Revsbech and Jorgensen 1986). With improvements and refinements of microelectrode techniques, routine measurements (at least in some laboratories) are now possible. The initial results (e.g., Lewandowski *et al.* 1991) showed that the models based on Equation 2.1 could be used, on a case-by-case basis, to calculate individual concentration profiles measured by the microelectrodes. However, concentration profiles observed in the same biofilm but only several micrometers apart often differed significantly in details. At one location, the nutrient concentration declined gradually with the diminishing distance to the substratum. At another position, nearby, the bulk liquid concentration extended almost to the substratum (see, for example, Figure 3 in De Beer *et al.* (1996)). This heterogeneity was the first indication that the commonly assumed conceptual image of a biofilm may be inadequate, and led to a dramatic re-evaluation of the biofilm structure and mass transfer mechanisms.

## 2.4 NEW EXPERIMENTAL TECHNIQUES

The re-evaluation was further accelerated by an almost simultaneous introduction of confocal laser microscopy and molecular probes to biofilm studies. With confocal microscopes (see Caldwell *et al.* 1992; Laurent *et al.* 1994; Surman *et al.* 1996) nondestructive, *in situ*, optical sectioning of biofilms became possible. Unlike previous methods, such as embedding and slicing, or microtomography, the artifacts were substantially diminished. Optical sectioning, combined with three-dimensional numerical reconstruction, allowed, for the first time, visualization of the structure of a biofilm in its (almost) undisturbed state. In a stark contrast to previous models depicting biofilms as flat homogeneous layers, these observations revealed a possibility of biofilm structures rich in forms, and heterogeneous on a scale as small as several micrometers. New terms such as 'mushrooms' and 'streamers' were applied to describe microscopic morphological forms of biomass (De Beer *et al.* 1994, 1996; Stewart *et al.* 1995). These chunks of biomass, while attached to the substratum, were found to be separated by voids, pores and channels. The flow of liquid through these voids was demonstrated using microphotography and microspheres (Stoodley *et al.* 1994), thus forcing recognition of advection as a mass transfer mechanism even at very small spatial resolutions.

At the same time, molecular techniques (especially fluorescent *in situ* hybridization: FISH) yielded images of biological diversity in biofilms (Poulsen *et al.* 1993). Although further refinements are still needed, the molecular probes are able to track the presence (and perhaps even activity) of various microbial taxa in the biofilm. In several instances, specific members of microbial populations were found organized in distinct agglomerations whereas in other cases, more uniform distributions were noted.

## 2.5 CHALLENGES IN MODELING

The images of biofilm structure and heterogeneity achieved through the advances in experimental techniques described above (microelectrodes, confocal microscopy, molecular probes) now require a new theoretical and modeling framework. For a modeler, two intertwined problems arose from the new observations: characterization or description of the complicated biofilm morphology and structure, and modeling of dynamic evolution of biofilm.

### 2.5.1 Characterization of morphology

Visually, two biofilms can often be seen as very different, but quantification of such differences (or similarities) poses a serious and still unresolved challenge. Without further developments of structure characterization, advances in the second area, the dynamic modeling, will also be limited. Although new models can be created, it is paramount that the predicted structures and morphologies are compared with real biofilms in a meaningful way. However, because a model cannot exactly reproduce all microscopic features of an individual biofilm fragment, the comparison between predictions and observation must be made in global, spatially averaged or statistical terms. Thus, the characterization techniques are essential tools for the development of realistic dynamic biofilm models. After all, if we do not have an agreed upon metric for model evaluation, we cannot assess the progress.

Although the need for a better characterization of biofilm structure was recognized (Christensen *et al.* 1989; Lewandowski *et al.* 1994, 1995) few advances have been made. For example, Zhang and Bishop (1994a,b) evaluated internal porosity and tortuosity of biofilm voids. Fractal geometry was employed for both pore space and biomass characterization (Zahid and Ganczarczyk 1994, Hermanowicz *et al.* 1995, 1996). For mature biofilms, the pore space had a fractal structure with the dimension lower than three (Zahid and Ganczarczyk 1994). For young biofilms, the voids did not have fractal structure but the biomass did (Hermanowicz et al. 1995, 1996). Using the fractal dimension, two types of biomass agglomerate and their scales were detected. In addition, possible flow effects on the formation of anisotropic biofilm morphology were recognized using the fractal analysis. However, these attempts are still in their early stages and must be further developed. It is likely that more than one parameter will be required for morphology characterization. Stanley (1986) listed twelve types of fractal dimension and noted that two objects may have some fractal dimensions identical and yet be very different visually. The ultimate test for characterization of biofilm morphology is whether a set of parameters derived for observations and measurements of a real biofilm can be used as an input to a model that produces a 'synthetic' biofilm indistinguishable, in some statistical sense, from a real one.

## 2.5.2 Description of microbial communities

Although characterization of biofilm *morphology* is important to understand their interactions with the surrounding environment (such as external mass transfer phenomena), description of biofilm *ecological structure* and its dynamic development are crucial for assessment of biofilm functions. In this context, morphology refers to the geometric, physical form of a biofilm. Structure includes the morphological features but also spatial distribution of different biofilm elements, such as various microbial populations, extracellular polymers and abiotic components (e.g., cellulose fibers or clay particles). The dynamics of biofilm development and the resulting structure depends on the processes of attachment (deposition), growth, death and detachment (for a seminal discussion, see Characklis (1984)). The rates of the processes and their relative importance may vary spatially and in time. Obviously, deposition and attachment are controlling processes in the evolution of an incipient biofilm, while their importance may diminish later as microbial growth becomes dominant. Early biofilm models often assumed an already existing biofilm (typically as a uniform flat layer) and considered only the penetration and transformations of nutrients (substrates) but not biofilm development. In addition, these biofilm models described biomass as a homogeneous medium consisting of a single microbial population despite a realization that in many applications biofilms contained multiple populations.

In some more advanced models, the distribution of different populations was assumed arbitrarily (e.g., Tanaka and Dunn 1982; Watanabe et al. 1985), but Kissel et al. (1984) and Wanner and Gujer (1985, 1986) were the first to consider *explicitly* microbial growth and detachment in their models. Although their respective models differed slightly, they both assumed that microorganisms grow inside the biofilm and displace other biofilm components. This process was characterized by a local biomass displacement velocity perpendicular to the substratum. The displacement velocity was proportional to a growth rate at each depth within the biofilm. Because the growth rate was a function of nutrient concentrations that diminished in deeper parts of a biofilm, the displacement velocity was also lower close to the substratum. If, in some parts of a biofilm, the growth rate was smaller then the microbial decay rate (typically because of low local nutrient concentrations), the displacement velocity could assume negative values and the biofilm or its portion could shrink. For growing biofilms the concept of biomass displacement led to an overall advective biomass flux through the biofilm surface. This flux was counterbalanced in the models by surface erosion of biofilm components. Depending on the balance between the erosion and the overall growth (expressed as the displacement velocity), the thickness of the modeled biofilm could either increase or decrease in time.

Although the models of Wanner and Gujer, and Kissel *et al.* could describe the growth of single population biofilms, their power lay in a possibility of simulating mixed populations in a biofilm. Each microbial population competed for space in the biofilm and its dominant location depended on the ratios of its displacement velocity to the overall displacement velocity. Because the velocities were related to local growth rates, which in turn were affected by nutrient concentrations, the model could simulate a large variety of microbial distributions along the biofilm depth. Although the growth and biomass displacement mechanism was, at that time, purely speculative, it constituted a major advancement in the understanding of the internal biofilm structure. It predicted a possibility of coexistence inside a biofilm of multiple microbial communities with diverse metabolisms and different growth rates. Such coexistence was possible when the growth rate (and thus the displacement velocity) of one microbial population was higher than the average in one part of the biofilm but lower in another part. Wanner and Gujer (1985) applied their model to a biofilm reactor (trickling filter) with mixed heterotrophic and nitrifying microbial populations. They supported the validity of their assumption by a good agreement between the predicted and observed fluxes of organics and ammonia into the biofilm at different locations in the reactor. The basic assumptions of the model were further generalized (Bryers 1992) and applied to specific biofilm populations (see, for example, Rittmann and Manem 1992). They were also extended to two- and three-dimensional modeling, where the displacement of growing biomass followed the gradients of nutrient concentrations (Gujer 1987; Noguera *et al.* 1999).

However, the agreement between observed and predicted overall fluxes of nutrients provided only an indirect support for model assumptions. Okabe *et al.* (1996) directly measured microbial population distributions inside a biofilm grown in a rotating biological contactor. They analyzed thin slices of a biofilm grown on a mixture of acetate and ammonium (as energy sources) and compared the observed densities of microbial populations with those predicted by the model. Similar experimental techniques were used in earlier studies (Masuda *et al.* 1991; Zhang *et al.* 1993) but the observations were not compared with predicted microbial population profiles. Okabe *et al.* (1996) found that nitrifiers and heterotrophs co-existed inside the biofilm, but their stratification was less pronounced then predicted. They attributed this discrepancy to 'the biofilm structural heterogeneity, such as water channels, which significantly affects mass transport'.

## 2.6 MODELING OF STRUCTURAL HETEROGENEITY

Modeling of 'structural heterogeneity' poses a significant challenge. Even in models that explicitly account for microbial growth, such as those based on the biomass displacement velocity (Kissel *et al.* 1984; Wanner and Gujer 1985), biomass grows in the direction of the concentration gradient. The resulting morphologies, even in two- and three-dimensional models, do not produce the full diversity observed in real biofilms like 'mushrooms' or 'streamers'. In addition, because the modeled growth rates of microbial populations vary smoothly in space, the simulated distributions of microbial populations are also continuous in the biofilm. Typically, predicted transitions between regions dominated by different populations are gradual, unlike many real biofilms where each population is concentrated in distinct clusters.

### 2.6.1 Discrete models

To model such heterotrophic biofilm structures, a new class of model is being developed. A common feature of these new models is a discrete representation of biomass (Hermanowicz 1997; Wimpenny and Colasanti 1997; Picioreanu *et al.* 1998). The new models use variants of cellular automata as a mathematical framework. As a result of both the discrete biomass representation and the application of cellular automata, a set of *local* rules is established for biomass development. No morphological form is assumed either explicitly or implicitly. Although these biofilm models are still in their infancy, they are able to reproduce a wide diversity of biofilm morphologies, as seen in Plate 2.2 (from Hermanowicz 1997). In the model used to generate these morphologies, only three very simple (if not simplistic) rules were used.

The *first* rule described the growth of biomass with a Monod-like function, with one growth-limiting nutrient supplied by molecular diffusion through the biomass clusters and the liquid boundary layer. The concentration in the bulk liquid, outside the clusters and the boundary layer, was kept constant. In real biofilm systems, nutrients outside the boundary layer are conveyed by advection and thus maintained at almost constant values (at least in a locally small neighborhood). This assumed mechanism of mass transfer yields nutrient concentration gradients that are consistent with the experimentally measured oxygen concentrations around biofilm clusters (Zhang and Bishop 1994c; De Beer *et al.* 1996). They observed steepest concentration gradients in a boundary layer that closely followed the contours of the biofilm surface.

The *second* rule in the model controlled the displacement of growing biomass. In its present form, the rule stipulated that the growing microbial cells either occupy an empty neighboring space or displace adjoining biomass in the direction of least mechanical resistance. Picioreanu *et al.* (1999) adopted a slightly different rule where biomass displacement occurred only when a critical biomass density inside a biofilm was reached. Adoption of such rules is quite speculative because virtually nothing is known about the mechanisms governing biomass 'movement' as a result of its growth in a biofilm.

Finally, the *third* rule characterized the detachment process. According to this rule, which is also speculative, biomass detachment rate was proportional to the hydrodynamic shear stress at the biofilm surface and inversely proportional to biofilm 'strength'. At present, this last parameter has no clear physical meaning, although it could be related to mechanical properties of the biomass.

The objective behind these three rules was to generalize current understanding of the processes governing the development of biofilms and to produce a universal, generic framework for their investigation. If a model, which includes only a small set of features, adequately represents the reality of a biofilm structure, we are more convinced that these features are common to many biofilms. Although complex models might perform better in describing individual cases, they require additional parameters to achieve a better fit. Then, they often lose predictive powers, because the choice of appropriate coefficients becomes bewildering. A similar approach to modeling of complex systems was advocated by Schweitzer (1997). He stated that, for a successful model, 'it is important to find a level of description which, on one hand, considers the specific features of the system and reflects the origination of new qualities, but on the other hand, is not flooded with microscopic details'. It is important to note that the rules of biofilm development were defined at a lowest, most local level, rather then imposed on a whole biofilm. Thus, whereas the response of each individual biofilm element can be calculated, the shape of the overall structure resulting from the *interactions* between the individual parts cannot be *a priori* predicted.

Based on the preliminary modeling results (Plate 2.2), both the nutrient concentration in the bulk liquid around the biomass ($C_s$ in Plate 2.2) and the external mass transfer limitations (characterized in Plate 2.2 by the dimensionless boundary layer thickness $\delta_B/x$) have a very large effect on the generated biofilm shapes. Unsurprisingly, as the bulk nutrient concentration increased, the simulated biofilm 'grew' more abundantly. However, the morphology was also clearly influenced by the external mass transfer limitations (i.e., by the thickness of the boundary layer). For a thin boundary layer, a compact, dense biofilm was generated. With a thicker boundary layer, biofilm morphology was more open, with channels appearing between biofilm aggregates. Such channels were observed in real biofilms (see Costerton *et al.* (1994) for a review). At even greater boundary layer thickness, the biofilm generated by this model formed 'stacks' or 'trees' towering over a very thin 'lawn' of a base biofilm. This morphology resembles biofilms observed by Keevil and co-workers and named 'heterogeneous mosaic' (Keevil and Walker 1992; Walker *et al.* 1995). Further modeling incorporated biofilm strength, although at this time in a quite simplistic manner (Hermanowicz 2001). The probability of detachment of a cell cluster was assumed to be a function of biofilm strength or cohesiveness, an as yet unspecified mechanical property. Even with such simple assumptions, the results of biofilm development simulations can lead to a hypothesis regarding biofilm structure. Similarly to the results presented in Plate 2.2, the morphology of the developing biofilm in Figure 2.3 depended on the thickness of the mass transfer boundary layer (expressed as the dimensionless ratio $\delta_B/x$)

The results suggest that the effects of bulk nutrient concentration (dimensionless $C_S/K$) are somewhat counterbalanced by the biofilm strength. For biofilms with low strength (i.e., higher probability of detachment), fully developed morphologies (shown in bottom rows of each panel) were similar to those with high strength (top rows) but required higher nutrient concentrations.

## 2.7 CONCLUSIONS AND FUTURE DIRECTIONS

The ability of this and similar models to yield a rich variety of different morphologies opens new avenues in biofilm research. The newly created framework offers many possibilities for numerical experimentations, formulations of new hypotheses and evaluation of feasible biofilm formation mechanisms. It is encouraging for the development of a universal theory, that the variety of morphologies can be generated by changing only a few global parameters.

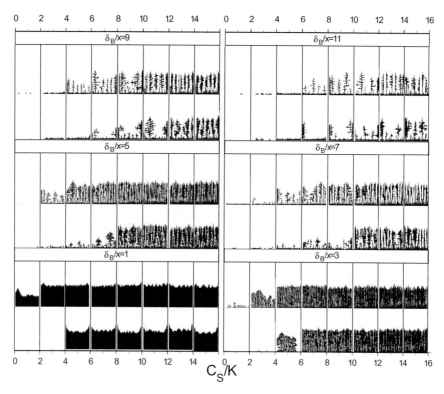

**Figure 2.3.** Simulated biofilm morphologies at different dimensionless nutrient concentrations $C_S/K$, boundary layer thickness $\delta_B/x$, and biofilm strength (top row in each panel, high strength; bottom row, low strength)

For example, the results suggest that the bulk nutrient concentration and the external mass transfer play very important roles in biofilm development, which postulates that the differences in biofilm strength can be compensated by changing nutrient concentrations. These hypotheses can be evaluated experimentally by using available techniques such as microelectrodes, microparticle tracing, nuclear magnetic resonance or laser Doppler velocimetry.

Using the presented framework, we can formulate a model that captures the essential aspects of biofilm development, and then we can play with the model until we discern 'how it works'. We can then re-evaluate the model with experimental data to see if our concepts need to be modified. We hope that each cycle, as seen in Figure 2.1, brings us further in the quest for better understanding of biofilm formation.

## 2.8 REFERENCES

Aris, R. (1957) On shape factors for irregular particles. 1. The steady state problem. Diffusion and reaction. *Chem. Eng. Sci.* **6**, 262–268.

Atkinson, B., Busch, A.W. and Dawkins, G.S. (1963) Recirculation, reaction kinetics, and effluent quality in a trickling filter flow model. *J. Wat. Poll. Control Fed.* **35**, 1307–1321.

Atkinson, B. and Swilley, E.L. (1963) A mathematical model for the trickling filter. In *Proceedings of the 18th Industrial Waste Conference, Purdue University, Lafayette, IN, 30 April – 2 May 1963*, pp. 706–737. Purdue University, Lafayette, Indiana.

Atkinson, B., Swilley, E.L., Busch, A.W. and Williams, D.A. (1967) Kinetics mass transfer and organism growth in a biological film reactor. *Trans. Inst. Chem. Engrs* **45**, 257–269.

Atkinson, B. and Davies, I.J. (1974) The overall rate of substrate uptake (reaction) by microbial films. *Trans. Instn Chem. Engrs* **52**, 248–259.

Bryers, J.D. (1992) Mixed Population Biofilms. In *Biofilms – Science and Technology*, (ed. L.F. Melo *et al.*), pp. 277–289. Kluwer, Dordrecht.

Caldwell D.E., Korber D.R. and Lawrence J.R. (1992) Imaging of bacterial cells by fluorescence exclusion using scanning confocal laser microscopy. *J. Microbiol. Methods* **15**, 249–261.

Characklis, W.G. (1984) Biofilm development: a process analysis. In *Microbial Adhesion and Aggregation* (ed. K.C. Marshall), pp. 137–158. Springer, Berlin.

Christensen, F.R., Kristensen, G.H. and Jansen, J.L.C. (1989) Biofilm structure – an important and neglected parameter in wastewater treatment. *Wat. Sci. Tech.* **21**(8/9), 805–814.

Costerton, J.W., Lewandowski, Z., Caldwell, D.E., Korber, D.R., De Beer, D. and James, G. (1994) Biofilms: the customize bioniche. *J. Bacteriol.* **176**, 2137–2142.

De Beer, D., Stoodley, P. and Lewandowski, Z. (1994) Liquid flow in heterogeneous biofilms. *Biotechnol. Bioengng* **44**, 636–641.

De Beer D., Stoodley, P. and Lewandowski, Z. (1996) Liquid flow and mass transport in heterogeneous biofilms. *Wat. Res.* **30**, 2761–2765.

Do, D.D. and Greenfield, P.F. (1981) A finite integral transform technique for solving the diffusion-reaction equation with Michaelis–Menten kinetics. *Mathl Biosci.* **54**, 31–47.

Gujer, W. (1987) The significance of segregation of biomass in biofilms. *Wat. Sci. Tech.* **19**(3/4), 495–504.

Gulevich, V., Renn, C.E. and Liebman, J.C. (1968) Role of diffusion in biological waste treatment. *Environ. Sci. Tech.* **2**, 113–119.

Harremoës, P. (1976) The significance of pore diffusion to filter denitrification. *J. Wat. Poll. Control Fed.* **48**, 377–388.

Hermanowicz, S.W., Schindler, U. and Wilderer, P. (1995) Fractal structure of biofilms: new tools for investigation of morphology. *Wat. Sci. Tech.* **32**(8), 99–105.

Hermanowicz, S.W., Schindler, U. and Wilderer, P. (1996) Anisotropic morphology and fractal dimensions of biofilms. *Wat. Res.* **30**, 753–755.

Hermanowicz, S.W. (1997) A model of two-dimensional biofilm morphology. In *Proceedings of the 2nd International Conference on Microorganisms in Activated Sludge and Biofilm Processes, 21–23 July 1997, University of California, Berkeley, CA* (ed. Jenkins, D. *et al.*). (Also published in *Wat. Sci. Tech.* **37**(4/5), 219–222 (1998).)

Hermanowicz, S.W. (2001) A simple 2D biofilm model yields a variety of morphological features. *Mathl Biosci.* **169**, 1–14.

Heukelekian, H. (1945) The relationship between accumulation, biochemical and biological characteristics of film and purification capacity of biofilter and standard filter. *Sewage Wks J.* **17**, 23–34.

Keevil, C.W. and Walker, J.T. (1992) Nomarski DIC microscopy and image analysis of biofilm. *Binary Comput. Microbiol.* **4**, 93–95.

Kissel, J.C., McCarty, P.L. and Street, R.L. (1984) Numerical simulations of mixed culture biofilms. *J. Env. Eng. ASCE* **110**, 393–405.

Knudsen, C.W., Roberts, G.W. and Satterfi, C.N. (1966) Effect of geometry on catalyst effectiveness factor — Langmuir-Hinshelwood kinetics. *Ind. Engng Chem. Fundamentals* **5**, 325–334.

Laurent, M., Johannin, G., Gilbert, N., Lucas, L., Cassio, D., Petit, P. and Fleury, A. (1994) Power and limits of laser scanning confocal microscopy. *Biol. Cell* **80**, 229–240.

Lewandowski, Z., Walser, G. and Characklis, W.G. (1991) Reaction kinetics in biofilms. *Biotechnol. Bioengng* **38**, 877–882.

Lewandowski, Z., Stoodley, P., Altobelli, S. and Fukushima, E. (1994) Hydrodynamics and kinetics in biofilm systems – recent advances and new problems. *Wat. Sci. Tech.* **29**(10/11), 223–229.

Lewandowski, Z., Stoodley, P. and Altobelli, S. (1995) Experimental and Conceptual Studies on Mass Transport in Biofilms. *Wat. Sci. Tech.* **31**(1), 153–162.

Masuda, S., Watanabe, Y. and Ishiguro, M. (1991) Biofilm properties and simultaneous nitrification and denitrification in aerobic rotating biological contactors. *Wat. Sci. Tech.* **23**(7/9), 1355–1363.

Moo-Young, M. and Kobayashi, T. (1972) Effectiveness factors for immobilized enzyme reactions. *Can. J. Chem. Engng* **50**, 162–167.

Noguera, D.R., Pizarro, G., Stahl, D.A. and Rittmann, B.E. (1999) Simulation of multispecies biofilm development in tree dimensions. *Wat. Sci. Tech.* **39**(7), 123–130.

Okabe, S., Hiratia, K., Ozawa, Y. and Watanabe, Y. (1996) Spatial microbial distributions of nitrifiers and heterotrophs in mixed-population biofilms. *Biotechnol. Bioengng* **50**, 24–35.

Picioreanu, C., van Loosdrecht, M.C.M. and Heijnen, J.J. (1998) Mathematical modeling of biofilm structure with a hybrid differential-discrete cellular automaton approach. *Biotechnol. Bioengng* **58**, 101–116.

Picioreanu, C., van Loosdrecht, M.C.M. and Heijnen, J.J. (1999) Discrete differential modeling of biofilm structure. *Wat. Sci. Tech.* **39**(7), 115–122.

Poulsen, L.K., Ballard, O. and Stahl, D.A. (1993) Use of rRNA fluorescence in situ hybridization for measuring the activity of single cells in young and established biofilms. *Appl. Env. Microbiol.* **59**, 1354–1360.

Revsbech, N.P. and Jorgensen, B.B. (1986) Microelectrodes: their use in microbial ecology. In *Advances in Microbial Ecology* (ed. K.C. Marshall), vol. 9, pp. 293–352. Plenum Press, New York.

Rittmann, B.E. and McCarty, P.L. (1980) Model of steady-state-biofilm kinetics. *Biotechnol. Bioengng* **22**, 2343–2357.

Rittmann, B.E. and Manem, J.A. (1992) Development and experimental evaluation of a steady-state multispecies biofilm model. *Biotechnol. Bioengng* **39**, 914–922.

Schulze, K. (1957) Experimental vertical screen trickling filters. *Sewage Indust. Wastes* **29**, 458–467.

Schulze, K. (1960) Load and efficiency of trickling filters. *Sewage Indust. Wastes* **32**, 245–258.

Schweitzer, F. (1997) Self-organization of complex structures. from individual to collective dynamics – some introductory remarks. In *Self-Organization of Complex Structures. From Individual to Collective Dynamics* (ed. F. Schweitzer), pp. xix–xxiv. Gordon and Breach Science Publishers, Amsterdam.

Stanley, H.E. (1986) Form: an introduction to self-similarity and fractal behavior. In *On Growth and Form* (ed. H.E. Stanley and N. Ostrowsky), pp. 21–53. Nijhoff, Boston.

Stewart, P.S., Murga, R., Srinivasan, R. and De Beer, D. (1995) Biofilm structural heterogeneity visualized by three microscopic methods. *Wat. Res.* **29**, 2006–2009.

Stewart, P.S. (1998) A review of experimental measurements of effective diffusive permeabilities and effective diffusion coefficients in biofilms. *Biotechnol. Bioengng* **59**, 261–272.

Stoodley, P., De Beer, D., Lewandowski, Z. (1994) Liquid flow in biofilm systems. *Appl. Env. Microbiol.* **60**, 2711–2716.

Surman, S.B., Walker, J.T., Goddard, D.T. and Morton, L.H.G. (1996) Comparison of microscope techniques for the examination of biofilms. *J. Microbiol. Methods* **25**, 57–70.

Swilley, E.L., Bryant, J.D. and Busch, A.W. (1964) Significance of transport phenomena in biological oxidation. In *Proceedings of the 19th Industrial Waste Conference, Purdue University, Lafayette, IN, 5–7 May 1964*, pp. 821–834. Purdue University, Lafayette, Indiana.

Tanaka, H. and Dunn, I.J. (1982) Kinetics of biofilm nitrification. *Biotechnol. Bioengng* **24**, 669–689.

Walker, J.T., Dowsett, A.B. and Rogers, J. (1995) Heterogeneous mosaic – a haven for waterborne pathogens. In *Microbial Biofilms* (ed. H.M. Lapin-Scott and J.W. Costerton), pp. 196–204, Cambridge University Press, Cambridge, UK.

Wanner, O. and Gujer, W. (1985) Competition in biofilms. *Wat. Sci. Tech.* **17**(2/3), 27–44.

Wanner, O. and Gujer, W. (1986) A multispecies biofilm model. *Biotechnol. Bioengng* **28**, 314–328.

Watanabe, Y., Masuda, S., Nishidome, K. and Wantawin, C. (1985) Mathematical model of simultaneous organic oxidation, nitrification and denitrification in rotating biological contactors. *Wat. Sci. Tech.* **17**(2/3), 385–397.

Whalen, W.J., Bungay, H.R. and Sanders, W.M. (1969) Microelectrode determination of oxygen profiles in microbial slime systems. *Environ. Sci. Tech.* **3**, 1297–1298.

Williamson, K. and McCarty, P.L. (1976) A model of substrate utilization by bacterial films. *J. Wat. Poll. Control Fed.* **48**, 9–24.

Wimpenny, J.W.T. and Colasanti, R. (1997) A unifying hypothesis for the structure of microbial biofilms based on cellular automaton models. *FEMS Microbiol. Ecol.* **22**, 1–16.

Zahid, W. and Ganczarczyk, J. (1994) Fractal properties of the RBC biofilm structure. *Wat. Sci. Tech.* **29**(10/11), 217–279.

Zhang, T.C., Fu, Y.C. and Bishop, P.L. (1993) Competition in biofilms. *Wat. Sci. Tech.* **29**(10/11), 263–270.

Zhang, T.C. and Bishop, P.L. (1994a) Evaluation of tortuosity factors and effective diffusivities in biofilms. *Wat. Res.* **28**, 2279–2287.

Zhang, T.C. and Bishop, P.L. (1994b) Density, porosity and pore structure of biofilms. *Wat. Res.* **28**, 2267–2277.

Zhang, T.C. and Bishop, P.L. (1994c) Experimental determination of the dissolved oxygen boundary layer and mass transfer resistance near the fluid–biofilm interface. *Wat. Sci. Tech.* **30**(11), 47–58.

# 3
# Towards new mathematical models for biofilms

*Volker Hösel and Volkmar Liebscher*

## 3.1 INTRODUCTION

In the past two decades great efforts were undertaken to develop a 'universal' biofilm model. The scientific community has used two basic approaches. The first is the classical model of Wanner and Gujer (1986), together with its extensions, which models the growth of the biofilm's entities with differential equations. It is therefore assumed that smooth functions, which result from averaging, describe the biofilm's relevant aspects. The other, more recent 'universal' model is proposed by Picioreanu *et al.* (1998). This model accounts for the complex heterogeneous structure observed in some biofilms and includes the concept of cellular automata. Other research groups have come up with refinements and additional ideas.

© 2003 IWA Publishing. *Biofilms in wastewater treatment*. Edited by S. Wuertz, P.L. Bishop and P.A. Wilderer. ISBN: 1 84339 007 8.

So, one may ask: which of the current models is superior in representing the most important factors determining biofilm architecture and hence transport processes within?

We believe that there is no general answer to this question because different environmental conditions produce a variety of structurally very different biofilms. This should imply that, depending on the specific biofilm, transport and architecture are governed by specific factors. A model can therefore be superior for specific types but not be suited to handle relevant effects in others.

With this in mind, we suggest to take a different, more general, approach and ask: which other classes of model could address interesting questions on biofilms; which are not answered by the established models? Therefore, we want to encourage scientists of different fields to contribute to the mathematical modeling of biofilms. Our hope is that this article will trigger a discussion between scientists working in the realms of biofilms and beyond.

The structure of this chapter is as follows. Starting from general methods of mathematical modeling, we discuss existing biofilm models. We then glimpse at some models from mathematical biology, which might be important in the area of biofilm modeling. Finally, we present possible stochastic models for biofilm heterogeneity. Because the essential ideas can be presented without recourse to mathematical formulae, we avoid these and refer the interested reader to relevant literature.

## 3.2 STRATEGIES OF MATHEMATICAL MODELING

Mathematical modeling of natural phenomena is mainly done for two purposes. First, to supply engineers, biologists and application-oriented scientists with estimates and predictions of processes to be investigated and controlled. Second, to start from basic principles and to supply an explanation of the processes from a more theoretical point of view. There is no sharp line distinguishing the models from these two classes, and the ideal model would be useful for both purposes.

Once there is an established model, one can study the modeled phenomena by analyzing the mathematical model. Different mathematical techniques are applied to receive quantitative or qualitative results. For example, assume that the growth of a bacteria species within the biofilm is modeled with differential equations. To obtain quantitative statements one has to determine the parameters of the model and to find solutions with numerical procedures. Alternatively one could vary the parameters and thus simulate different environmental conditions. If the model is complex, the computational effort may be large and sometimes not feasible (compare with Chapter 4). In any case, mathematical analysis focusing on the model's qualitative aspects may be helpful. With mathematical analysis one can discuss models with very general assumptions on the parameters and functions involved. Parameter dependence and sensitivity issues can be addressed with all the available tools.

Often a 'primitive' model, which is amenable to rigorous analysis, is very useful as a first step in understanding cause-and-effect relationships in complex situations. In contrast to mathematical analysis, simulations cannot be done without specifically defined functions and rules.

To continue the example above, one may be interested in the mechanisms controlling the growth of a bacterial population. Typical qualitative statements are that the species will thrive if a specific parameter increases or that another parameter may cause extinction if it decreases beyond a certain threshold.

Qualitative and quantitative techniques complement each other and are both applicable to the different models built for control or explanation purposes.

As it is impossible to represent all aspects of a complex system in one 'universal model', one has first to choose relevant questions that should be addressed. Therefore, a close cooperation between the mathematicians and the other involved scientists is necessary. Often non-mathematicians shy away from abstract calculus and only participate with the question: 'Which parameters do you need for 'your' model?'. In our opinion this is not a promising attitude for modeling. Instead, mathematicians and scientists from other fields should discuss together topics like: what is the purpose of the model to be developed, which aspects of the research object are relevant for a certain question, are the explicit or implicit assumptions for a concrete model plausible, etc.

As already mentioned, one encounters models built for different purposes. Those models designed with the aim to improve the understanding of the system, should be simple and include only a few, but relevant, parameters. This makes the handling of the models feasible and allows easy interpretation and verification. In the field of pure research, these should be the favorable models. Models connected to control aspects are more important in the field of engineering. These models are often complex, with many parameters involved. Thus simulation techniques are sometimes the only possibility to study performance of the models themselves. Improved understanding of the entire system might also help to improve the modeling of control aspects.

After model development, verification has to be done. It is less involving to compare the more qualitative (simple) models with the experiment. Owing to the many parameters in complex models, these can be fitted to almost every given (quantitative) data set. The problem here is the robustness of the fit (reliability of data and estimated parameters). Thus, one could state the general rule: the simpler the model, the easier the verification. Table 3.1 roughly summarizes the above discussion.

**Table 3.1.** Comparison of features usually attributed to simple and to complex models

| Feature | Simple model | Complex model |
| --- | --- | --- |
| Number of parameters | Small | Large |
| Results derived | Qualitative | Quantitative |
| Model investigation | Mathematical analysis | Simulation |
| Model verification | Easy | Difficult |
| Fitting data | Weak | Strong |
| Prediction power | Strong | Weak |
| Purpose | Understanding | Control |

## 3.3 EXAMPLES OF MATHEMATICAL MODELS

We now want to introduce some general approaches, which can also be found in the modeling of biofilm. Our choice represents no complete or systematic classification of mathematical models. We only want to illustrate some of the prevailing philosophies in modeling. Often, different strategies are correlated to different scientific disciplines.

### 3.3.1 Physical approach

An important paradigm in physics is that to each question on the object there correspond certain scales of space and time relevant for the model. Depending on these scales, there exist certain first principles (conservation laws, etc.) yielding reasonable models with reduced complexity.

In the modeling of biofilms, relevant space scales might be represented by continuum mechanics in the form of Navier–Stokes equations or by models related to smaller constituents like bacteria, polysaccharides, and so forth.

The timescale aspect is reflected in the models' static or dynamic properties. Static models are not time dependent. They correspond to small or to large timescales in the real world. Small timescales mean that we consider local equilibria, which are not stable but rather are meta-stable. One could, for example, regard a fully developed biofilm as static if only a short time interval of the growth process is considered. Large timescales correspond to asymptotic equilibria. Such an equilibrium, which is reached after a long time, may represent a species successful in the competition for limited supplies of nutrients and space.

Dynamic models are typically correlated to a larger range on the timescale. As an example, one could regard the growth of fresh biofilm in bio-reactors. This time-dependent process deserves a dynamic model.

If there is a dynamic model for a phenomenon, the corresponding static model for local or asymptotic equilibria is more easily developed.

### 3.3.2 Introducing randomness

One way to model a complex part of a system is to replace this part by a random variable. Then, the information is coded in the distributions of the variables. The corresponding model parameters we are interested in describe collective aspects or mean values of the system.

As an example, let us consider the division of a cell. It is obvious that the complex biochemical processes involved cannot be modeled for every cell in a deterministic way. One therefore models the division as a random process. In this case, the most important parameter is the growth rate of the population, which can then be estimated easily.

To keep track of the randomness, further parameters like scale of fluctuations or correlation of noise appear in stochastic models. During parameter determination and verification of the model, statistical methods are involved which have to be suitable for the specific model. Statistics also provides a methodology for deciding between different models.

### 3.3.3 Models from system theory

Models coming from system theory are built from a few constituents and allow for a complex structure of interactions. Typical examples are predator–prey models or models for rivaling species. In mathematical terms, one encounters dynamic systems. The idea is to represent the interaction of a few constituents with a set of coupled nonlinear differential equations. Examples are Lotka–Volterra systems. An application of dynamic systems to develop scenarios explaining the breakdown of a bio-reactor is demonstrated by Hösel and Walcher (2000). Another example, mathematical modeling in epidemiology, is often quite simple and covers only a few aspects of a complex phenomenon; yet it proved useful, for example in the design of vaccination strategies (see, for example, Kermack and McKendrick 1927; Bailey 1975).

### 3.3.4 Compound models

Compound models are hybrids of models belonging to different classes. For example, there could be parts of the model that are stochastic and others that are purely deterministic. Compound models are very flexible and may cover a variety of phenomena. But, one should be aware that the interaction of the different compounds may increase the difficulty of mathematical analysis and simulation. Recently, compound models for biofilms have been published by Picioreanu *et al.* (1998, 1999).

## 3.4 ESTABLISHED BIOFILM MODELS

Most of the biofilm models used so far presume a continuum concept. In this case, the components of the biofilm, like microbial species, are not characterized by exact location, size or shape, but by quantities resulting from averages over small biofilm volume elements. For example, the density of the biofilm at some locus $p$ is defined as mass of biofilm in a volume element with center $p$ divided by the volume of the element.

Local analysis of the resulting averages gives the distribution and the time dependence of the biofilm components. Relevant physical and biochemical processes are here modeled with differential equations.

The one-dimensional biofilm models assume that the biofilm is homogeneous in layers parallel to the substratum. Thus an additional averaging over a 'representative' area $A$ of such a layer results in spatial dependencies from $z$ only, $z$ being the distance to the substratum surface. Wanner and Gujer (1986) presented a rather general concept for modeling mixed-culture biofilms with a one-dimensional approach. Since then, their model and its extensions (Wanner and Reichert 1996; Reichert and Wanner 1997) became the standard model for many real biofilms (see, for example, Horn and Hempel 1997). Of course, these models are suitable only for questions that can be addressed by neglecting the complex spatial structure of the biofilm. Good results, for example, are provided for biofilms in fast flow conditions (Horn and Hempel 1997).

In the 1990s, new techniques like confocal laser scanning microscopy (CLSM) were developed, which allow non-destructive *in situ* investigations of real biofilms. The findings are that biofilm structures vary from simple homogeneous layers to complex mushroom- or tulip-like structures. The continuum model does not seem adequate to describe biofilms, which are heterogeneous with channels and voids in them. To model these types of biofilm, one needs to characterize the complex morphology. There have been some attempts to quantify the morphology (see, for example, Zhang and Bishop 1994; Hermanowicz *et al.* 1995) and more recent approaches are discussed in Chapters 6 and 7.

In the classical biofilm models, one has to impose the structure of the biofilm before solving differential equations. To avoid this, several research groups recently proposed to model the growth of biofilms with a cellular automaton approach. These models originated with Conway's Game of Life (see, for example, Berlekamp *et al.* 1992). The entities of such an automaton are grid units, which may represent the basic units of a biofilm model. These grid units have states like 'occupied' or 'empty' which are updated in discrete iterative steps. The change of a unit's state is controlled by rules, which depend on local conditions like the state of the unit and its neighbors. In biofilm modeling these rules especially involve the nutrient supply of the biofilm units. The possibility of interpreting the growth of such model biofilms as self-organizing owing to local 'rules of behavior' seems to make these concepts attractive to biologists and ecologists.

Most recent developments are hybrid models combining cellular automaton approaches with differential equations for the flow. The discrete-differential model of Picioreanu *et al.* (1998, 1999) is an example. A more detailed survey of these and other models can be found in Hösel and Walcher (1999).

## 3.5 MODELS FROM MATHEMATICAL BIOLOGY

There are several classical and recent works in mathematical biology that are highly relevant for basic questions associated with the modeling and description of biofilms, as well as other complex systems in biology, ecology and medicine. These models are small models, in the sense that they are only concerned with a few of the many mechanisms and aspects that have to be considered in the 'full picture'.

As an early example with some relevance to biofilm modeling we mention the classical paper by Turing (1952), in which he first described and analyzed a mechanism that produces spatially heterogeneous patterns from homogeneous starting conditions. Studying such mechanisms may be helpful for an understanding of the heterogeneity of biofilms.

There are several recent works that are obviously important for biofilm modeling, such as the work by Ballyk and Smith (1999), who investigated (single-species) microbial growth with wall attachment in a plug flow reactor. The principal application the authors have in mind is concerned with the human large intestine, but clearly the model is basically applicable to biofilms in water pipes, etc. The model equations are analyzed in a rigorous manner, including existence, uniqueness and boundedness results. The steady states of the system and their stability properties are determined. (One of these steady states describes complete washout of microbes, whereas microbes persist in the other.) Interestingly, although the stability analysis is basically an infinite-dimensional matter, there are only a few parameters (which can be linked to physically or biologically meaningful quantities) that actually decide stability.

Currently, small mathematical models like the above are certainly important for the qualitative understanding of biological phenomena. Recent trends also seem to point towards the use of such models in quantitative investigations.

## 3.6 IDEAS FOR STOCHASTIC MODELS

Another way to reduce complexity is, as mentioned above, the introduction of randomness. All cellular automata models presented up to now neglect the fine structure of the biofilm for complexity reasons. But, in reality, there are different bacterial populations inside the biofilm forming different geometric patterns (balls, filaments). Therefore, the *fine structure* of a biofilm may be modeled by random geometric objects. In terms of mathematics, we enter the field of stochastic geometry. Even if we neglect the geometry and consider the centers of the different objects, we obtain a (random) point configuration, a point process. The simplest distributions of such point configurations, corresponding to complete randomness, are (homogeneous) Poisson processes. There is only one parameter, the intensity $\lambda$, i.e., the expected number of points in a unit area (volume), which is easy to estimate. Of course, such a model does not show the desired heterogeneity like real biofilm images (see Figure 3.1), but there are several feasible modifications, as follows.

- Cluster processes. The Poisson model is only assumed to generate the centers of clusters. Any center point generates a cluster, a set of points. The whole configuration (which is observed) is constituted by the superposition of all generated clusters. Such models definitely show much more heterogeneity than the Poisson process model, but the corresponding statistics are much more of an issue here (see Figure 2.1).
- Gibbsian-type probability measures. Here the key ingredient is an energy function $H(x)$ modeling the 'interaction' of the objects leading to the Gibbs measure as equilibrium distribution. The probability of a realization $x$ is proportional to $e^{H(x)}$, so large $H(x)$ means repulsion and small $H(x)$ means attraction. Even small models of this type can describe such attraction–repulsion relations well (e.g., Särkkä and Högmander 1998). A connection to existing (dynamic) models is established by the fact that the Gibbs measures are in equilibrium under certain birth- and death-processes, which are very similar to the stochastic cellular automata models (Preston 1975).
- Marked point processes. Each point carries a mark, which may describe several properties of an assumed object at that point. This may be population type, activity, shape of the objects, etc. Such an approach is quite flexible in connection with the above-mentioned possibilities, e.g. for multi-species problems. If the mark corresponds to the shape of the object, one arrives at germ models in stochastic geometry (Stoyan *et al.* 1995).

**Figure 3.1.** Simulations of a Poisson process (left) and two cluster processes (middle, right) with increasing tendency to build agglomerations.

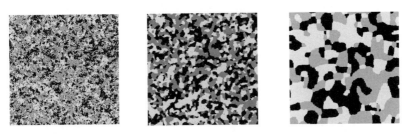

**Figure 3.2.** Simulations from Potts models (from left to right increasing $\beta$).

A different approach, much nearer to the data provided by CLSM, would be to use statistical models for, say, binary values on a rectangular two- or three-dimensional grid. The simplest model with apparent heterogeneity is provided by the Ising model (well-known to physicists), which is again of Gibbsian type. Now the energy function is the number of neighboring pixels, with different values multiplied by a parameter $\beta$. Thereby, small $\beta$ forces large clusters, and there are relations to other geometric parameters (see Georgii and Haeggstroem 1996 and Chayes et al. 1995). Concerning the statistics, there are a lot of results (see, for example, Frigessi and Piccioni 1990 and Guyon 1995). Thus, this seems a good opportunity to code the geometry of a CLSM image into a few parameters with good interpretation (provided such a model is verified). Again, it is possible to enrich the model by considering more than two values, which leads to so-called Potts models. Different values could then represent different species. Figure 3.2 shows examples of such model biofilms.

All these possibilities may contribute to modeling the state of a piece of biofilm, and there are many well-established statistical procedures to verify a certain model on a data set. As far as the dynamics of a biofilm culture is concerned, the above-mentioned birth and death processes, perhaps in a refined form, may be suitable. At the moment, two- or three-dimensional screenings of a fixed biofilm culture over a certain amount of time do not seem to be available. Therefore, at present, we do not regard it as feasible to develop detailed models for the dynamics of biofilm structure, as there is insufficient possibility to verify the models.

## 3.7 CONCLUDING REMARKS

There is a large spectrum of real biofilms as well as possible (data-driven) biofilm models. Which models one should favor depends on the type of biofilm and the questions one wants to address. Scientists from different fields should investigate such interesting questions in an interdisciplinary approach. Therefore, we postulate that there should be a lively ongoing discussion between mathematicians and non-mathematicians regarding the modeling of biofilms.

The different modeling approaches lead to different mathematical techniques. Some of the mathematical concepts not frequently used in biofilm modeling may play an important role in future developments. Thus, dynamic systems can describe the competition of species, and stochastic geometry could be used to describe the complex heterogeneous structure of biofilms. In general, it seems reasonable to study various types of simple mathematical model with few parameters to improve our understanding of partial aspects of the complex system biofilm.

## 3.8 REFERENCES

Bailey, N.T.J. (1975) *The Mathematical Theory of Infectious Diseases and its Applications*, 2nd edn. Charles Griffin, London and High Wycombe.
Ballyk, M. and Smith, H. (1999) A model of microbial growth in a plug flow reactor with wall attachment. *Mathl Biosci.* **158**, 95–126.
Berlekamp, E.R., Conway, J.H. and Guy, R.K. (1992) *Winning Ways for Your Mathematical Plays*. Academic Press, New York.
Charaklis, W.G. and Marshall, K.C. (1990) *Biofilms*. Wiley-Interscience, New York.
Chayes, J.T., Chayes, L. and Kotecky, R. (1995) The analysis of the Widom–Rowlinson model by stochastic geometric methods. *Commun. Math. Phys.* **172**(3), 551–569.
Frigessi, A. and Piccioni, M. (1990) Parameter estimation for two-dimensional Ising fields corrupted by noise. *Stochastic Processes Appl.* **34**(2), 297–311.
Georgii, H.-O. and Haeggstroem, O. (1996) Phase transition in continuum Potts models. *Commun. Math. Phys.* **181**(2), 507–528.
Guyon, X. (1995) *Random Fields on a Network. Modeling, Statistics, and Applications*. Springer, New York.
Hermanowicz, S.W., Schindler, U. and Wilderer, P.A. (1995) Anisotropic morphology and fractal dimension of biofilms. *Wat. Res.* **30**, 753–755.

Horn, H. and Hempel, D.C. (1997) Substrate utilization and mass transfer in an autotrophic biofilm system: Experimental results and numerical simulation. *Biotechnol. Bioengng* **53**(4), 363–371.

Hösel, V. and Walcher, S. (2000) A generalization of the single-species chemostat. Preprint. Institute of Biomathematics and Biometry, GSF Neuherberg, Germany

Hösel, V. and Walcher, S. (1999) On the mathematical modeling of biofilms. GSF Report 10/99 Neuherberg, Germany.

Kermack, W..O. and McKendrick, A.G. (1927) A contribution to the mathematical theory of epidemics. *Proc. R. Soc. Lond.* A **115**, 700–721

Picioreanu, C., van Loosdrecht, M.C.M. and Heijnen, J.J. (1998) Mathematical modeling of biofilm structure with a hybrid differential discrete cellular automaton approach. *Biotechnol. Bioengng* **58**(1), 101–116.

Picioreanu, C., van Loosdrecht, M.C.M. and Heijnen, J.J. (1999) Discrete-differential modeling of biofilm structure. *Wat. Sci. Tech.* **39**(7), 115–122.

Preston, C. (1975) Spatial birth-and-death processes. *Bull. Int. Stat. Inst.* **46**(2), 371–391.

Reichert, P. and Wanner, O. (1997) Movement of solids in biofilms: Significance of liquid phase transport. *Wat. Sci. Tech.* **36**(1), 321–328.

Särkkä, A. and Högmander, H. (1999) Multitype spatial point patterns with hierarchical interactions. *Biometrics* **55**, 1051–1058.

Stoyan, D., Kendall, W.S. and Mecke, J. (1995) *Stochastic Geometry and its Applications*. Wiley, Chichester.

Turing, A.M. (1952) The chemical basis of morphogenesis. *Phil. Trans. R. Soc. Lond.* B **237**, 37–72.

Wanner, O. and Reichert, P. (1996) Mathematical modeling of mixed-culture biofilm. *Biotechnol. Bioengng* **49**, 172–184.

Wanner, O. and Gujer, W. (1986) A multi-species biofilm model. *Biotechnol. Bioengng* **28**, 314–328.

Zhang, T.C. and Bishop, P.L. (1994) Density, porosity and pore structure of biofilms. *Wat. Res.* **28**, 2267–2277.

# 4

# Beyond models: requirements and chances of computational biofilms

*Hans-Joachim Bungartz and Miriam Mehl*

## 4.1 INTRODUCTION: SIMULATING BIOFILM SYSTEMS

The breathtaking progress of both computer hardware and software to exploit its potential has led to an increasing attractiveness of computer-based simulations (numerical as well as non-numerical ones); that is, of a *computational* approach besides theoretical analysis and experiments. This development has shown a tremendous impact on (mathematical) modeling, because now the derivation of models no longer seems to be of interest for theorists only, but can be valuable for practitioners, too. Hence, in many fields of science and engineering, modeling and simulation have entered the center of research efforts, and it can be considered as typical that one of the chapters of this book is dedicated to simulation.

© 2003 IWA Publishing. *Biofilms in wastewater treatment.* Edited by S. Wuertz, P.L. Bishop and P.A. Wilderer. ISBN: 1 84339 007 8.

# Beyond models: requirements and chances of computational biofilms

Complementary to the other contributions to this book, which focus more on modeling aspects of biofilms or on concrete results obtained by simulations, we emphasize the computational part of the story; that is, to show that numerical simulation of biofilms not only means to look for better models and to apply existing software, but also to enhance the algorithms and methods currently used and to design and implement new ones. Consequently, our contribution addresses this section's scientific question about the most important factors determining biofilm architecture and transport processes in an indirect way: we point out what has to be done from the point of view of numerical analysis and computer science so as to provide biofilm experts with efficient high-performance simulation tools that are well suited to run detailed studies resolving both time and space, and which represent the state-of-the-art of simulation technology. Thus, it is our aim to enable them to give a profound answer to the above question – which, in our opinion, is not possible without a detailed study and understanding of the processes on the microscale.

Although computational physics or computational fluid mechanics have become established fields of research, the term 'computational biofilms', chosen for the title of this contribution, still sounds slightly strange. In fact, discussions with biofilm researchers often reveal a significant skepticism about numerical simulations. Whereas the engineers among them do need real-time simulations for the operation of reactors or plants and, thus, only question the benefit of approaches going beyond PC-based one-dimensional computations, the reservations of biologists seem to be of a more principal nature. The first reason is certainly the larger distance between biology on the one hand and mathematics and computer science on the other, compared with physics or engineering, which entails the necessity to consult experts from outside the biofilm community if serious numerical simulations are to be tackled.

Furthermore, the physical and the biological complexity and heterogeneity of biofilm systems turn out to be important hindrances, because they are responsible for a permanent dissatisfaction of biologists with biofilm models. On the one hand, purely biological models based on the occurrence and competition of species, for example, can hardly reflect the variety of physical effects that can be observed in biofilms. On the other hand, however, the computational researcher who concentrates first on the more principal physical or physicochemical processes like flow, diffusive and advective transport, or reaction, sees himself or herself immediately confronted with some kind of prayer wheel uttering phrases like 'what about biology?' or 'do you have any idea what is really going on in a biofilm?' Although, generally, there will be neither a complete picture nor a perfect model, things must get started!

Hence, because even taking into account at least the most important effects with a full resolution of space and time is not possible for the foreseeable future, the benefits of extended computational (that is, supercomputing) efforts are doubted, and attention is often restricted to simple one-dimensional or steady-state models and simulations on the system level or macroscale. Finally, the omnipresent skepticism about the computational approach stems from the fact that biofilm models that have been derived in recent years are not simplifying condensations of a comprehensive understanding of the underlying morphology and processes to allow simulations, but serve, rather, just to improve the current and fragmentary knowledge of biofilms and are, therefore, unsatisfactory, because they reflect this incomplete understanding.

As a consequence, though spatial resolution found its way into biofilm *models* in the past decade, most biofilm *simulations* are still restricted to the one-dimensional world of programs like AQUASIM (Reichert 1994). Today, AQUASIM is widespread and accepted in the scientific community, and its underlying models are constantly extended and refined. Many questions, especially those of engineers concerning the stable operation of bioreactors, can be answered with the help of this program or comparable tools in a satisfying way. However, owing to the lack of a fully three-dimensional discretization of the considered domain, its scope is limited, and it does not allow one to dive into spatial details. On the other hand, novel analytical and investigative methods such as fluorescence *in situ* hybridization (FISH), confocal laser scanning microscopy (CLSM) or microelectrodes direct the scientific community's attention to exactly those details on the microscale that have been hidden so far, and they increase interest in corresponding three-dimensional numerical simulations that are far more challenging from the computational point of view.

Hence, if we are both willing and able to invest more on the computational battlefield, the scope of systematic numerical simulations is much wider than the one of programs of the above-mentioned type. For these investments, there is a series of challenging tasks that involve the expertise of applied mathematicians, computer scientists, and biofilm experts. The mathematical model is the starting point of each computational approach, but it is just one ingredient on the way to useful numerical simulations. It is important to note that, to exploit the full computational potential, improved models are essential, but not sufficient. Rather, there is an obvious and crucial need for progress in each of the following steps.

## Mathematical modeling

The first step of each simulation has to be the derivation or selection of a suitable mathematical model of the processes to be studied. For biofilms, phenomenological or heuristic models derived by biofilm expert groups like the IWA Activated Sludge Models are dominant, whereas strictly formal approaches from applied mathematics leading to more compact and manageable models with a better theoretical foundation still play only a minor part. There is no one correct or best model, but the appropriate choice usually depends on the level of observation and, thus, on the accuracy to be obtained. Apart from the classical mathematical requirements for such a model to be as simple as possible to provide the desired precision and to be (uniquely) solvable, the computational aspect, additionally, calls for the model to be well suited for a later numerical treatment.

## Numerical treatment

After the model has been established, which is, in fact, quite a theoretical step and not at all computational, the actual numerical simulation can start. For that, both the respective algebraic or differential equations and the underlying geometry have to be discretized, that is, restricted to a finite number of relations involving a finite number of unknowns. Furthermore, for the computer-aided approximate solution of the resulting systems of discretized equations, efficient numerical methods have to be applied or even developed.

## Implementation

Next, the obtained numerical algorithms have to be implemented on a computer. Owing to the complexity of biofilm systems, this will be typically a supercomputer like a parallel or a vector-parallel computer, or a network of computers. Against the background of scarce resources concerning both storage and computing time, the choice of the programming language and of data structures, as well as parallelization and vectorization of the code, are the crucial issues here. Finally, along with the impressive size of modern simulation codes, a systematic program design and other aspects of modern software engineering enter the game, too.

## Embedding

Because numerical simulations are based on a broad variety of input data, and because the results and insight obtained from computations allow a direct feedback to the experts of the respective application, we need efficient interfaces for embedding the computations. For example, automated techniques to transform measured data into data representations suitable for the available simulation tools or to produce visualizations based directly upon the calculated values are of crucial importance.

### Visualization

To interpret the data resulting from a numerical simulation (as well as measured data), sophisticated methods for the visual representation of huge sets of data, (i.e., for their visualization) are necessary. This holds especially for a complete three-dimensional resolution of space, where straightforward techniques, like drawing an arrow at each grid point to visualize a flow field, fail.

### Validation

Last but not least, validation of the computed results is a good time for a review by the experts of the respective field of application. Because computational people often tend to be fond of any kind of computed results, the expertise and measurements of their experimental colleagues are imperative, if we want to do some calibration or parameter fitting of our model and code, and if we thus intend to 'separate the wheat from the chaff'.

In the remainder of the chapter, we discuss the requirements of the above steps of a numerical simulation against the special background of biofilm systems, and we present our approach for a comprehensive and flexible simulation of flow and transport in and growth of biofilms on a microscale, which is in a small section of some defined biofilm that can be observed by modern microscopic techniques like CLSM.

## 4.2 MATHEMATICAL MODELS

As there are many biofilm specialists among the authors of this book, and as the first three chapters focus on aspects of biofilm modeling, we omit the otherwise mandatory paragraph on the essence and characteristics of biofilms and limit ourselves to some characteristics that are essential for numerical simulation:

(1) Concerning the underlying geometry, there is no typical architecture of biofilm systems, but a broad variety of appearances (Zhang and Bishop 1994; Hermanowicz *et al.* 1995; Bishop 1997; Wimpenny and Colasanti 1997; Eberl *et al.* 2000). However, the geometric leitmotif of the actual biofilm (that is, the layer between bulk liquid and substratum) is complexity and heterogeneity.

(2) Concerning dimensionality, it has to be stated that biofilm systems, in general, are essentially three-dimensional. Nevertheless, some types can be described as stacks of rather homogeneous porous layers perpendicular to the surface of the substratum (Charaklis and Marshall 1990; Wanner and Reichert 1996; Horn and Hempel 1997a; Reichert and Wanner 1997).

(3) There are a multitude of factors influencing the growth of the biomass, the relative importance of which is far from being undisputed. Among these factors are physical or physicochemical ones such as the characteristics of the surrounding fluid and flow, the diffusive and convective transport of dissolved substances as well as the resulting nutrient level, chemical reactions, shear forces, and an eventual detachment (van Loosdrecht et al. 1997; Wimpenny and Colasanti 1997; Hermanowicz 1999) and biological ones including the types, properties and numbers of microorganisms and their activities (Horn and Hempel 1997b; Noguera et al. 1999). Obviously, at first glance, this distinction looks rather arbitrary, but the somewhat strange rivalry between physics and biology can be found in many discussions on the subject.

Against this background, several decisive questions have to be answered before or during the design of an appropriate biofilm model. The first issue is dimensionality of space. Shall the model reflect space and be able to represent three-dimensional effects (heterogeneity, anisotropy or spatial phenomena such as turbulence), or shall it be based on a reduced dimensionality (2-D via symmetry of rotation in a cylindrical reactor, or 1-D as in the widespread standard layer models)? Another topic of interest is the treatment of time. Do we want to model all processes in a fully transient way, or do we look for a mixed model with time-dependent processes, but with stationary boundary conditions or a fixed geometry, for example?

Moreover, it is important to think about the scale on which our simulations shall take place, because the available computing resources impose certain upper bounds for the resolution of both space and time. For example, if these bounds are $1000 \times 1000 \times 1000$ grid points and 1000 time steps (which means $10^{12}$ different discrete values per physical variable like pressure), it is obvious that a simulation of a biofilm reactor that is 1 m high during a period of one day cannot be expected to explicitly take into account processes that need less time than a second to run or geometric details that are smaller than 1 $mm^3$. Especially if we take a macroscopic point of view and work with a rather coarse resolution, the question arises how the influence of important microscopic effects, nevertheless, can be taken into account on the coarse scale, for example by averaging techniques in the sense of representative elementary volumes (Korber et al. 1992) or by homogenization methods (Tartar 1980), that is, limit processes. These considerations are of crucial importance, because biofilm systems show extreme multiscale characteristics in both space and time. Concerning space, geometric details on the microscopic level are in the range of several micrometers, whereas whole reactors can reach lengths of several meters. Concerning time, differences in the time scales of up to eight orders of magnitude, or even more, can occur (Picioreanu *et al.* 1999).

Finally, the question of relevance of the different influences and effects has to be dealt with, to allow a reasonable selection of what shall be integrated into the model. As the brainstorming at the IAWQ Biofilm Workshop in Garching, 1998 (where the idea for writing this book arose) revealed once more, it is a very difficult job to get any kind of agreement on that point. Therefore, because of our objective to obtain simulation software of some durability that does not become useless when some experts' opinions change, a hierarchical approach is advantageous, as it allows us to start with only a rough model and to extend it when new knowledge, better numerical algorithms, or larger computing facilities are available.

How do the existing models answer these questions? The attempt to get some rough classification of biofilm models leads us to at least three different approaches:

(1) The first class of models is based upon algebraic and differential equations expressing relationships between physical quantities. In this sense, they are *equation-based* or *physical* models. Depending on the amount of computational work to be invested, and depending on the objective the model shall serve, they range from simple one-dimensional layer models (Charaklis and Marshall 1990; Wanner and Reichert 1996; Horn and Hempel 1997a; Reichert and Wanner 1997) to three-dimensional spatial models. The latter involve Navier–Stokes equations, convection–diffusion-reaction equations, and so forth (Griebel *et al.* 1997; Picioreanu *et al.* 1998) representing the basic laws of continuum mechanics. Even in the higher-dimensional case, the resolution of space is not necessarily very fine, because averaging techniques may allow one to avoid studying things in spatial detail. Concerning the timescale, such models are primarily suited for describing flow, transport, and reaction in a stationary geometry. The (usually much slower) processes of biomass growth, movement, and detachment, however, are hard to represent on this level.

(2) To overcome this drawback has been the main reason for the development of the second class of models, which try to reflect the metabolic activity of the different species present and to take into account these processes by deriving rules for the behavior of the biomass as a whole. Such *rule-based* or *biological* models have lost contact with the details of the physical level, but they allow a more straightforward access to biological paradigms like

competition and, hence, to biomass growth and movement. Therefore, they seem sometimes to be more attractive from a phenomenological point of view. The most important representatives of this approach are the models that are based on cellular automata (see Wimpenny and Colasanti 1997; Hermanowicz 1999), but even fractal models (Hermanowicz et al. 1995, Hermanowicz et al. 1996; Lewandowski et al. 1999) have been developed.

(3) Finally, there are hybrid models that try to bring together the physical and the biological world by combining continuum mechanics and cellular automata (Picioreanu et al. 1998, 1999; Noguera et al. 1999; Eberl et al. 2000). The chance of profiting from the advantages of both approaches is very promising. However, the problems resulting from the huge differences in the occurring scales are not yet solved in a satisfactory way.

Obviously, the definition or choice of an appropriate biofilm model depends on which phenomena shall be described and studied with its help. For our purpose of a microscopic view of fluid flow, transport, and reactions in a biofilm, we start with studying continuum mechanics by a three-dimensional resolution of space, which is also done by other authors (Picioreanu et al. 1998; 1999; Eberl et al. 2000) in the continuous part of their hybrid model.

The fluid flow is given by the three-dimensional velocity $\mathbf{u} = (u_1, u_2, u_3)$, which is determined by the incompressible Navier–Stokes equations (Griebel et al. 1997; Eberl et al. 2000):

$$\frac{\partial u_i}{\partial t} = \nu \sum_{j=1}^{3} \frac{\partial^2 u_i}{\partial x_j^2} - \sum_{j=1}^{3} u_j \frac{\partial u_i}{\partial x_j} - \frac{1}{\rho}\frac{\partial p}{\partial x_i} + g_i, \qquad i=1,2,3, \tag{4.1}$$

$$0 = \sum_{j=1}^{3} \frac{\partial u_j}{\partial x_j}, \tag{4.2}$$

or, in a more compact notation,

$$\mathbf{u}_t = \nu \Delta \mathbf{u} - (\mathbf{u} \cdot \mathrm{grad})\mathbf{u} - \frac{1}{\rho}\mathrm{grad}\ p + \mathbf{g}, \qquad \mathrm{div}\ \mathbf{u} = 0. \tag{4.3}$$

Here, $\mathbf{x} = (x_1, x_2, x_3)$ is the vector of the space coordinates, $\mathbf{g}$ denotes gravity, $\rho$ and $\nu$ are density and kinematic viscosity, respectively, and $p$ is the pressure in the fluid.

The concentration $c$ of some substrate in the fluid and in the biofilm is described by a convection–diffusion equation in the fluid domain $\Omega_f$, by a diffusion-reaction equation in the biofilm $\Omega_b$ itself, and by Fick's Law (according to Eberl et al. 2000) describing sorption at the interface $\Gamma$ of $\Omega_f$ and $\Omega_b$:

$$\frac{\partial c}{\partial t} = D_f \sum_{j=1}^{3} \frac{\partial^2 c}{\partial x_j^2} - \sum_{j=1}^{3} u_j \frac{\partial c}{\partial x_j} \quad \text{in } \Omega_f, \tag{4.4}$$

$$\frac{\partial c}{\partial t} = D_b \sum_{j=1}^{3} \frac{\partial^2 c}{\partial x_j^2} + r(c) \quad \text{in } \Omega_b, \tag{4.5}$$

$$D_f \left. \frac{\partial c}{\partial \mathbf{n}} \right|_{\Omega_f} = D_b \left. \frac{\partial c}{\partial \mathbf{n}} \right|_{\Omega_b} \quad \text{at } \Gamma. \tag{4.6}$$

Again, a more compact notation of Equations 4.4 and 4.5 reads

$$c_t = D_f \Delta c - \mathbf{u} \cdot \text{grad } c \text{ in } \Omega_f, \quad c_t = D_b \Delta c + r(c) \text{ in } \Omega_b. \tag{4.7}$$

In Equation 4.6 $(\partial c)/(\partial \mathbf{n})|_{\Omega_f}$ is the first normal derivative of the concentration with respect to the interface $\Gamma$ at the fluid side of $\Gamma$, and $(\partial c)/(\partial \mathbf{n})|_{\Omega_b}$ is defined in an analogous way as the first normal derivative at the biofilm side of $\Gamma$. $D_f$ and $D_b$ are the diffusion coefficients in the fluid and in the biofilm, respectively, and $r$ denotes the reaction term within the biofilm. Equation 4.6 causes the normal derivative of $c$ to be discontinuous at the interface $\Gamma$ if $D_f \neq D_b$. Equations 4.4–4.6 only provide a frame where the various parameters like $D_f$, $D_b$, or $r$ need some experimental input or modeling, themselves. However, as already mentioned, the focus of this contribution is put on the computational part, and we, hence, restrict ourselves to a qualitative modeling step and to simulations with somewhat artificial parameters and quantities. Furthermore, note that the above proceeding can be easily generalized to $n > 1$ substrates by just introducing quantities $c_j$, $r_j$, $D_{f,j}$, and $D_{b,j}$, $1 \leq j \leq n$, and by taking into account the above relations (Equations 4.4–4.6) for each substrate.

Concerning the dynamics of a biofilm, we need an equation that describes the relations between the concentration(s) $c_j$ and the developing biofilm, that is, growth or decay of the biomass $m$, respectively. Here, van Loosdrecht and his colleagues (Picioreanu et al. 1998, 1999; Eberl et al. 2000) use one ordinary differential equation

$$\frac{\partial m}{\partial t}(t,x) = r_m(m, c_1, \ldots, c_n), \tag{4.8}$$

and base the propagation of the biomass on a simple cellular automaton. As far as we know, the possibly more precise, but by far more complicated, way to base the propagation of biomass and, especially, the dynamics of the shape of the boundary between bulk fluid and biofilm on mechanisms like they are used for crystal growth or related phenomena, has not been considered so far.

Finally, note that, owing to the extremely different timescales occurring, it is impossible to represent both the fast (flow) and the slow (growth) processes within the same simulation without essential simplifications. Therefore, a quasi-stationary approach with the continuous part taking place in a fixed geometry and the discrete part providing an update in longer time intervals is used. At that point, there is an important need for additional modeling efforts. But now, let us turn to the first central aspect of this paper, the efficient numerical treatment of model equations.

## 4.3 NUMERICAL METHODS

Models use some mathematical formalism. For example, they provide or consist of algebraic or differential equations, which describe the relations of the quantities we are interested in and which are, thus, suited to determine them. In our scenario discussed in the previous section, we have the Navier–Stokes Equations 4.1 and 4.2 for the velocity and the pressure, and convection–diffusion-reaction equations 4.4, 4.5 and 4.6 for the substrate concentration(s). The problem is that, here as well as in general, the model equations cannot be solved analytically. Hence, we have to treat the model numerically, that is, we must approximate the different quantities with the help of a *finite* number of values in a *finite* number of suitable computational steps.

To this end, in a first step, the continuous model has to be transformed into a set of discrete equations. Afterwards, we need efficient numerical algorithms for the fast solution of these systems of discretized equations. The first part, the so-called *discretization*, can be subdivided into two major steps: the discretization of the underlying domain and the discretization of the equations.

### 4.3.1 Discretization of the domain

If we deal with technical objects, the domain of interest is, usually, available as a digital geometric model generated with the help of some standard tool from computer-aided design and based upon, for example, construction sketches. For 'natural' objects, as we are confronted with in the context of biofilms, the geometric information concerning the domain of interest is more diffuse. Here, the input is not a precise description, but, rather, consists of some photos, digital images, or image stacks stemming from microscopy or some other measurement technology.

Next, this input has to be transformed into a discrete representation that is adapted to the further numerical treatment, but that reflects, nevertheless, the original as exactly as possible. Most numerical methods need simple geometries or, at least, geometries that can be made up of simple basic elements like rectangles and triangles in 2-D or cuboids and tetrahedra in 3-D, respectively. In some cases, a mathematical mapping step, as illustrated in Figure 4.1, can help to get simple computational domains like squares or cubes.

**Figure 4.1.** Mapping of a 2-D domain onto a quadratic computational domain.

However, note that this mapping has to be applied to the model equations, too, which may make things even more difficult. Anyway, for porous media like biofilms, such a proceeding is inapplicable, because suitable mappings that can be handled in a reasonable way do not exist.

For the discretization step itself, that is, for the introduction of a discrete set of *grid points* and of a corresponding finite set of *computational cells* or *elements*, two principal strategies are available:

(1) The geometry can be approximated by an *unstructured* grid consisting of elements of a specific type (tetrahedra or cuboids), but of quite arbitrary shape (see Figure 4.2). Owing to the obvious lack of any systematic order or structure of the resulting grid, the position and relations of both grid points and cells have to be stored explicitly, which leads to a significant administrative overhead. However, the big advantage of unstructured grids is their flexibility with respect to even very complicated geometries.

(2) The geometry can be approximated by a *structured* grid consisting of elements of a specific type and of a specific shape (see Figure 4.3). Now, because of the inherent structure, the handling of the grid is very simple and can be done in very efficient ways, for example, by matrices or quad-/octrees (Samet 1984; Breitling *et al.* 1999). In particular, the treatment of varying or moving geometries is very straightforward. Note that the term 'structured' does not mean that the size of all cells is the same. A local refinement where a certain cube is subdivided into eight congruent subcubes does not destroy the structure of the grid.

**Figure 4.2.** Unstructured surface and volume grids around a wing/slat area (from http://ad-www.larc.nasa.gov/cab/People/ SP/Web/HTML/eet.html).

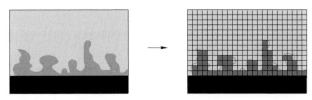

**Figure 4.3.** Approximation of a 2-D biofilm–fluid domain by a regular structured grid consisting of squares.

At first glance, the restrictions and the loss of information resulting from the discretization of the domain seem to be an important drawback to realistic biofilm simulations. However, one has to keep in mind that the input data themselves (for example, confocal laser scanning images) are restricted concerning their resolution and, thus, are error-prone, too. Consequently, if we manage to deal with this high resolution – which is our aim for biofilms in small spatial scales – we do not lose any information.

For our biofilm simulations, we choose a structured approach of orthogonal or Cartesian grids. First, this discretization is closer to the digitized images we use as input. Second, the handling of varying geometries is much simpler (just change the fluid–biofilm or fluid–biofilm–EPS (extra polymeric substances) characteristics of the single computational cells). Finally, structured grids are the appropriate choice for our finite difference discretization of the model equations to be discussed in the next subsection.

## 4.3.2 Discretization of the equations

In the second discretization step, all functions and operators occurring in the model equations have to be discretized in a way that is compatible with the discretization of the domain, that is, with the chosen grid. Again, there is a wide variety of possible strategies, for example finite differences (Jordan 1965), finite elements (Braess 2001; Bungartz 1998), finite volumes (Patankar 1980; Versteeg and Malalasekera 1995), or cellular (Berlekamp et al. 1982; Green 1990) and lattice automata (Babovsky 1998; Chen and Doolen 1998).

The *finite difference* method, typically used with structured regular grids, directly approximates a function $f$ via a discrete set of values assigned to the discrete grid points $x_i$, $1 \leq i \leq n$, in the underlying domain $\Omega$:

$$f(x) \xrightarrow{\text{discretization}} \{f_i : f_i \approx f(x_i), x_i \in \Omega, i = 1,\ldots,n\}. \tag{4.9}$$

The differential operators in the model equations are replaced by difference quotients, which leads to one discrete equation for each unknown value in each grid point. For example, the first derivative $f'(x)$ of a function $f(x)$ is replaced by a difference quotient like $[(f(x+h)-f(x))/h]$ or $[(f(x+h)-f(x-h))/2h]$. Applied to the discrete form of the function $f(x)$ from Equation 4.9, we obtain, for example, the approximation

$$\frac{f_{i+1} - f_i}{x_{i+1} - x_i} \approx f'(x_i) \tag{4.10}$$

in each grid point $x_i$.

In contrast to that, the *finite element* method approximates some function via finite linear combinations of basis functions associated with the discrete grid points, for example piecewise linear functions or piecewise polynomials. These basis functions may be arranged in a nodal or a hierarchical basis (see Figure 4.4), where the latter typically provides an improved convergence behavior. Instead of the model equations themselves, the finite element solution (which minimizes some suitable energy error function) only has to fulfil some weak form of the partial differential equations derived via Galerkin's approach, which finally leads to the desired set of discrete equations.

In the *finite volume* method, the original differential equations are integrated, using some piecewise constant or piecewise linear approximation of all functions in each cell (volume). Here, the governing principle is the conservation of physical quantities in each finite cell.

# Beyond models: requirements and chances of computational biofilms

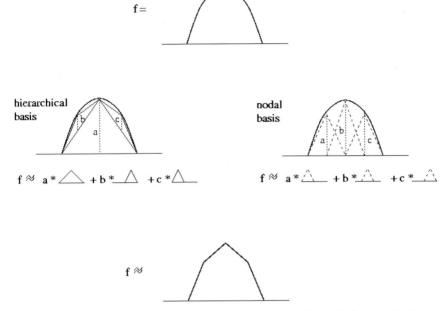

**Figure 4.4.** Approximation of a function with nodal and hierarchical piecewise linear basis functions in 1-D.

In contrast to the methods above, *cellular* and *lattice automata* do not directly tackle the model equations from continuum mechanics introduced in section 4.2, but approximate the kinetic behavior of the system via a finite set of local rules or by studying local probability distributions of (virtual) particles. Note that the most enhanced automata approaches, the so-called *Lattice Boltzmann Automata*, can be interpreted as an approximation to a continuum Boltzmann equation but have, nevertheless, close relations to the Navier–Stokes equations by the statistical moments of the regarded random variables. Their main advantage is the easy handling of complicated geometries. On the other hand, up to now, their numerical potential is quite limited.

For our microscale simulations of fluid flow and transport in biofilms on Cartesian grids (Bungartz *et al.* 2000), we use a finite difference discretization. Thus, as mentioned above, the approximations to continuous functions are only defined in the discrete grid points. Note that, for numerical reasons, we work with a so-called *staggered grid* (Griebel *et al.* 1997; see Figure 4.5) instead of a *cell-centered* or *collocated* discretization. That is, there are different sets of grid

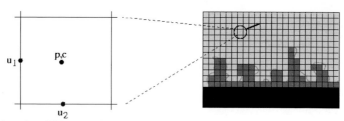

points for the different physical quantities (pressure, substrate concentration, and velocity components) that we want to compute.

**Figure 4.5.** Staggered grid for velocity components, pressure, and substrate concentration (2-D for reasons of simplicity): The velocity $u_1$ in $x_1$-direction (horizontal) is assigned to the midpoint of the left edge of the cell, the velocity $u_2$ in $x_2$-direction (vertical) to the midpoint of the lower edge, the pressure p and the substrate concentration c to the midpoint of the cell itself.

Based on this discretization of the occurring functions like velocity and pressure, we can now turn to the discretization of the differential operators in our model equations. Because this is a very important aspect from the point of view of numerical mathematics, but a rather technical matter, the reader who is not interested in numerical details may skip the following paragraphs.

We can distinguish two classes of differential operators in Equations 4.1, 4.2, 4.4 and 4.5: time derivatives and spatial derivatives. For the latter in Equation 4.1 – in particular for the convective (first derivative) terms $u_j\,(\partial u_i)/(\partial x_j)$ – the choice of suitable difference quotients is not straightforward. The obvious way would be to use central differences for the first derivatives, that is, to take into account the two respective neighbors of the actual grid point:

$$f'(x_m) \approx \frac{f_{m+1} - f_{m-1}}{x_{m+1} - x_{m-1}} = \frac{f_{m+1} - f_{m-1}}{2\delta x}. \tag{4.11}$$

Unfortunately, this leads to stability problems and oscillations in the solution of the discrete Navier-Stokes equations. For the convective terms of Equation 4.1, the Donor-Cell scheme (Griebel et al. 1997) is a well-known remedy. It needs staggered grids and stabilizes the discrete equations by consideration of the direction of the flow, but worsens the quality of operator approximation. Hence, we actually take a weighted mean of central differences and of the Donor-Cell scheme and, thus, can combine the better order of the first with the stabilizing effect of the latter scheme.

The discretization of all other spatial derivatives is straightforward. For example, for the Laplacian, the usual five-point stencil is used:

$$f''(x_m) \approx \frac{f_{m+1} - 2 \cdot f_m + f_{m-1}}{\delta x^2}. \qquad (4.12)$$

All first derivatives apart from the convective terms discussed above are discretized by central differences. The discretization of the time derivatives is a very important point for the performance of a numerical method. For an explicit discretization due to

$$\frac{\partial f}{\partial t} \approx \frac{f^{n+1} - f^n}{t^{n+1} - t^n} \qquad (4.13)$$

with $f^n \approx f(t^n)$, where all other terms in the equations are evaluated at the (old) time $t^n$, the function values at the (new) time $t^{n+1}$ can be computed directly from the old ones. At first glance, this is very cheap and comfortable. However, explicit methods usually require very small time steps and, thus, increase computing time significantly. On the other hand, a fully implicit time discretization with all spatial derivatives to be evaluated at $t^{n+1}$ would result in a whole system of equations – in our case even nonlinear ones – to be solved in each time step. Therefore, for the Navier–Stokes equations, we apply a mixture – explicit in the velocities and implicit with respect to the pressure:

$$u_i^{n+1} = \underbrace{u_i^n + \delta t \cdot \left( v \sum_{j=1}^{3} \left[ \frac{\partial^2 u_i}{\partial x_j^2} \right]^n - \sum_{j=1}^{3} u_j^n \left[ \frac{\partial u_i}{\partial x_j} \right]^n + g_i \right)}_{=:F_i^n} - \delta t \cdot \left[ \frac{\partial p}{\partial x_i} \right]^{n+1}, \qquad (4.14)$$

$$0 = \sum_{j=1}^{3} \left[ \frac{\partial u_j}{\partial x_j} \right]^{n+1}, \quad i = 1,2,3 . \qquad (4.15)$$

where $[EX]^n$ stands for the discretization in space of the respective expression $EX$ (as described above in Equations 4.11 and 4.12 for example). A short calculation transforms Equation 4.15 to

$$[\Delta p]^{n+1} = \frac{1}{\delta t} \sum_{j=1}^{3} \left[ \frac{\partial F_j}{\partial x_j} \right]^n. \qquad (4.16)$$

Thus, performing one time step ($t^n \to t^{n+1}$) requires, first, to compute the $F_i^n$ and to approximate the $\partial F_i^n / \partial x_i$ for $i = 1,2,3$, second, to solve the Poisson Equation 4.16 (see the following subsection) and, finally, to update the velocities according to Equation 4.14. Concerning the transport equations, the choice of the scheme for time discretization is not that critical. Explicit methods are the easiest ones, and the results we obtained up to now do not suggest that an implicit strategy may be worthwhile.

And what about biomass growth? From a numerical point of view, the coupling of a discretized version of Equation 4.8 to the discretized Navier–Stokes and transport equations is no problem. However, biomass growth, decay, or spreading are processes that run much slower than do flow and diffusive or convective transport. Actually, differences up to eight orders of magnitude can be observed (Picioreanu *et al.* 1999). Because, obviously, it is too time-consuming to perform $10^8$ time steps before any change in biofilm geometry occurs, we must simplify the treatment of time. In Eberl *et al.* (2000), steady-state solutions of the fast processes (flow, convection, diffusion, reaction) are coupled with the equations for the slow ones (biomass growth and spreading) just by solving them alternately. Though the method seems to provide reasonable results, it is not satisfactory from the modeling point of view. What could be done is the derivation of some scheme for a homogenization in time so as to average and integrate the influence of the fast processes in the computation of the slow ones. This requires some additional mathematical modeling efforts.

No matter which type of discretization we choose, the discretization step finally leads to finite systems of linear or linearized equations which have to be solved to determine the discrete approximation to our model equations. This solution process will be discussed in the next subsection.

### 4.3.3 Solving the discrete equations

As we use explicit or semi-implicit time discretization, we only have to deal with systems of linear equations. The development of the performance of numerical algorithms for the iterative solution of such systems during recent decades is more than impressive. Actually, the improvements that have been made on the algorithmic side are comparable to the increase in performance due to better hardware components, for example, processors (Figures 4.6 and 4.7). It is only by this interaction of algorithmic and technical progress that, today, we are able to tackle the numerical solution of the Navier–Stokes equations in 3-D complicated domains.

Direct solvers for systems of linear equations like Gaussian elimination cannot be used because of the size of the problems and resulting matrices. For example, for steady-state calculations at a spatial resolution of $512 \times 512 \times 60$ computational cells, we need more than 78 million unknowns to represent velocity, pressure, and just one substrate. Because the computational effort to solve a system of $N$ unknowns grows like $N^3$, we have to look for efficient *iterative* (that is, approximate) solvers for large, and typically sparse, systems of linear equations.

Beyond models: requirements and chances of computational biofilms 77

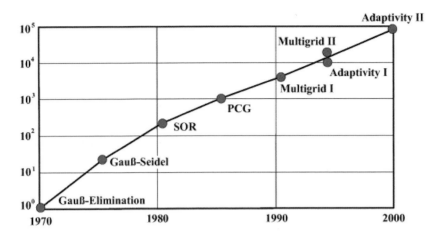

**Figure 4.6.** Increase in performance due to enhanced algorithms for the numerical solution of systems of linear equations (from standard direct solvers through basic iterative schemes to sophisticated adaptive multilevel methods).

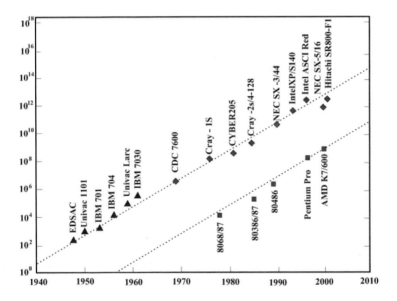

**Figure 4.7.** Increase in performance due to enhanced hardware components.

Here, once more, the high complexity of the biofilm's geometry is responsible for many difficulties. Standard multigrid or multilevel methods (Hackbusch 1985), which are the most sophisticated and fastest solvers available that involve only $O(N)$ arithmetic operations to solve a system of $N$ linear equations, are often inapplicable, because the coarse levels or grids cannot represent the (nevertheless important) geometric details of the fine levels. Hence, a special multigrid method (Mehl 2001; Bungartz et al. 2003) had to be developed to handle such complex geometries. It uses operator-dependent interpolation to transfer results from coarse to fine grids and Galerkin operators on the coarse levels. Thus, interpolation and coarse grid operators depend on the geometric structure of the actual grid cell. Computing times per time step were substantially reduced with this method (Mehl 2001; Bungartz et al. 2003) compared with the successive over-relaxation (SOR) method (see Griebel et al. 1997) used before in the code *Nast++* discussed in the next section.

## 4.4 IMPLEMENTATION

Simulation codes never are finished. They reflect the expertise of mathematicians, computer scientists, and experts from the respective field of application; they require years of designing and producing code; they are subject to permanent updates and enhancements owing to progress on the numerical or modeling side, to extensions of functionality, or to improved numerical modules to replace the present ones; they are expected to be run on computers that were not there when the work on the code started; and they are and will be used by people who often know close to nothing about the history or inner life of the code. Therefore, implementation is a permanent process, during which we are faced with both software and hardware aspects and with requirements from the user. Because the reader interested in biofilms is probably not that interested in these aspects, we do not go into detail here. However, we want to emphasize that the kind of job that has to be done is much more than a programmer in the usual sense does.

On the software side, especially when we speak of large program packages, the focus has shifted from the mere act of programming to *design* and *specification* of a code. Here, important aspects that result from the characteristics of simulation codes listed above are easy maintenance and flexibility. In our situation of biofilm systems, the latter holds especially against the background of a hardly satisfactory or finished modeling situation, which makes the fast implementation or integration of enhanced models very important. Hence, a modular and object-oriented design and structure are crucial to ensure flexibility. Furthermore, other aspects like good debugging features (to save time and to avoid unnoticed errors) are also very important.

On the hardware side, there is the whole range of computer architectures from laptops to supercomputers. Test runs during prototyping are typically made on PCs or workstations (usually based on simple script languages and on a 'quick-and-dirty' type programming style). The final version for the large or realistic examples with their resulting huge storage and runtime requirements (which are far beyond the potential of the above-mentioned machines), however, has to be written in a language that leads to fast machine codes, and it must be tailored to the respective (super-) computer architecture. Today's supercomputers are typically vector, vector-parallel, or parallel machines. Thus, the implementations of the various numerical algorithms, as well as the organization of the whole program package, have to be well suited for parallelization and vectorization, if we want to come close to the peak performance of those 'number crunchers'.

Finally, the user wishes for a platform-independent code, that is, a code that can easily be compiled and run on almost any kind of computer and under almost any kind of operating system. Hence, the code must fulfil standards that are supported by the common compilers. In addition, it might be helpful to provide a tool for the automatic selection of a compiler compatible with the respective computer architecture.

For our biofilm simulations, we use the program package *Nast++* (Brück 1998), which is an object-oriented advanced version and extension of the older flow simulation program *NaSt* (Griebel *et al.* 1997). In *Nast++*, which is written in C++, the discretized equations described in the previous section are implemented.

To ensure flexibility, not only inside the code but also at its interfaces, *Nast++* is actually a library. Thus, the various modules of *Nast++* can be accessed from other codes too, by inclusion of the respective library. For instance, we can use *Nast++* to compute flow and substrate concentrations in a biofilm system, and handle biofilm growth, detachment, etc. with the help of another program. As *Nast++* is strictly object-oriented, new facilities like improved solvers (on the numerical side) as well as modified reaction terms or transport equations for further substrates (on the model side) can be easily added. There are two versions of the *Nast++* library available: a first slow version with numerous error detecting tests (Brück 1998) for debugging purposes; and a second runtime-optimized version without these tests for the actual simulations.

To guarantee platform independence as far as possible, *Nast++* fulfils the standard ISO 14882 established in 1998, which determines the programming language C++ and the corresponding libraries. Some features of this standard are not yet supported by compilers in a sufficient way and, therefore, are not used in *Nast++* (Brück 1998).

Finally, there exists a parallel version of *Nast++* based upon the *MPI-II* standard (Pögl 1999). Up to now, the parallelized code has been run both on a cluster of workstations and PCs and on a Fujitsu VPP700 vector-parallel supercomputer. For flow computations on a *Sphingomonas* sp. monoculture biofilm, a parallel efficiency of 88% was reached on a PC-cluster with eight processors, for example (Mehl 2001).

## 4.5 EMBEDDING

To be a helpful tool for researchers in the biofilm field (or, at least, to become one in the future), numerical simulation has to be embedded into their daily work. To this end, it is not sufficient to develop computer programs for the respective simulation itself but, rather, simple and efficient interfaces between experimental and computational data are of great importance, too. On the one hand, this means that, before simulation starts, measured data have to be processed automatically to serve as an input for the computations, and, on the other hand, at the end, computational results must be processed to allow easy comparisons with experimental data and interpretations by biofilm experts.

Concerning the necessary *input* for any computation, we can distinguish three types of data: information about the concrete values of parameters in the model (for example, think of the viscosity of the actual fluid used, or think of the values of the diffusion coefficients $D_f$ and $D_b$ in the transport part of our model, cf. section 4.2), information about the computational domain or geometry of interest (what does the section of the biofilm to be investigated look like?), and information about initial or boundary values (which are typical nutrient levels at the boundary of our computational domain?). Although several parts of this information can be characterized and handed over using few numerical entities like the Reynolds number, the overwhelming part consists of spatially resolved data that must be generated with the help of statistical methods or obtained from suitable experimental studies.

Although the synthetic generation of input data like biofilm geometries based upon statistical or fractal processes is possible, the direct coupling of experimental and computational methods seems to be more promising. In our situation of microscale simulations, the input data can be obtained with the help of CLSM (Kuehn *et al.* 1998). Like other tomographic approaches, CLSM and the subsequent digital image analysis provide stacks of 2-D digital images, each of which represents a certain horizontal layer or slice of a defined vertical position in the respective 3-D domain. Depending on the focus of the actual experiment, the 3-D image stack may contain spatially resolved information about the section's morphology (simple gray-scale pictures distinguishing just between biomass and fluid, or more sophisticated formats detecting specific substances) or about certain physical quantities (the velocity field, for example).

With the help of an automated interface and based upon their brightness or color information, these 3-D image data are now transformed into the specific input format of our code *Nast++* to generate the discretized computational geometry and to provide the code with additional information contained in the image data. Note that, owing to the marker-and-cell (MAC) discretization of *Nast++* described in the previous sections, this representation of the geometry, again, consists of an array of three-dimensional data (see Plate 4.1). However, the resolution of the image data from CLSM (given as a stack of 2-D pixel arrays) is not necessarily the same as the resolution of the computational data (now available as a 3-D voxel array).

Although it is our aim to compute at the same level of resolution at which the measurements are done, owing to storage restrictions, this may be impossible without more sophisticated hierarchical and adaptive data structures like octrees (see, for example, Samet 1984; Breitling *et al.* 1999), which can handle a varying resolution of the computational geometry. Thus, at least a slight reduction of resolution may be inevitable. Plate 4.1, for instance, shows a single CLSM image of $512 \times 512$ pixels, whereas the resolution of the corresponding slice of the computational geometry is only of a $256 \times 256$ resolution. Taking into account that the respective stack of CLSM pictures consisted of 60 horizontal slices altogether, this still leads to a total of nearly forty million discrete values (per physical quantity to be computed!).

The discussion of the embedding and processing of the *output data* of a numerical simulation leads us to the next step of a complete simulation cycle, the visualization of simulation data.

## 4.6 VISUALIZATION: HOT AIR OR CATALYST?

Striving for higher and higher resolutions in their simulations, computational researchers have to cope with a problem following simulation: how can we analyze and interpret the results of a numerical simulation without having to look at millions of numbers? This question has to be addressed, especially if the desired result cannot be compressed into a single or a few characteristic numbers (like the $c_w$ value in aerodynamics, for example), but rather consists of the overall data set as a whole.

When studying fluid flow and transport in biofilm systems on a microscale we are, in fact, confronted with this problem, because we are interested in details of the flow field, in local minima or maxima of substrate concentration, or in gradients of concentration. Here, because of the enormous bandwidth of visual perception compared with the other human senses, graphical processing and representation or *visualization* of computed data can help and, consequently, has tremendously gained in importance in recent years. Note that the same holds for experimental methods, where even non-optical techniques like radar-based methods usually end up with colored images, because simple visualization modules are already integrated in modern microscopes or tomographs.

Although the visualization of 2-D data is quite straightforward, the spatial resolution, again, causes some additional difficulties. Obviously, it is hardly helpful to visualize the velocity field in a flow channel with small arrows (as is often done in two space dimensions) – we would not be able to 'see the wood for the trees'. Therefore, the usual approach consists of a reduction of dimensionality, where we restrict ourselves to the visualization of a lower dimensional section of the overall data. For instance, one might think of isosurfaces as shown in Plate 4.2, (ortho-) slices as given in Plate 4.3, streamlines (see Plate 4.4), or particle tracing.

Finally, apart from visualizing the values of certain quantities, color can be used to distinguish between different species or between the biomass itself and EPS in the same image (see Plate 4.5). In addition to the potential on the computational side, such possibilities are of interest in a merely experimental context, too, since they allow one to combine several CLSM image stacks (resulting from different experiments in the same scenario, but with different species marked) and to get an impression of the overall composition of the biofilm system considered.

Although the visualization techniques shown in Plates 4.2–4.5 are helpful and widely used for the interpretation or post-processing of numerical simulations, the drawback of their inherent low dimensionality is still present. Without the help of different colors, the geometric structure of the biofilm studied in Plate 4.2 is hard to detect, because we only see a lot of flocks, but cannot really recognize the 3-D structure. Consider the streamlines in Plate 4.4 that depict the flow field where they are visible, but which can hardly be interpreted in subdomains hidden by the biomass.

Hence, to profit from the human capability of spatial (that is, really 3-D) perception, we need stereoscopic visualizations. In this field, the perhaps best-known approach is to use anaglyphes that are based upon the superposition of red and green perspective views of the same scene, corresponding to the two eyes of the viewer. By comparison, polarization-based methods, which are well known from the IMAX cinemas, for example, and which provide a much higher

Pseudo-homogenous model
(after Atkinson *et al.* 1967)

Biofilm image (confocal laser scanning
microscope) (Stoodley, P., DeBeer, D.,
Lewandowski, Z. ASM Biofilm Collection;
www.asmusa.org)

Cellular automata model
(see Hermanowicz 1997)

**Plate 2.1** Similarities of biofilm morphologies.

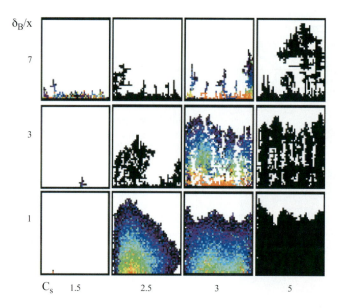

**Plate 2.2.** Examples of biofilm morphologies generated by a discrete model. $C_s$, dimensionless bulk nutrient concentration, $\delta_B/x$, dimensionless boundary layer thickness (adapted from Hermanowicz 1997).

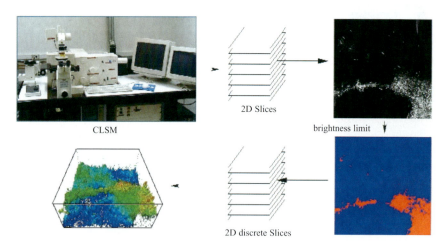

**Plate 4.1.** Automated generation of a discrete computational geometry from a real biofilm with the help of CLSM and image processing as used by Kuehn *et al.* (1998).

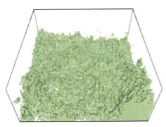 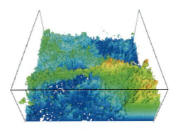

**Plate 4.2.** Visualization of a biofilm surface with an isosurface, that is, a surface representing all points of a fixed value of some function (on the right-hand side, the color indicates the height measured between substratum and biofilm surface; blue stands for low, red for high layers).

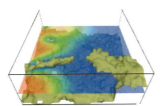 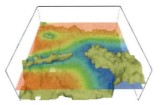

**Plate 4.3.** Visualization of substrate flow into a biofilm via an isosurface for the biofilm's morphology and an orthoslice colored according to substrate concentration (red means high, blue means low concentration).

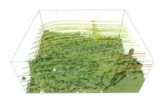

**Plate 4.4.** Visualization of the velocity field in a biofilm with an isosurface for the morphology and with streamlines for the fluid flow (here, the color indicates the absolute value of the velocity: blue means slow, red stands for fast flow).

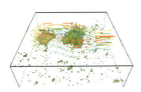 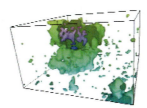

**Plate 4.6.** Fluid-obstacle geometry obtained from different CLSM images of a two-species biofilm with corresponding flow field (left) and computational biomass-EPS geometry (green: *Sphingomonas* sp. strain LB 126; pink: EPS).

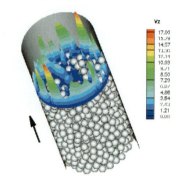

Distribution of Axial Component of Velocity on the Top-Plane in Packedbed

**Plate 5.1.** Channeling effect in the near wall region of a packed-bed reactor, where $V_z$ is the velocity component along the z-axis of the reactor.

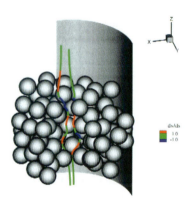

**Plate 5.2.** The elongation of fluid elements. $dv/ds$ is the elongation rate in $[1/s]$.

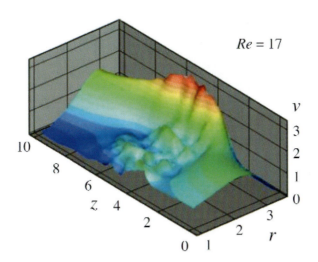

**Plate 5.3.** Qualitative carpet plot of the axial velocity distribution in a reactor. $v$ is the value of the velocity, $r$ is the radial coordinate from the reactor center to the wall, and $z$ is the axial coordinate of the reactor height. Red color indicates a high velocity, blue color a low velocity.

**Plate 5.4.** See over.

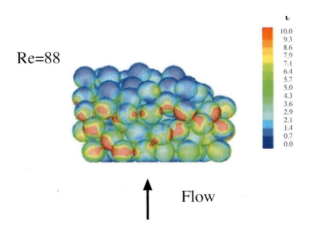

**Plate 5.5.** Tangential stress distribution on the biofilm carriers for a given Reynolds number, where $\tau$ is the shear stress.

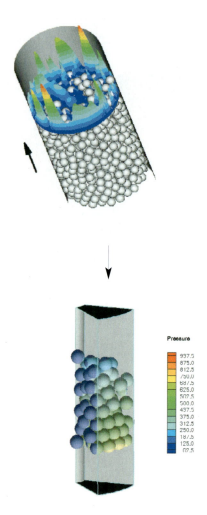

**Plate 5.4.** Reducing the number of particles by 'cutting' a full reactor into slices.

quality with respect to 3-D impression, colored representations, and independence of the viewer's position, are not very widespread in simulation and applications because of the high costs accompanying them and to the restriction to specially equipped projection rooms.

Therefore, we developed a mobile and low-budget method based on two simple and portable LCD video projectors, a specially coated screen, and polarization filters and glasses (for the setup, see http://www5.informatik.tu-muenchen.de/forschung/visualisierung/3dbeam.de). This approach combines the high quality of stereoscopy by polarization with the straightforward handling of simpler techniques.

A second problem – particularly in complex geometries that biofilms confront us with – is that single images are not sufficient for a comprehensive examination of a whole 3-D scene. For example, if we want to investigate the distribution of EPS in the given domain (Kuehn et al. 2001), we must be able to explore the whole structure, i.e., we need different perspective views (as they can be provided by a film instead of single images), and we need to look into hidden subregions like pores or channels, too. Hence, possibilities of interactive virtual walks or flights through the respective domain are of great importance. Modern visualization packages like AVS/Express (Advanced Visual System, http://www.avs.com/products/expovr.htm) provide the necessary tools; they can be used to generate the respective sequences of perspective image twins simply by moving the mouse, and the combination with our stereoscopic approach results in films that are really three-dimensional.

Of course, one must never forget that the beauty or the aesthetics of a visualization are not at all related to the correctness of the underlying data (Globus and Raible 1992): Even data that are completely false can be used for pretty pictures! Nevertheless, serious visualization is far from populism: The respective fields of application can and do profit a lot from methods that have been developed in the fun- and consumer-driven field of computer graphics, as both experimental and computational results become accessible and easily understandable.

## 4.7 VALIDATION

Having derived an apparently reasonable model and having developed or tuned numerical algorithms that converge to an apparently reasonable solution of the model equations, we still cannot be sure that our computed results are correct in the sense that they represent the phenomena we want to investigate at least approximately, for there are several potential sources of errors.

(1) Are there any problems with inexact computer arithmetic (for example, rounding errors)?
(2) Do our numerical algorithms really converge to the solution of the discretized model equations?
(3) Is the discretization adequate, or do our computations neglect information on finer scales of space or time that are important for the overall result?
(4) Is the model too coarse, that is, does the model already neglect certain scales or phenomena which must be taken into account in order to reflect the real processes with the desired accuracy?

These considerations show that some kind of validation of our results is inevitable. However, especially for the validation of the model equations, it is not at all evident how to do this. Actually, this is one of the most delicate aspects of numerical simulation. One might think of an experimental validation, but is it not one of the aims of simulation to provide insight where experiments are impossible (as it is in astrophysics, for example, when we want to simulate the life cycle of galaxies)? Moreover, measurements are incomplete and error-prone as are numerical simulations. For instance, in our code, we can trace the path of one single virtual particle through the flow field. On the other hand, up to now, CLSM can only track large sets of microbeads. Because of the tomographic approach, we see several particles in each horizontal slice at a time. Thus, the experimental way can provide flow rates, but it cannot be used to validate our (virtual) particle's path. So who wants to validate whose results?

Of course, these problems must not be used as an excuse for both sides to stop cooperation, but rather they show the importance of a tight coupling. At least to some extent, the comparison and mutual control of results can increase the reliability of both experimental and computational data.

## 4.8 CONCLUDING REMARKS

With this contribution, we want to lessen the gap between 'classical' biofilm research and state-of-the-art simulation techniques. For that, it is important that numerical simulation is seen as a tool that can provide insight in several situations, but which, by now, is far from being able to answer all open questions. Furthermore, we think that it is essential to understand simulation as an interdisciplinary challenge that goes far beyond working with some (commercially) available programs of limited scope.

First, it has to be pointed out that the scope of any simulation is limited by the quality of the underlying model. Because, in the case of biofilms, modeling is still an open field, all computational results are, of course, limited to this knowledge. Second, biofilms are highly complex systems, which makes their modeling and simulation much harder compared with other applications like computational fluid dynamics in technical applications or semiconductor technology where computational techniques are quite widespread. In contrast to those fields, here we have flow aspects, transport aspects (diffusion and convection in the bulk fluid phase, diffusion and reaction within the biofilm, and adsorption at the interface), structural aspects (shear forces, detachment), chemical aspects (reactions), biological aspects (metabolism of different classes of species), and even electrostatic aspects due to ionization. Although it is quite clear what to do in each single field, their combination is a tough challenge, especially against the background of limited resources of computing time and storage.

Furthermore, the highly complicated and heterogeneous geometries enforce a high resolution of the discrete grids, and they cause a lot of difficulties in exploiting the advantages of modern and efficient numerical algorithms like multilevel methods.

Finally, we have highly different scales in time and in space. A single simulation can only tackle a certain range of scales. One cannot simulate the operation of a whole wastewater plant during several months and expect, at the same time, to still see what each single microbe does during a certain second. Nevertheless, it is one objective of microscale experiments and simulations to learn about the influence of the single microbe's behavior on the operation of the plant, which shows the necessity of a coupling of, and information flow between, simulations on the different scales.

As a summary, we emphasize (and we hope that we could, at least to some extent, convince the reader) that numerical simulations will gain in importance for biofilm research, as we have already seen in other disciplines of science and engineering. If applied in close contact with experimental investigations, simulation can help to accelerate the process of learning and understanding biofilms. However, the expectations should not be exaggerated – there are and there will always be limitations. To keep this balance is the main challenge. Always think of the weather forecast: we need it, we want it, we get it, and we mistrust it!

## 4.9 REFERENCES

Babovsky, H. (1998) *Die Boltzmann-Gleichung: Modellbildung – Numerik – Anwendungen.* B. G. Teubner, Stuttgart & Leipzig.
Berlekamp, E.R., Conway, J.H. and Guy, R.K. (1982) *Winning Ways for your Mathematical Plays.* Academic Press, New York.
Bishop, P.L. (1997) Biofilm structure and kinetics. *Wat. Sci. Tech.* **36**(1), 287–194.
Braess, D. (2001) *Finite Elements. Theory, Fast Solvers and Applications in Solid Mechanics.* Cambridge University Press.

Breitling, P., Bungartz, H.-J. and Frank, A. (1999) Hierarchical concepts for improved interfaces between modeling, simulation, and visualization. In *Proc. Vision, Modeling, and Visualization* (ed. B. Girod, H. Niemann and H.-P. Seidel), pp. 269–276. infix, St. Augustin. Germany.

Brück, B. (1998) Ein objektorientiertes Framework zur modularen Strömungssimulation. Diploma thesis, Institut für Informatik, TU München.

Bungartz, H.-J. (1998) *Finite Elements of Higher Order on Sparse Grids.* Berichte aus der Informatik, Shaker Verlag, Aachen.

Bungartz, H.-J., Kuehn, M., Mehl, M., Hausner, M. and Wuertz, S. (2000) Fluid flow and transport in defined biofilms: experiments and numerical simulations on a microscale. *Wat. Sci. Tech.* **41**(4–5), 331–338.

Bungartz, H.-J., Kuehn, M., Mehl, M. and Wuertz, S. (2003) Space- and time-resolved simulations of processes in biofilms on a microscale. In *Polymer and Cell Dynamics – Multiscale Modeling and Numerical Simulations*, pp. 175–188. Birkhäuser-Verlag, Basel.

Charaklis, W.G. and Marshall, K.C. (1990) *Biofilms.* Wiley-Interscience, New York.

Chen, S. and Doolen, G.D. (1998) Lattice Boltzmann method for fluid flow. *Annu. Rev. Fluid Mech.* **30**, 329–364.

Eberl, H., Picioreanu C. and van Loosdrecht, M.C.M. (2000) Modeling geometrical heterogeneity in biofilms. In *High Performance Computing Systems and Applications* (ed. A. Pollard *et al.*). Kluwer Acadamic, Dordrecht, The Netherlands.

Globus, A. and Raible, E. (1992) *13 Ways to Say Nothing with Scientific Visualization.* NASA Ames Research Center, Report RNR-92-006.

Green, D.G. (1990) Cellular automata models in biology. *Math. Comp. Mod.* **13**, 69–74.

Griebel, M., Dornseifer, T. and Neunhoeffer, T. (1997) *Numerical Simulation in Fluid Dynamics – a Practical Introduction.* SIAM, Philadelphia.

Hackbusch, W. (1985) *Multigrid Methods and Applications.* Springer, Heidelberg.

Hermanowicz, S.W. (1999) Two-dimensional simulations of biofilm development: effects of external environmental conditions. *Wat. Sci. Tech.* **39**(7), 107–114.

Hermanowicz, S.W., Schindler, U. and Wilderer, P. (1995) Anisotrophic morphology and fractal dimension of biofilms. *Wat. Res.* **30**, 753–755.

Hermanowicz, S.W., Schindler, U. and Wilderer, P.A. (1996) Fractal structure of biofilms: new tools for investigation of morphology. *Wat. Sci. Tech.* **32**(8), 99–105.

Horn, H. and Hempel, D.C. (1997a) Substrate utilization and mass transfer in an autotrophic biofilm system: experimental results and simulation. *Biotech. Bioengng* **53**(4), 363–371.

Horn, H. and Hempel, D.C. (1997b) Growth and decay in an auto-/heterotrophic biofilm. *Wat. Res.* **31**(9), 2243–2252.

Jordan, C. (1965) *Calculus of Finite Differences.* Chelsea Publ. Co., New York.

Korber, D.R., Lawrence, J.R., Hendry, M.J. and Caldwell, D.E. (1992) Programs for determining statistically representative areas of microbial biofilms. *Binary* **4**, 204–210.

Kuehn, M., Hausner, M., Bungartz, H.-J., Wagner, M., Wilderer, P.A. and Wuertz, S. (1998) Automated confocal laser scanning microscopy and semiautomated image processing for analysis of biofilms. *Appl. Env. Microbiol.* **64**(11), 4115–4127.

Kuehn, M., Mehl, M., Hausner, M., Bungartz, H.-J. and Wuertz, S. (2001) Time-resolved study of biofilm architecture and transport processes using experimental and simulation techniques: the role of EPS. *Wat. Sci. Tech.* **43**(8), 143–151.

Lewandowski, Z., Webb, D., Hamilton, M. and Harkin, G. (1999) Quantifying biofilm structure. *Wat. Sci. Tech.* **39**(7), 71–76.

Mehl, M. (2001) Ein interdisziplinärer Ansatz zur dreidimensionalen numerischen Simulation von Strömung, Stofftransport und Wachstum in Biofilmsystemen auf der Mikroskala. Ph.D. thesis, Institut für Informatik, TU München.

Noguera, D.R., Pizarro, G., Stahl, D.A. and Rittman, B.E. (1999) Simulation of multispecies biofilm development in three dimensions. *Wat. Sci. Tech.* **39**(7), 123–130.

Patankar, S. (1980) *Numerical Heat Transfer and Fluid Flow*. McGraw-Hill, New York.

Picioreanu, C., van Loosdrecht, M.C.M. and Heijnen, J.J. (1998) A new combined differential-discrete cellular automaton approach for biofilm modeling: application for growth in gel beads. *Biotech. Bioengng* **57**(6).

Picioreanu, C., van Loosdrecht, M.C.M. and Heijnen, J.J. (1999) Discrete-differential modeling of biofilm structure. *Wat. Sci. Tech.* **39**(7), 115–122.

Pögl, M. (1999) Erweiterungen von Nast++: Drei Raumdimensionen, Geometriebeschreibung und Parallelisierung. Systementwicklungsprojekt, Institut für Informatik, TU München.

Reichert, P. (1994) Aquasim – a tool for simulation and data analysis of aquatic systems. *Wat. Sci. Tech.* **30**(2), 21–30.

Reichert, P. and Wanner, O. (1997) Movement of solids in biofilms: significance of liquid phase transport. *Wat. Sci. Tech.* **36**(1), 321–328.

Samet, H. (1984) The Quadtree and related hierarchical data structures. *ACM Computing Surveys* **16**(2), 187–260.

Tartar, L. (1980) Incompressible fluid flow in a porous medium – convergence of the homogenization process. *Lecture Notes Phys.* **129**, 368–377.

van Loosdrecht, M.C.M., Picioreanu, C. and Heijnen, J.J. (1997) A more unifying hypothesis for biofilm structures. *FEMS Microbiol. Ecol.* **24**, 181–183.

Versteeg, H.K. and Malalasekera, W. (1995) *An Introduction to Computational Fluid Dynamics – the Finite Volume Method*. Longman Scientific & Technical, Harlow, Essex, UK.

Wanner, O. and Reichert, P. (1996) Mathematical modeling of mixed-culture biofilms. *Biotech. Bioengng* **49**, 172–184.

Wimpenny, J. W. T. and Colasanti, R. (1997) A unifying hypothesis for the structure of microbial biofilms based on cellular automaton models. *FEMS Microbiol. Ecol.* **22**(1), 1–16.

Zhang, T.C. and Bishop, P.L. (1994) Density, porosity and pore structure of biofilms. *Wat. Res.* **28**, 2267–2277.

# 5
# On the influence of fluid flow in a packed-bed biofilm reactor

*Stefan Esterl, Christoph Hartmann and Antonio Delgado*

## 5.1 INTRODUCTION

Numerical simulation has become a very powerful tool in diagnosing and prognosticating the behavior of complex processes in nature and technology. This chapter deals with the numerical simulation of the flow behavior in packed-bed biofilm reactors for wastewater cleaning. The studies and considerations presented should contribute to a better understanding of the fundamentals of bioreactors at different scales to improve the data on which design and operation of the bioreactor are based. Most relevant global data for typical reactors have been compiled in Table 5.1.

© 2003 IWA Publishing. *Biofilms in wastewater treatment*. Edited by S. Wuertz, P.L. Bishop and P.A. Wilderer. ISBN: 1 84339 007 8.

It is well accepted that the diameter of granules $d$ and the flow rate averaged velocity $u_m$ are the characteristic length and velocity scales, respectively. Friction is characterized by the kinematic viscosity $v$ of the wastewater. Thus, the Reynolds number $Re$ represents the ratio between convective momentum transport and diffusive transport due to friction and is defined as $Re = u_m d/v$.

**Table 5.1.** Tangential stress for different reactor types and different Reynolds numbers. The simulation data (which cover the $Re$ domain up to about $Re = 20$) are for $Re = 10$. The diameter ratio is the ratio of the diameter of the container to that of a single particle.

| Parameter | Lab-scale reactor | Pilot-scale reactor | Simulation |
|---|---|---|---|
| No. of particles | ca. 70,000 | ca. 44,000,000 | 1,000 |
| Particle Reynolds number $Re$ | 35 | 12 | 10 |
| Diameter ratio | 33 | 370 | 10 |
| Superficial flow velocity | 0.02 m/s | 0.007 m/s | 0.006 m/s |
| Max. tangential stress $\tau$ | 6 N/m² | 2.2 N/m² | 2 N/m² |

From Table 5.1 it is evident that it is possible to realize laboratory-scale reactors in which the fluid dynamic parameters, such as velocity and tangential stress, are very similar to those in full-scale reactors. Of course, the main differences between lab-scale reactors and pilot-scale reactors are present in all quantities, which correlate to the reactor dimensions. Most of the consequences of this fact will be discussed later.

Furthermore, Table 5.1 gives a first estimation of the potential and limitations of numerical simulation. Similar to the comments made above, the dynamic parameters in laboratory and even full-scale reactors can be assessed by numerical simulation. The limitations are due to the large deviations in reactor dimensions.

When talking about the design and operation of packed-bed biofilm reactors, the role of fluid mechanics aspects on the local flow structure is not always fully appreciated. This might be a result of the apparent problem in gaining information about the flow structure in a reactor. Owing to the difficulties in placing measurement instruments inside such a reactor, without disturbing or destroying the aggregates inside, knowledge of the local physical quantities, e.g. pressure or velocity, is very limited.

There are additional requirements to be considered when designing and operating a bioreactor, compared with well-studied chemical reactors,. The fluid carries energy, nutrients and pollutants to the cultivated organisms. Furthermore, it exerts mechanical stresses on the biofilm, which is, to a greater or lesser extent, sensitive to this impact. The superficial velocity determines whether the biological processes are diffusion or reaction limited and provides the boundary conditions for the transport processes taking place inside the biofilm.

### 5.1.1 Basic questions and considerations

The mechanical stresses acting on the biofilm can be subdivided into normal and tangential stresses. The effects of the latter, e.g. shear stress, have been studied in many publications (see, for example, Mersmann *et al.* 1990; Hoffmann 1995; Edwards 1989). The pressure, extensional or elongational stresses in shear-free flows and normal stress components, which are generated in a wide range of rheological substances during shearing, belong to the category of normal stresses. As discussed by Nirschl and Delgado (1997), the role of elongational effects on biotechnological processes has been poorly investigated. Recent findings elucidate that, in packed-beds, extensional stress can substantially exceed the magnitude of shear stress (Debus 1997; Debus *et al.* 1998). This agrees well with the findings of Durst *et al.* (1987), which provided strong evidence that the influence of pressure, and especially of extensional stress, must be considered, too.

There are further questions regarding both fluid mechanical and biological aspects of the processes in bioreactors (Papoutsakis 1991):
- What kinds of forces affect cells in a flow environment?
- Are the effects on organisms due to the intensity and/or the frequency of the forces?
- What kinds of interactions have the most impact on the cells?
- Do fluid mechanical stresses cause cell death or injury, or do they reduce the cell growth?
- Do fluid mechanical forces affect the physiology, molecular processes, product expression and morphology of cells and biofilm, and how?
- Do cells respond, and adapt in response, to fluid mechanical forces?

These questions are usually posed in the context of the cultivation of sensitive animal cells. However, some of them can be adopted directly to treat problems concerning the generation process and the mechanical stability of biofilms. In the process of wastewater treatment, the biotic and abiotic matter experiences a variety of flow environments during the different operation periods. They may be classified in the following categories:
- The bulk liquid whose flow can be assumed to be laminar, (see also section 5.1.2 and Table 5.1). It is responsible for the interactions between fluid and carrier material such as beads. It exerts the stresses and delivers nutrients to the biofilm.
- The flow and transport processes inside the biofilm caused, for example, by concentration gradients or active movements of invertebrates.
- Fluid flows near the biofilm surface induced by higher organisms. It is known that such organisms, with a size of $10^{-4}$ m (e.g. nematodes or

ciliates), have the capability of active movement and therefore influence the flow field around the biofilm on a more microscopic scale.
- Flow induced collisions of carrier beads between one another or between bead and bioreactor. This environment contributes mostly to damage and abrasion of the biofilm during backwashing.

It is not the goal of this chapter to discuss all processes taking place in the operation of a reactor in a conclusive and detailed manner. This is mainly due to their complexity and variety. One example is the influence of invertebrate motion on the biofilm. The process is not yet well understood. Modeling and simulating the local flow field generated by these organisms will require very extensive efforts, which must be reserved for future activities. The transport processes in the biofilm are treated elsewhere in this book (see Chapters 6 and 7). Substantial progress in simulating the backwashing can only be achieved with a better understanding of cohesion and adhesion forces and the response (deformation, abrasion and fracture) of biomaterials on mechanical loads. We consider this one of the biggest challenges in the future.

In the next sections, an attempt is made to find some answers and conclusions to the questions outlined above.

## 5.1.2 Conclusions from the 'state-of-the-art'

Biological wastewater treatment suffers from the fact that real progress can only be achieved when the fundamental processes are studied from different points of view. However, the number of disciplines involved and, consequently, the number of publications, is very extensive. Therefore, it seems to be reasonable in this context to focus on the aspects of most relevance, disregarding any acclaims for completeness. But even when doing so, it is of crucial importance to include knowledge well established in surrounding fields. In accordance with this, the following survey gives an overview about the literature concerning the hydrodynamics of liquid flow in packed-beds and biological reactors.

The overview is ordered in the following manner:
- Interactions between flow and microbiological structures;
- Resistance against damage due to flow induced forces and the cohesion of biological material;
- Hydrodynamics of liquid flow in packed-bed biofilm reactors;
- Influence of 'imperfections' of the spherical carriers, such as particle roughness.

It appears to be convenient to start the discussion of the 'state of the art' with some statements about biofilm and biofilm architecture. The physical and physicochemical properties of a biofilm are mainly due to extracellular polymeric substances (EPS) (Flemming 1994; see also Chapter 8). The EPS matrix is very heterogeneous and its components can vary strongly according to the hydrodynamics, nutrient conditions and included organisms. This implies that the biofilm 'reacts' to the environmental conditions and changes its architecture. By changing the quantity and composition of EPS, the mechanical stability of biofilm and microbial aggregates concerning cohesion and adhesion on the substratum is altered. In biofilm reactors this alteration (which sometimes causes 'sloughing off' of biofilms) can have a major influence on the stability of operating conditions and maintenance. The crucial question, which has to be answered, is: how do the hydrodynamics influence the structure of EPS and hence the architecture of the biofilm?

Azeredo and Oliveira (2000) studied the effect of the amount of extracellular polymers produced by bacteria in biofilm formation and composition in a reactor under different conditions (see also Chapter 9). They cultivated two thin biofilms (< 200 µm) of different mutants of *Sphingomonas paucimobilis*, a high and a low producer of gellan (a polysaccharide). The biofilms were formed around four glass cylinders immersed in the reactor. After 12 days' cultivation, the authors performed shear experiments by rotating the cylinders at 500, 1000 and 2000 r.p.m. and found a generally higher removal rate with increasing shear rate for both mutants. Biofilm formed by the mutant capable of producing a high amount of gellan was shown to have stronger cohesion. This result is very important when studying the influence of flow on the biofilm architecture. It strongly indicates that mutants that produce more gellan can have a higher chance of survival. In phases of high stress, the result is a changing EPS composition and biofilm architecture.

Picioreanu and van Loosdrecht (1998) investigated the problem of adopting biofilm architecture by simulating a growing film with a cellular automation model (refer to Chapters 1, 2 and 4). The model allows the characterization of the biofilm surface shape, roughness and porosity as a result of microbial growth in different environmental conditions. The crucial environmental conditions are substrate transport/growth rate and convective flow together with biomass detachment. Porous biofilms, with many channels and voids, with 'finger like' colonies were obtained in a substrate-limited régime. Compact and dense biofilms resulted at high substrate-transfer rates. In this case diffusion has a minor influence and the biofilm development is limited only by the growth rate.

These results show that the biofilm has the ability to adapt to hydrodynamic conditions, and the composition of the biofilm is changed. The biofilm architectures have an important influence on the stable operation of plants. Therefore, it seems to be possible by manipulating the abiotic and biotic structure of biofilm reactors to improve their efficiency for advanced wastewater treatment.

In this context, further comments on the hydrodynamic considerations given in the literature are necessary. Firstly, most of the results found are based on plane flow geometry (Picioreanu and van Loosdrecht 1998). Here, flow and concentration boundary layers are described. This is done even when the characteristic dimensionless numbers do not indicate that the basic conditions for the existence of boundary layers are fulfilled. Furthermore, the results obtained for plain geometry are assumed to be valid for packed-bed reactors. But this appears to be more than questionable as even the pressure, stress and flow field from a single granule strongly deviates from that of a plane plate. Additionally, the wastewater in a packed bed is accelerated and decelerated due to the local changes of porosity and, therefore, the biofilm experiences different mechanical effects.

Many different authors address the flow régime. This is due to the fact that the interactions between flow and biological material are expected to be totally different in the case of laminar or turbulent flow. Here, no efforts have been made to check the possibility of the presence of rheological properties of the wastewater.

Dybbs and Edwards (1995) tried to determine the flow régimes with tracer experiments in a lucent container filled with glass rods in a packed bed analogous configuration. They found the existence of four flow régimes, based on the particle Reynolds number $Re$.

The régimes found by Dybbs and Edwards (1995), in addition to a creeping flow ($Re < 1$), are a stationary laminar flow ($1 < Re < 70$), a non-stationary laminar flow ($70 < Re < 130$) and a turbulent flow ($Re > 130$).

Further insight into the flow field and structure of a packed bed was provided by Rottschäfer (1996). He determined the porosity function with an experimental set-up and measured the velocities with a laser in a Perspex container filled with Perspex spheres and a refraction-index-adopted medium.

An important result, which can be transferred to biofilm reactors, is that of the flow régime. Usually, the range of $Re$ available in lab-scale as well as in pilot-scale reactors is substantially smaller then $Re=100$ (see Table 5.1). The flow is laminar in this $Re$ domain (Rottschäfer 1996), in agreement with glass rod experiments (Dybbs and Edwards 1995).

In principle, flow separation and laminar wakes can occur because of the existence of divergence channels in the space between the granules and at the outflow region as a consequence of positive pressure gradients. But the nuclear magnetic resonance (NMR) measurements by Götz (1997) showed that there are no areas of recirculation inside the reactor in the $Re$ domain of relevance. The author explained this as due to the narrow packing structure. When a wake develops at the back of a sphere, its further dispersion will be inhibited by the other particles.

All these points lead to the conclusion that the flow in biofilm reactors with moderate velocities is of 'good nature'. In this context the question arises whether this flow has an influence on the architecture of the biofilm. To find the right answer, the mechanisms of microbial damage due to fluid mechanical impact have to be elucidated.

Whether the influence of shear forces has a damaging effect strongly depends on the structure of the exposed material. Not knowing the exact structure and cohesion forces, it is very difficult to make statements about stability. It must be emphasized that the effects of mechanical forces not only depend on their magnitude but also on the period of exposure, and when flow is concerned, on the dynamics. Unfortunately, these facts are often ignored in literature.

Brück (1998) and Nirschl and Delgado (1997) describe models used to explain the destruction of fragile biological aggregates due to shear and normal stresses. At a given shear rate, if the shear stress $\tau$ exceeds a critical value, the diameter of the aggregate decreases exponentially with the stress duration. This implies that even low mechanical stress can have a crucial impact on the agglomerate size if the duration is long enough. The considerations given by Brück agree well with results presented by Horwatt et al. (1992). They describe the break up of flocks using an Eden and a Liner-Trajectory-Aggregation Model (LTA), respectively. In case of the LTA model, the critical shear force is independent of the flock size. Small aggregates experience lower hydrodynamic forces, but these forces have to break up lower binding strengths to destroy the aggregates. If the particle growth is simulated by the Eden model, there exists a relatively dense aggregate core, and this leads to a density higher by a factor of ten compared with the LTA Model.

Information about the maximal stresses microorganisms can endure and the interaction between biofilm architecture and fluid flow is very limited and mostly refers to effects caused by shearing. Mersmann et al. (1990) give a survey of the maximal stresses bacteria, eukaryotic yeasts, plant and animal cells can withstand. In a compilation presented by the authors, cells of bacteria and higher microorganisms sustain shear stresses up to $10^8$ N/m$^2$ when they are exposed to an extremely short-term mechanical load (see also Table 5.2). The tolerance of plant and animal cells given in Table 5.2 is much lower, at $2 \times 10^6$ N/m$^2$. These values refer to suspended organisms, not those embedded in a biofilm. Organisms protected by EPS seem to withstand even more stress.

When talking about stability of biological materials, a further form of stress has to be taken into account. Durst et al. (1987) showed analytically that the pressure drop in a packed-bed reactor is 60% caused by elongational effects. In this context, elongational flow can damage biological material more effectively than pure shear flow (Nirschl and Delgado 1997). Figure 5.1 shows mechanically deformed yeast cells extruded through a nozzle.

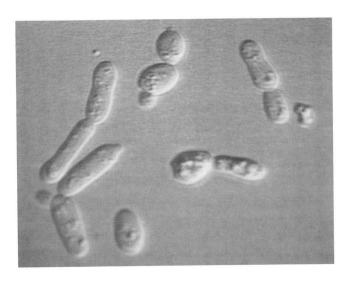

**Figure 5.1.** Yeast cells deformed in an elongational flow.

This photograph shows the shape of the cells just after passing the nozzle. Usually the shape of yeast cells corresponds to that of an ellipsoid. It is evident that they suffered an elongation. As it was not the subject of the investigation, no statement could be made about a possibly irreversible extension or even an inactivation. But it must be expected that at least a substantial alteration in the metabolism of the yeast cell occurred.

The main reason for deviations of the behavior of biomaterials in shear and elongational flow is the different kinematics of both flow forms, which could generate time-dependent mechanical loads on the biotechnological substances. A pure shear flow induces a rotation of the suspended biomaterial (Arnold and Delgado 1998; Guo et al. 1998). In perfect spherical particles, the rotation occurs with a constant angular velocity. In contrast to this, the angular velocity is time-dependent for non-spherical particles and biological materials such as microbial cells. As a result of this, the surface of the cells is impacted periodically, whereas the maximum range of normal and shear stress generated does not depend on time.

In elongational flow, a rotation of particles or biological material only occurs in the non-spherical case. This rotation is restricted to a small time period in which the particle takes the position of highest stability regarding the direction of elongation. This means, for example, that yeast cells (which can be described approximately as an ellipsoid, as mentioned above) are oriented preferably in such a way that the long axis of the ellipsoid indicates the axis of elongation.

It is of crucial importance that the mechanical load on the surface of a cell increases dramatically with time during extension. In mono-axial elongation, the increase is exponential with time. Thus, the magnitudes of fluid mechanical forces appearing during elongation can substantially exceed those produced by shear flows.

For completeness, the literature has been consulted for a further theme of interest in connection with the investigations carried out. The question is whether and in what manner the properties of the granule surface can affect the generation of the biofilm and its mechanical stability against damage and detachment.

Different from the removal of thick biofilms in wastewater reactors, the detachment of biomaterial from a carrier is of high relevance in food sciences, cleaning processes and medicine. In most of the cases, an inert or chemically reactive fluid is employed to generate mechanical forces on the surface of the biofilm and to transport the removed biomaterial. It is well established that the mechanical impact needed for lifting up a biofilm differs by many orders of magnitude, depending on the physicochemical properties of the biofilm and the surface of the carrier.

As mentioned above, in practice the carrier substratum for the biofilm cannot be assumed to be perfectly spherical and slick. The beads show small channels and a certain surface roughness. Visser (1995) gives a detailed review on particle adhesion and removal. Experiments on the aggregation kinetics of *Saccharomyces cerevisiae* on solid surfaces revealed a crucial influence of the roughness on the aggregation rate. At slick polished surfaces with a roughness lower than 1 µm, the effect of adhesion dominates the formation of a biofilm. The cells are positioned flat on the surface and have the lowest possible distance to the surface. In this case the responsible forces for adhesion (Lifshitz – van der Waals and electrostatic forces) are maximal (see also Chapter 9). In case of a higher roughness, between 1–6 µm, the aggregation rate reaches a minimum. Owing to the higher distance between particle and surface, the adhesion forces are diminished and the hydrodynamic effects, like shear and elongation, lead to an abrasion of biological material. At a roughness higher than 6 µm, the hydrodynamic effects are still present, but not large enough to detach the cells. In this case the cells are mostly located in channels and valleys, and therefore protected from direct fluid-mechanical impact.

## 5.2 SOME COMMENTS ON SPECIFIC TOPICS

The following comments are intended to elucidate the potential, and also the limits, of numerical simulation. There exists a large discrepancy among engineers and scientists studying wastewater in their understanding of this topic. Thus, this section is provided to contribute to the declared aim of the present book, i.e. to achieve a better exchange of information among the different disciplines. Another goal is the discussion of fluid mechanical terms such as boundary layers, which are often used in very different ways in the literature.

### 5.2.1 Elephant versus microorganism: the different scales to simulate

There are many powerful methods in microbiology suitable for investigating the behavior of microbial cells. If one postulates that the volume of a cell contains about $10^{-18}$ m$^3$, the number of cells required to compose an elephant would amount to approximately $10^{19}$ cells. Assuming, furthermore, a biologist will investigate each cell to describe the whole 'system' of an elephant, and under the assumption that the investigation of each cell lasts no longer than a millisecond, it will take about $3 \times 10^8$ years to complete.

This is a tremendously long time. But the situation is more tedious when studying developments in time. Coming back to the example of the elephant: if for any reason it would be of interest to study the kinetics of the cells for a time of 1 week, and assuming that a characteristic inner time of cells is of the order of magnitude of 0.1 s, the number of individual investigations will increase to the order of magnitude of $10^{25}$.

These data should illustrate why it is not possible to simulate all flow processes that can contribute or influence the generation of biofilms in wastewater reactors, even when using the most powerful computer available. This is due to the enormous ranges in the characteristic time and spatial scales. The spatial scale in a bioreactor starts from $10^{-6}$ m for bacteria, $10^{-5}$ m for the biofilm, $10^{-4}$ m for higher organisms and ends up at $10^1$ m or higher for pilot scale and full-scale reactors (Figure 5.2).

At present, it is not possible to simulate the behavior of a complete reactor with a spatial resolution of $10^{-6}$ m. In this chapter, simulations are based on a reference spatial scale in the order of magnitude of the diameter of the support granules. A simulation on a smaller spatial scale, by looking at the secondary flows caused by higher organisms, would reveal further insight into the flow and transport processes and should be considered in future activities.

Working on the scale of support granules allows access to some essential questions above. How good is the performance of transport processes to the biofilm? How large are the mechanical stresses acting on the biofilm? How do the mechanical stresses vary in a biofilm reactor? Is it proper to assume the flows in biofilm reactors have boundary layer characteristics? Is it a turbulent or laminar flow? Is it possible to transfer results obtained for plane transport processes to packed-bed reactors? Which are the effects of flow separation? Must a biofilm be considered to be compressible or incompressible?

For the time resolution achievable, it should be stated that the 'characteristic inner time' of the process depends on the point of view. Biochemical reactions can proceed within a time of 1 ms or even lower. The inner time $\tau$ scale for the momentum transport in the flow field can be shown to be proportional to $(d^2/\nu)$. For clean water and a granule diameter of $5 \times 10^{-3}$ m, the value of $\tau$ is about 25 s. From this it can be concluded that the characteristic time for the generation of the biofilm respective to the operation period for wastewater treatment is much higher (about 1 week) than $\tau$. Thus, it appears justified to treat the flow in the reactor as quasi-steady.

## 5.2.2 On fields, layers, flow régimes, pressure gradients, and their importance in packed-bed biofilm reactors

As mentioned above, there is some confusion in the literature about the momentum and mass transport in so-called layers. Furthermore, the existence of local flow zones containing whirls, i.e. eddies, is often interpreted as an indication that the flow is turbulent. Last but not least, the existence of substantial pressure gradients in flow configurations other than plane flows is often ignored, although the geometry has a decisive influence on the flow and pressure distribution.

In fluid mechanics the knowledge that in certain flow processes the exchange of momentum occurs in a thin layer is considered to be the decisive step towards a comprehensive understanding of flow phenomena and the unification of theoretical and experimental findings that have been well established, but poorly accessible, before.

The concept of boundary layers is based on the following ideas: The momentum transport (convective and due to friction) occurs in a layer of thickness $\delta$ that is much thinner than the characteristic length $L$ in the main flow direction $z$. Outside the boundary layer, the fluid flow can be considered as shear free, i.e. the velocity and pressure field can be calculated from a potential theory. To fulfil the basic assumption of a very thin boundary layer, the Reynolds number must take very high values, as the ratio of $\delta/L$ is of the order

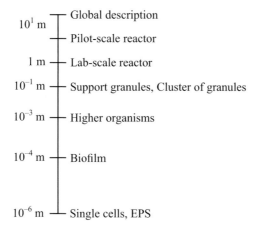

**Figure 5.2.** The different spatial scales in a packed bed biofilm reactor.

of magnitude of $Re^{-1/2}$. In practice the Reynolds number should be at least $10^3$. As a consequence of this, it can be stated that the velocity in the main flow direction $z$ is much higher than that in the transverse direction $y$ (distance from the surface of the body studied). Furthermore, the gradients of all kinematical quantities in the $z$-direction are much smaller than those in the $y$-direction.

A further consequence is that the pressure does not depend on $y$, i.e. for a given $z$ it is constant within the boundary layer. Thus, the pressure adopts the same value in the boundary layer as in the outer, potential field. In the case of a flat plane plate oriented in the direction of the undisturbed outer flow field, the pressure in the outer field and, therefore in the boundary layer, does not depend on $z$. This is due to the fact that the streamlines do not suffer any distortion. In contrast to this, thick, curved bodies and geometries orientated under a certain angle of attack relative to the direction of the undisturbed flow induce a pressure gradient in the $z$ direction. Consequently, it is not possible to assume that the results obtained for a biofilm on a flat surface are valid for packed bed reactors. It must be expected that the transport of microorganisms needed for the primary adhesion, as well as the transport of nutrients to the colonies, occurs with diverse mechanisms, and thus a different shape of geometry might lead to a different biofilm architecture when comparing flat and curved structures (such as granules).

The thickness δ of the boundary layer and its stability depend strongly on the pressure distribution. In the case of negative pressure gradients (accelerated flow), the flow is stabilized. In contrast, an increase of the pressure in the $z$ direction could lead to boundary layer separation. This generates detached eddies, i.e. dead water zones. Thus, in packed bed biofilm reactors, not only forces due to acceleration or deceleration, but also due to secondary flows, may be present. The latter inhibit the transport of biological material to the wall. This could also have consequences on the biofilm architecture, e.g. a non-uniform thickness on the granules.

The stability of the flow depends also on the flow régime. At high Reynolds numbers, the flow in general becomes turbulent. As it carries more energy, the momentum transfer between different fluid layers is enhanced, and the boundary layer is consequently more stable than in the laminar case. In the literature, some considerations concerning the turbulent flow case can be found. The results obtained for plane flows are often transferred to complicated geometrical systems. In this context, it must be emphasized that, on a flat plane, the turbulence generation begins a certain distance from the leading edge. This results from the fact that the flow field is characterized by the axial co-ordinate. In contrast to this, in biofilm reactors, the granule diameter is the characteristic length scale. Thus, the flow does not become turbulent in the reactor due to the small diameter of the granule and the velocities available (see Table 5.1).

In the literature (e.g. Picioreanu *et al.* 2000; Horn and Hempel 1995) there are some attempts to apply the basic ideas of boundary layer theory to cases where it is very difficult to determine an exact boundary layer thickness and use the results in complicated bioreactor systems. When talking about mass transfer from the surrounding fluid to microorganisms, one often speaks about the existence of 'layers', neglecting that the Reynolds number (based on the particle diameter) of most of the wastewater in the reactor is too low for the generation of a boundary layer. Thus, the dimension of the 'layer' is absolutely arbitrary. This is of decisive importance. From a theoretical point of view, when the Reynolds number is small, local flow processes are influenced by the whole flow field. At higher Reynolds numbers the flow develops along a characteristic axis and the dimensions of a boundary layer can be uniquely defined.

When thinking about reactors, it must be emphasized that the employment of the boundary layer concept appears to be more than questionable, even when assuming the Reynolds number to be sufficiently high. This is due to different reasons. A more strict definition of the fundamental boundary layer concept says that it is not valid when the local Reynolds number $Re$ has a value lower than a minimal value (for example 1000). In contrast to the definition of the Reynolds number in the introduction, the Reynolds number in the classical boundary layer theory is defined with the length of a flat plate as the reference length. This means that the boundary layer concept is only valid at a certain distance of a characteristic point where the boundary layer generation begins (for example the stagnation point). In the case of a bioreactor it is impossible to find areas where a boundary layer with a specific thickness exists. Owing to the narrow packing structure of the carriers, eventually evolving layers are disturbed by neighboring particles. The thickness of the layer is not solely determined by the fluid velocity but mainly by the geometrical arrangement of the particles. As the boundary layer thickness has a crucial influence on the exact determination of the mass transfer, theoretical and experimental approaches to determine the thickness are limited to biofilms growing on a flat geometry or on single particles at high Reynolds numbers.

## 5.3 CURRENT INVESTIGATIONS

The following sections give an overview about the hydrodynamic patterns in a biofilm reactor. The major points of interest will be the transport of biological materials and nutrients, the distribution of forces and mechanical stresses, as well as the problem of the complete description and scaling-up of a reactor by means of numerical simulation.

### 5.3.1 Numerical method

Numerical methods represent the basis for accessing data necessary to get a better insight into the influence of local flow on biofilm generation. The method used here is described very shortly. The basic numerical method and grid technique (Nirschl *et al.* 1993, 1995; Nirschl 1994) was adopted and the code extended to calculate the flow within packed-bed reactors (Debus *et al.* 1998; Esterl *et al.* 1998). This extension included the parallelization of the code, as well as the ability to calculate particle numbers up to 1000. The code is paralleled with the message passing library PVM and is based on a modified master–slave concept, where the master does the message handling and additionally calculates the grid of the container. The slave processes calculate the flow field around the carriers of the biofilm. After each iteration, there is an exchange of information between the processes.

The computer code solves the three-dimensional, incompressible Navier–Stokes and energy equations using an implicit, second-order, finite-volume approach. The governing flow equations are the continuity and momentum equations in integral form:

$$\iint_A \rho \vec{v}\, dV = 0,$$

$$\iiint_V \frac{\partial \vec{v}}{\partial t}\, dV + \iiint_V \vec{v}\cdot\nabla\vec{v}\, dV = -\frac{1}{\rho}\iint_A p\, d\vec{A} + \nu \iint_A \nabla\vec{v}\, d\vec{A}.$$

Here $\vec{v}$ denotes the velocity vector, $\rho$ the density of the fluid, $t$ the time, $p$ the hydrostatic pressure, $V$ the volume and $\vec{A}$ the surface vector of a control volume.

The equations are made dimensionless with characteristic values of the given system. All geometric lengths have been made dimensionless with the granule diameter $D_p$. The velocity vector $\vec{v}$ is related to the superficial flow velocity $U_\infty$, and the pressure $p$ and the shear stress $\tau$ is related to the stagnation point pressure generated by a flow of the velocity $U_\infty$.

The symbol $*$ denotes the dimensionless variables. For convenience, the radii

$$x* = \frac{x}{D_p}; \quad y* = \frac{y}{D_p}; \quad z* = \frac{z}{D_p}; \quad \vec{v}* = \frac{\vec{v}}{U_\infty};$$

$$p* = \frac{p - p_\infty}{\rho U_\infty^2}; \quad \tau* = \frac{\tau}{\rho U_\infty^2}.$$

have also been made dimensionless by using $D_p$.

The three velocity components are marched in time using an implicit predictor/corrector scheme along alternating co-ordinate directions. The pressure correction consists of solving a Poisson equation derived from the continuity equation. The solving of the Poisson equation is accelerated by a generalized minimal residual algorithm (GMRES). Second-order central differences are used for approximating the time and space derivatives. The physical assumptions for the numerical studies are: Newtonian, incompressible fluid; steady state, laminar flow; isothermal flow, non-reactive fixed particles. The flow pattern in a biological reactor can be characterized by the Reynolds number, which is defined above.

Owing to the high gradients in the narrow interspaces between the carriers, the upper range of realizable Reynolds numbers is limited to 20 (based on the particle diameter). According to Table 5.1, this appears to be sufficient, as this is the typical range of Reynolds numbers occurring in biofilm reactors.

It is possible to make the code more stable for dealing with reactors at higher Reynolds numbers. In this case an artificial viscosity is implemented. The artificial viscosity damps the high gradients, especially at the beginning of the iteration, and is reduced to a 'normal' level during the iteration.

## 5.3.2 A comparison between laboratory and pilot reactor or the scale-up problem

It has often been found that different populations of bacteria and higher microorganisms can be detected in reactors of different sizes. It is very difficult to ascertain the cause of this deviation, as it is not possible to realize similar environmental conditions. There are deviations in the temperature, composition and quantity of nutrients and wastewater, as well as in hydrodynamic aspects.

As discussed in the introduction (compare Table 5.1), from the point of view of hydrodynamics and under the assumption that both reactors are operated at the same particle Reynolds numbers, the major differences between a lab-scale and pilot-scale reactor occur due to reactor dimensions. In this context, it is evident that the characteristics of the regions influenced by the cylindrical container wall and the core region will be substantially different in laboratory and full-scale reactors.

In general, it is convenient to divide a reactor into four different zones (Figure 5.3): an inflow and outflow region, a core, and a near wall zone. Within the zones the physics of the flow is characterized by very different mechanisms.

At the inflow (within the first three to four layers of carriers), the flow develops and high radial velocity components can be observed. The non-uniform mass flow from the center to the near wall region is due to the high porosity in this area. Figure 5.4 represents computationally determined porosity profiles of a packed-bed filled with monodisperse or polydisperse carrier material, respectively. For monodisperse material the porosity profile from the wall (left) to the center of a packed-bed filled is represented with a black line. It is apparent that the expansion of the high porosity region near the wall is much more distinctive than in the polydisperse case shown with a gray line in the figure. The porosity profile has the form of a damped oscillation. The flow strictly follows the porosity profile. At a distance of three to four particle diameters from the wall (for the particle shown in Figure 5.4), the amplitude of the porosity is widely damped. Thus, the flow kinematics will change in the order of magnitude of the damped porosity.

Compared with this, the profile on the polydisperse packed-bed (Figure 5.4) shows that the oscillation region lasts only about one particle diameter into the packing zone. Consequently, the resulting flow field is more balanced in the wall region.

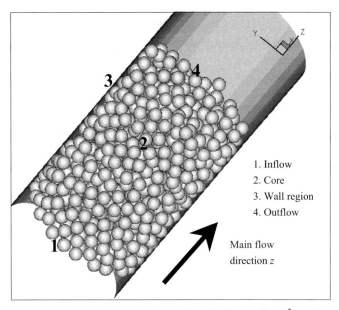

**Figure 5.3.** Different zones in a reactor of the dimensionless radius $R_c^* = R_c/D_p$.

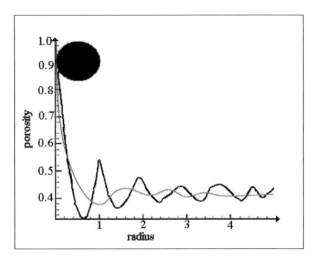

**Figure 5.4.** Porosity ($\varepsilon$) profiles of a packed-bed with monodisperse (black line) and polydisperse (gray line) support granules with an average particle diameter $D_p$.

The extension of the near wall region is independent of the reactor size (assuming the reactor diameter to be much larger that the particle diameter) and depends only on the particle size distribution. Having particles with a small diameter enlarges the probability to fill up empty spaces near the wall, diminishing the ordering influence of the walls.

For a small reactor with a low volume and a low diameter ratio of *reactor/granule,* the volume of the region with high porosity is much more predominant than for a large reactor. Esterl *et al.* (1998) showed in a model that the high porosity region covers up to 20% of the total volume of a lab-scale reactor. In contrast to this, the region of high porosity can be neglected for reactors with a large diameter ratio of *reactor/granule.*

Especially for reactors filled with monodisperse carriers and a low diameter ratio *reactor/granule,* the assumption of plug-flow is inaccurate. As mentioned above, the fluid strictly follows the porosity profile and this leads to regional distinctions in the axial velocity components. The change in velocities between wall and core region ranges on an order of magnitude of 5–10. Plate 5.1 shows the typical channeling of axial velocity components. The simulation was performed with 1000 particles, a particle Reynolds number of 10, and an average porosity of $\varepsilon = 0.45$. Red colors indicate high axial velocity.

The fourth region can be classified as outflow. It starts with the last layer of spheres and extends further downstream. In contrast to the mass flow from the core to the wall region at the inflow, the radial velocity components head in the other direction. Thus the inhomogeneous distribution of axial velocity components induced by the channeling effect will be balanced out. The velocity peaks near the wall fade away, depending on the Reynolds number. The dominance of viscous forces leads to a slow asymptotic fading; high inertial forces cause a fast decrease of the peaks. In earlier measurements of flow velocities at the outflow region (Rottschäfer 1996), an attempt was made to draw direct conclusions from the flow inside the bed. This procedure allows realistic implications only when the Reynolds number is low enough. The subdivision of the biofilm reactor into four zones provokes not only fluid mechanical considerations, but also raises new questions about biology and process engineering.

It is well known that the biofilm thickness varies with the position within the reactor. Biofilms are much thicker in the inflow region than in the outflow. In general two different mechanisms can be suggested to be responsible for this effect. One is the varying concentration of nutrients along the reactor axis; the other is the influence of mechanical stress formed by shear and normal stresses. From the results of the simulation, it can be deduced that the shear effects are of similar magnitude along the reactor axis. Thus it appears unlikely that tangential stress is responsible for a varying vertical biofilm thickness. In contrast to this, the normal stresses vary along the reactor axis. In the inflow, the fluid gets decelerated (stagnation points) and strongly redistributed; in the outflow, the fluid is accelerated and homogenized. But, apart from this, the different kinematics do not change the absolute value of normal stress. It is of the same magnitude as shown qualitatively in Plate 5.2. Thus mechanical stress seems not to be the cause of the differing biofilm thickness.

Consequently, the formation of a biofilm is predominantly influenced by the hydrodynamic transport of microorganisms and, after formation of a biofilm, the transport of nutrients to the carriers. The probability of adhesion of microorganisms in the inflow is much higher than in the outflow. This is caused by the long residence time due to the stagnation points. Furthermore, the organisms in the outflow experience limited nutrients caused by microbial substrate conversion in the layers below. So, from both reasons, it is obvious that the biofilm formation is inhibited in the outflow region. The formation of different biofilm architectures and the selection of microbial species may be due to increasing or decreasing nutrient concentrations in the reactor caused by changing hydrodynamic conditions (e.g. channeling effect). Currently this hypothesis cannot be proved because experimental data are missing and the regular backwashing frequently changes the position of the particles.

Plate 5.3 gives a complete overview of the demarcation of the different zones. It is based on a simulation of 120 particles and a Reynolds number of 17, and shows the axial velocity component. The velocity was found by averaging the velocities along certain radii at a fixed axial position in the reactor. When this procedure is done for many radii, a 'line' of the average velocity distribution from the center to the wall results. Repetition of this process for many axial positions results in a 'carpet' describing the velocity distribution in the whole reactor.

The flow direction is in the $z$ component from lower left to upper right. The first steep rise and the lower 'hills' beneath indicate the beginning of the packing zone and coincide remarkably with the porosity profile shown in Figure 5.4. When the small 'hills' disappear and the high velocity peak indicated by the red color asymptotically slows down, the packing zone ends and the flow balances to normal tube flow.

### 5.3.3 Numerical simulation of lab- and pilot-scale bioreactors

The simulation of lab-scale and pilot-scale bioreactors raises several problems. To give as realistic a picture as possible, the simulation claims realistic configurations. But computational performance limits the realizable particle number. With currently available equipment, it is possible to perform simulations up to about 1000 particles. This is relatively low compared with real conditions. Pilot-scale reactors, with an average particle diameter of 6.2 mm and a fill volume of 10 m³, contain about 44,000,000 particles; even smaller lab reactors with a volume of $2 \times 10^{-2}$ m³ contain around 90,000 particles. So direct comparisons between simulations with 1000 particles and a real plant need special attention to avoid scale-up problems (see also section 5.3.2).

Two basic approaches offer a solution. One represents the four different zones by using the physics of the flow. This requires enough particles to represent all zones. By simulating a reactor with a diameter ratio *reactor/granule* of about 20, this requirement can be fulfilled. The reactor then consists of a core region with 10 particle diameters, a wall region of five particle diameters into the core, an inflow region with three to four layers of carriers and 10 layers atop the inflow region to assure fully developed flow. To realize this, a simulation with about 1000 particles has to be performed. This is in the range of the computational facilities and allows direct conclusions on the behavior of lab-scale reactors.

The other approach is to use the statistically periodic configuration of a packed bed, simulating only one small slice and transferring the results to a full reactor simulation. This approach is a matter of current work, and the first results are promising. A basic goal is to reach a number of up to 10,000 particles, which is very close to a lab-scale reactor. Plate 5.4 shows the basic idea of splitting the whole reactor into slices and calculating the flow field in only one slice.

Comparisons of this approach with pressure drop equations from the literature show that the pressure field and distribution are represented correctly. Figure 5.5 shows a comparison between the theoretical approaches by Ergun (1952) or Molerus (1977) and the result of the simulation. In this case a slice of 45° was taken. This reduced the number of particles to be calculated from 120 to 12. In Figure 5.5 it is shown that, for the two simulation cases, the radius ratio of cylinder to sphere is 7. The data in the Figure are recalculated from the dimensionless numerical results, which assumed that the sphere diameter is 5 mm and the fluid is water at 20 °C.

## 5.3.4 Considerations of the effect of mechanical stress

From the statements above it is obvious that the hydrodynamic conditions are one of the important factors in determining the reactor behavior. Among them the most important are the overall flow rate, the pressure drop, as well as the velocity and pressure field. With these values, the relevant stresses acting on the biofilm can be formulated. As mentioned above, this topic is discussed differently in the literature (e.g. Picoreanu *et al.* 2000; George and Chellapandain 1996; Daka and Laidler 1978; 1988) and the results are often obtained from simple flow geometries such as plate or two-dimensional channels.

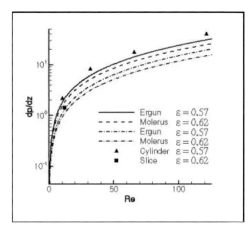

**Figure 5.5.** Comparison between pressure drop equations from literature (Ergun's and Molerus' approaches with different porosities, ε) and a simulation of a 'reactor slice' and a whole cylinder reactor. The quantity d$p$/d$z$ is the axial pressure gradient in the reactor.

When speaking of the effects of mechanical forces, it should be stated first that the considerations given here refer to the recirculation phase of the biofilm reactor. Furthermore, it should be emphasized that, even when being able to determine the local mechanical stresses on the biofilm, it is extremely difficult to determine the exact response of the biofilm to the mechanical load. This problem is present in all fields of biotechnology. The mechanical properties of biological materials are poorly understood and investigated. Thus a prediction of deformations, erosive processes or even fracture is not possible at the moment.

On the other hand, many observations on the behavior of different biological materials exposed to mechanical load have been done. Thus, it appears acceptable, in reference to other materials, to conclude whether the flow induced mechanical stress has the potential to alter or destroy the architecture of a biofilm.

Table 5.2 lists lethal shear stresses for cells (see Mersmann *et al.* 1990) as well as critical shear stresses inhibiting the agglomeration of bacteria on a surface (Duddgridge and Kent 1982) or leading to a droplet break up (Hinrichs 1993). These different values are intended as an overview over the wide range of forces acting on biological material. In Mersmann *et al.* (1990), the lethal shear rate for bacteria is relatively high compared to the values found elsewhere (Brück 1998). This appears to be due to the exposure time of the cells to the stresses due to high pressure extrusion. If the duration of the mechanical impact is very short, the cells can stand this stress for a certain period (compare also with section 5.1.1.). A similar behavior is found when both biotic and abiotic materials are investigated (Nirschl and Delgado 1997).

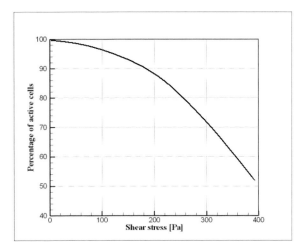

**Figure 5.6.** Deactivation of yeast cells exposed to shear stress for a period of 5 minutes.

**Table 5.2.** Critical shear stress values.

| Biological material | Critical shear stress (N/m$^2$) | Reference |
|---|---|---|
| Bacteria (high pressure extrusion) | $10^7$–$10^8$ | Mersmann et al. (1990) |
| Bacteria (agglomeration on flat plate) | $10^1$ | Duddridge and Kent (1982) |
| Yeast | $8 \times 10^7$ | Mersmann et al. (1990) |
| Plant cells | $2 \times 10^6$ | Mersmann et al. (1990) |
| Animal cells | $5 \times 10^{-3}$ – $5 \times 10^2$ | Mersmann et al. (1990) |
| Hot wort | $5 \times 10^1$ | Brück (1998) |
| Milk fat droplets | $1.9 \times 10^1$ | Hinrichs (1993) |

In contrast to this, Figure 5.6 shows the influence of shear stress on the deactivation of yeast cells after 5 minutes of shearing. The deactivation of *Saccharomyces cerevisiae* started when the shear rate exceeded a relatively low value, about 50 Pa. When the forces do not exceed a certain critical value, the time of effect can be infinite, without any damaging influence.

Compared with the stresses shown in Figure 5.6, the stresses occurring in biological reactors are relatively low. Plate 5.5 shows the dimensionless shear stress distribution in a reactor. From Table 5.1 it can be seen that the maximum value at the given Reynolds number reaches a value of about 2 N/m$^2$. It is clear that the organisms and investigating methods cannot be compared directly; however, the results indicate that the microorganisms in the biofilm are not damaged or destroyed by the low shear and normal forces, although they are exposed to stress for a long time.

Flemming (1994) states that, for shear stresses below 13 N/m$^2$, it is not even possible to prevent 'free' microorganisms from adhering on a membrane surface. So it seems very unlikely that the shear stress available in common reactors during the recirculation phase substantially destroys or alters the architecture of a biofilm (cells are protected by EPS and growing on rough surfaces).

Owing to the wide range of shear forces, which can be seen in Plate 5.5, the biofilm might have a different architecture in the different regions of the reactor. But the predominant influence on the biofilm architecture and abrasion is certainly the backwashing process. Detachment experiments by Arnz (2001) showed that the more intensive the recirculation the stronger is the adhesion of the biofilm on the carriers.

## 5.4 INFLUENCE OF FLUID FLOW AND SUBSTRATE CONCENTRATION ON BIOFILM ARCHITECTURE

To give an answer to the basic question 'what is the most important factor determining the biofilm architecture and hence transport processes within', we must consider the kinetics of the conversion in a biofilm. In general, the architecture of the biofilm is affected by two fluid-dynamic mechanisms. These are the flow-induced mechanical stress and the flow-induced transport of nutrients, which are represented by the Reynolds number and the Damköhler number, respectively. The Reynolds number was introduced in one of the preceding sections. The Damköhler number,

$$Da = \frac{V_{Max}}{K_S C_0},$$

is defined as the ratio between the maximum conversion rate $V_{Max}$ of the biofilm and the maximum mass transfer rate $K_S*C_0$. Herein $K_S$ denotes the mass transfer coefficient, and $C_0$ is the inflow substrate concentration. The influence of the Damköhler number can be explained by the illustration of two extreme cases: $Da \gg 1$ and $Da \ll 1$.

If $Da$ is much larger than one, the conversion capacity of the biofilm significantly exceeds the transfer rate of substrate to the biofilm. In this case we talk about a transport-limited régime, which means that the substrate transport by convection and diffusion is the conversion limiting process. In this régime, the biofilm growth rate and the biofilm architecture strongly depend on the Reynolds number. The growth rate increases with increasing Reynolds number, because the mass transfer coefficient gets larger. Furthermore, the transfer can also be enhanced by enlarging the inflow concentration $C_0$, as the maximum mass transfer rate $K_S*C_0$ depends on the inflow concentration. Owing to substrate limitation, the kinetics of the biofilm will always be of mixed- or first order.

The architecture depends on the flow induced mechanical stresses onto the biofilm. The biofilm architecture in the case $Da \gg 1$ can be described as a structure of scarcity. We conjecture that the microbial population tries to enhance the substrate transfer, for example by enlarging the surface area of the biofilm and building a loose and branched architecture if the Reynolds number is small enough. If the Reynolds number has a higher value but the mass transfer rate is still low (e.g. because of a very low substrate concentration), the biofilm also tends to build a branched architecture until the flow-induced forces become too large and biofilm is sheared off. An erosion of biofilms will start at Reynolds numbers of a few hundred. The erosion does not exist in a packed-bed biofilm reactor, because the Reynolds number is too low; the dominating mechanism of biofilm loss in a reactor is due to the abrasion during the backwashing phases. Thus, the biofilm growing in a packed-bed reactor is highly sensitive to mechanical load as it is not adapted to high shear and elongational impact. This hypothesis is supported by the fact that, at the low fluid velocities, it is more likely to have a transport limited régime in the reactor and, therefore, the biofilm builds a branched and loose architecture.

The second extreme case that should be discussed is at a Damköhler number much smaller than 1. Here, the maximum mass transfer rate exceeds the maximum conversion rate of the biofilm. This means that the organisms always have enough substrate for their metabolism and their growth rate does not depend on the Reynolds number.

The biofilm structure is affected by the flow-induced load, developing into a very loose and slack architecture at low Reynolds numbers. At higher Reynolds numbers, its architecture is very compact and dense.

## 5.5 CONCLUSIONS AND OUTLOOK

This chapter deals with the effect of the flow in packed-bed biofilm reactors. The aim was not only to elucidate aspects of hydrodynamics, but also to address some new questions more concretely. A further goal was to facilitate the exchange of information between the different disciplines involved in biofilm reactor research. Most emphasis has been devoted to the results obtained by our own numerical simulations. To make deductions for our own results, these statements are complemented with a discussion of data from different disciplines in the literature.

When simulating biotechnological processes, the expectations of investigators from different disciplines involved could differ enormously. Therefore, the potential of numerical simulation has been illustrated heuristically so as to answer one of the questions addressed in this chapter. It has been shown that numerical simulation must be restricted to a certain range of length and time scales. In the present contribution, the length scale considered extends from the diameter of a single carrier particle to the diameter of the full-scale reactor. Thus, no attempt has been made to study the transport processes within the biofilm (these are treated in subsequent chapters in this book), or even the fluid movement induced by single microorganisms.

For the pure fluid mechanical aspects, it must be stated that, at least during the recirculating phase, the flow in a packed-bed reactor is laminar. Furthermore, for the typical values of the Reynolds number ($Re < 50$) available in laboratory, pilot-scale and full-scale reactors, the simulation shows that the flow is steady laminar and free of secondary flows, in agreement with experimental results (Dybbs and Edwards 1995; Rottschäfer 1996). Locally, the wastewater is accelerated and decelerated and, therefore, not only tangential stresses but also normal stresses are present. The flow characteristics violated essential requirements of boundary layer theory and, therefore, it is not conclusive to consider the transport of nutrients and energy or mass transfer within layers, as is often done in the literature. Furthermore, it seems to be inadmissible to transfer the results obtained in simple geometry, such as flat channels, when describing the processes in packed-bed biofilm reactors.

About the question concerning the functions and effects of the fluid flow in the reactor, the numerical results show that there are several different aspects to be considered. In general, the flow transports nutrients and pollutants to an established biofilm. Furthermore, it carries microorganisms to the carrier material during the phase of biofilm formation. Finally, the momentum exchange between the wastewater and the biofilm results in mechanical stress, which is assumed to be able to inhibit biofilm formation, to alter the behavior of the cultivated microorganisms, or even to inactivate them.

A comparison of the numerical results and some literature data for different biological materials gives strong evidence that the shear stresses present in a packed-bed biofilm reactor (unfortunately, in the literature there are no investigations regarding the effect of normal stress on biological materials, as yet!) are too small to be a major influence on biofilm formation or damage of micro-organisms.

As a consequence of this, the main effects of the flow are the transport of microorganisms during the biofilm formation phase and the transport of nutrients, pollutants and energy to the biofilm in later stages. It also appears reasonable that the local flow conditions will decisively influence the architecture of the biofilm. The most evident observation supporting this hypothesis is that the thickness of the biofilm is much higher in the inflow than in the outflow region of the reactor. The major reason for this is certainly the higher concentration of nutrients at the inflow, but there are also some hydrodynamic aspects contributing to the different thickness.

In the inflow zone, the flow is redistributed and a large number of stagnation points are formed. As a consequence of this, the time available for the adhesion of microorganisms to the carrier material and, therefore, the adhesion probability, is large. Similarly, the residence time of pollutants and nutrients in the neighborhood of an established biofilm is large. Thus, biochemical reactions occur without limitation. In contrast to this, the environmental conditions in the outflow zone reduce the generation of biofilm materials.

Data from literature and the results of our own numerical calculations show that the existence of the container wall leads to a high porosity region within a distance of about five sphere diameters. In this zone the local flow velocity is substantially higher than in the average channeling zone. In contrast to this, the porosity in a very large core region has a fixed mean value, which does not depend on the radial or axial position in the reactor. This results in a local flow, which does not have any significant characteristics, but depends on the random structure in the neighboring region.

Unfortunately, no specific microbiological and biochemical biofilm data for the core zone and the channeling region are available. But it is expected that the flow conditions will influence these sectors in a similar way as in the inflow and outflow zones. In this context, it is of crucial importance to look for a conclusive answer to the question of whether the flow in a laboratory apparatus is somewhat comparable to that present in the full-scale reactor. The results obtained numerically to date give only partial information on this topic. This is because of the very high numerical efforts required. Thus, most of the data calculated with a reasonable computing time are only adequate for approximating the flow processes in a laboratory reactor.

For simulating the reactors up to full-scale, different methods were suggested and tested in this chapter. The results found are very promising but also very sparse. Therefore, this question must remain unanswered.

No answers are available for the many additional questions. What is the interference between the flow forces and the adhesion forces (electrostatic, van der Waals forces, etc.)? Do the microorganisms adapt to mechanical forces even when they are relative low as found here? Do the microorganisms react to flow fluctuations? May fatigue of biological material play a role when describing possible effects of fluid mechanics on the biofilm? In what way does EPS protect the microorganisms in the biofilm? From the literature, it is known that shear flows are not as effective in homogenization as elongational flows and thus, what is the influence of the elongational character of the flow between the carrier particles on the nutrient transport and even on biological conversion? Can surface roughness influence the wastewater flow and, thus, the biofilm formation? In what way does flow influence the biofilm?

The blocking of certain reactor regions and the enlarged drag coefficient of a single carrier, as well as of the whole packing zone, is of major interest, as this crucially influences the operation. Nicolella and van Loosdrecht (1999) performed settling experiments with biofilm-coated particles and found an approximate 1.6 times higher drag coefficient compared with normal spherical particles. Projected to the expansion of the whole reactor, this indicates an enormous increase of the fluid-mechanical resistance during the growth of the biofilm during recirculation. The authors attributed the enlarged resistance of one particle to the surface roughness, a point that is certainly true for one particle, but when describing the whole reactor the increasing thickness of the biofilm must be included in the consideration. Therefore, one of the major challenges for simulation is to focus on the influence of the surface roughness and its effect, as well as the change of flow resistance, and on the formation of stagnant regions with no fluid flow due to excessive biofilm growth.

## 5.6 REFERENCES

Arnold, S. and Delgado, A. (1998) Untersuchungen zum Bewegungsverhalten einzelner Partikel in einer monoaxialen Dehnströmung unter kompensierter Gravitationswirkung. GAMM – Tagung, 6–9 April, 1998, Bremen.

Arnz, P. (2001) Biological nutrient removal from municipal wastewater in sequencing batch biofilm reactors. Ph.D. thesis, Institute of Water Quality Control and Wastewater Management, Technical Univ. Munich. (*Berichte aus Wassergüte- und Abfallwirtschaft* **164**.)

Azeredo, J. and Oliveira, R. (2000) The role of exopolymers produced by *Sphingomonas paucimobilis* in biofilm formation and composition. *Biofouling* **16**, 17–27.

Brück, D. (1998) Einfluß mechanischer Belastungen auf dispergierte, empfindliche Inhaltsstoffe von Flüssigkeiten in Leitungssystemen und Behälterströmungen, Lehrstuhl für Fluidmechanik und Prozessautomation, Technical Univ. Munich.

Daka, N. and Laidler, K.J. (1978) Flow kinetics of lactate dehydrogenase chemically attached to nylon tubing. *Can. J. Biochem.* **56**, 774–779.

Daka, J.N. and Laidler, J.K. (1988) Immobilisation and kinetics of lactate dehydrogenase at a rotating nylon disk. *Biotech. Bioengng* **32**, 213–219.

Debus, K. (1997) Numerische Untersuchung zur Kugelhaufendurchströmung – Ansätze zur Berechnung strömungsbedingter Deformation verformbarer Körper, Ph. D. thesis, Lehrstuhl für Fluidmechanik und Prozessautomation, Technical Univ. Munich.

Debus, K., Nirschl, H., Delgado, A. and Denk, V. (1998) Numerische Simulation des lokalen Impulsaustausches in Kugelschüttungen. *Chemie Ingenieur Technik* **70**(4), 415–418.

Duddridge J.E. and Kent, C.A. (1982) Effect of surface shear stress on the attachment of *Pseudomonas fluorescens* to stainless steel under defined flow conditions, *Biotech. Bioengng* **24**, 153–164.
Durst, F., Haas, R. and Interthal, W. (1987) The nature of flows through porous media. *J. Non-Newtonian Fluid Mech.* **22**, 169–189.
Dybbs A. and Edwards, R.V. (1995) A new look at porous media fluid mechanics – Darcy to turbulent. In *NATO ASI Ser.* **82**(1), 199–210.
Edwards, N. (1989) A novel device for the assessment of shear effects on suspended microbial cultures. *Appl. Microbiol. Biotech.* **30**, 190–195.
Ergun, S. (1952) Fluid flow through packed beds. *Chem. Eng. Prog.* **48**, 89–94.
Esterl, S., Nirschl, H. and Delgado A. (1998) Three dimensional calculations for the flow through packed beds. In *Proc. of the Fourth ECCOMAS Computational Fluid Dynamics Conference, Athens, 07.09-11.09, 1998.*
Flemming, H.-C. (1994) Biofilme, Biofouling und mikrobielle Schädigung von Werkstoffen, Habilitationsschrift, Forschungs- und Entwicklungsinstitut für Industrie- und Siedlungswasserw irtschaft sowie Abfallwirtschaft E. V. Stuttgart, Kommissionsverlag R. Oldenburg, München, Stuttgarter Berichte zur Siedlungswasserwirtschaft, Vol. 129.
George, S. and Chellapandian, M. (1996) Flow rate dependent kinetics of urease immobilized onto diverse matrices. *Bioprocess Engng* **15**, 311–315.
Götz, J. (1997) Visualisierung der Strömungsverhältnisse in schüttgefüllten Festbettreaktoren mit Hilfe der Kernspintomographie zur Bestimmung der lokalen Porositäten, der Geschwindigkeiten sowie der axialen und radialen Dispersionskonstanten, University of Karlsruhe, unpublished.
Guo, X., Nirschl, H. and Delgado A. (1998) Numerical 3-D-simulation of the behavior of aggregates in elongational flow. GAMM – Tagung, 6–9 April, 1998, Bremen.
Hinrichs, J. (1993) Die mechanische Stabilitat von Fettkugeln mit kristallinem Fettanteil im Srömungsfeld. Lehrstuhl für Fluidmechanik und Prozessautomation, Technical Univ. Munich.
Hoffmann, J. (1995) Ermittlung von maximalen Scherspannungen in Rührbehältern, *Chemie Ingenieur Technik* **67**(2), 210–214.
Horn H. and Hempel D.C. (1995) Mass transfer coefficients for an autotrophic and a heterotrophic biofilm system. *Wat Sci. Tech.* **32**(8) 199–204.
Horwatt, W.S. (1992) Simulation of the breakup of dense agglomerates in simple shear flows. *Rubber Chem. Tech.* **65**, 805–821.
Mersmann A., Schneider, G. and Voit, H. (1990) Selection and design of aerobic bioreactors, *Chem. Engng Tech.* **13**, 357–370.
Molerus, O. (1977) Druckverlustgleichung für die Durchströmung von Kugelschüttungen im laminaren und Übergangsbereich. *Chemie Ingenieur Technik.* **49**, 657–664.
Nicolella, C. and van Loosdrecht, M.C. (1999) Terminal settling velocity and bed-expansion characteristics of biofilm-coated particles. *Biotech. Bioengng* **62**, 62–70.
Nirschl, H. (1994) Mikrofluidmechanik – Numerische und experimentelle Untersuchungen zur Umströmung kleiner Körper. Ph.D. thesis, Lehrstuhl für Fluidmechanik und Prozessautomation, Technical Univ. Munich.
Nirschl, H. and Delgado, A. (1997) Simulation of particle behavior in elongational flow. In *Proc. 1st Intern. Symp. Food Rheol. Struct., Zürich, 16–21 March.*

Nirschl, H. Dwyer, H.A. and Denk, V. (1993) Three dimensional chimera grid scheme for the calculation of particle-wall interactions. In *Proc. of the 5th Int. Symp. of Computational Fluid Dynamics, Sendai Intern. Center. Aug.31- Sept.3, Tokyo, 1993*, Vol. II, pp. 357–362.

Nirschl, H., Dwyer, H.A. and Denk, V. (1995) Three dimensional calculations of the simple shear flow around a single particle between two moving walls. *J. Fluid Mech.* **283**, 273–285.

Papoutsakis, E. (1991) Fluid-mechanical damage of animal cells in bioreactors. *Trends Biotech.* **9**, 427–437.

Picioreanu, C. and van Loosdrecht, C. M. (1998) Mathematical modeling of biofilm structure with a hybrid differential-discrete cellular automaton approach, *Biotech. Bioengng* **58**, 101–116.

Picioreanu, C., van Loosdrecht, M. and Heinen, J.A. (2000) Theoretical study on the effect of surface roughness on mass transport and transformation in biofilms. *Biotech. Bioengng* **68**, 355–369.

Rottschäfer, K. (1996) Geschwindigkeitsverteilungen in durchströmten Füllkörperschüttungen. Ph.D. thesis, Lehrstuhl für Fluidmechanik und Prozessautomation, Technical Univ. Munich.

Visser, J. (1995) Particle adhesion and removal. *Particulate Sci. Tech.* **13**, 169–196.

# Modeling and simulation: Conclusions

## *Paul L. Bishop*

Wastewater biofilms are very complex structures. The architecture that a biofilm takes during growth is governed by a multitude of factors, such as water quality characteristics, types and amounts of substrates and nutrients, types and populations of microorganisms present, system hydrodynamics, biofilm age, type of support material, etc. Microscopic examinations of many biofilms have shown that biofilms can grow in a myriad of forms including as dense mats, in mushroom shapes, and in finger shapes.

The scientific question posed for this section of the book on Biofilm Modeling and Simulation was: *What is the most important factor determining biofilm architecture and hence transport processes within?* As can be seen by the preceding five chapters, major improvements in biofilm modeling and simulation approaches have been made in recent years, but in most cases they are still not sufficient to accurately predict the biofilm architecture resulting from a given set of environmental conditions. In many cases, the models can be tweaked to produce biofilm architectures that mimic one of those found in nature, and can then be used to evaluate impacts of changes of growth systems on the architecture. But what is really needed is the opposite: using the major determining biofilm architecture factors to develop the models. Before models can be used to adequately determine a resulting biofilm structure, though, we must first decide on what are the major controlling factors for biofilm growth and structure.

The first fact that must be accepted is that there is no, and probably never will be, single model that can describe biofilm growth processes. A single model that could account for all impacts on formation of biofilm architecture would be far too complex to be of practical use. It would be analogous to

---

© 2003 IWA Publishing. *Biofilms in wastewater treatment*. Edited by S. Wuertz, P.L. Bishop and P.A. Wilderer. ISBN: 1 84339 007 8.

attempting to predict global weather events resulting from a minor disturbance over the Atlantic Ocean. Instead, we should be developing simpler biofilm models that are relevant to a particular application.

The second general conclusion from this section is that, because of the many operations occurring in the biofilm and the complexity of the necessary models, it is essential that modeling and interpretation of model output be done by an interdisciplinary team of researchers, rather than by individuals. The team should consist of environmental engineers, microbiologists and chemists knowledgeable of biofilms, as well as mathematicians with strong modeling backgrounds.

Numerous approaches to biofilm modeling can be taken. In the past, most biofilm models used a one-dimensional continuum approach that used average spatial properties and did not predict exact changes at any given location. These were often sufficient for the uses they were put to. For example, they could quite adequately describe average nutrient profiles with depth in a biofilm. These models were soon expanded to two, and in some cases three, dimensions. As researchers began to ask more detailed questions, though, such as whether there are channels in a biofilm that could affect transport processes, these models were found to be inadequate. Models that described the physical architecture of the biofilm were needed.

Most biofilms models now being developed are based on use of cellular automata. These assume a grid structure, where the grid units represent spaces in a biofilm. They can be occupied by microorganisms or can be empty. By applying rules that depend on local conditions in the biofilm, such as the nutrient supply or hydrodynamics, the biofilm can be allowed to 'grow' (occupy new spaces in the grid) in iterative steps that meet the rules. Great strides have been made in cellular automaton biofilm modeling over the past five years, and they can now be used to depict the shape of biofilm structures seen in nature under a variety of growth conditions. However, they still have their shortcomings. For one thing, they only represent the microorganisms in the biofilm, not the biofilm itself. They do not yet describe the extracellular polymeric substances (EPS) surrounding the microbial cells. Also, it is relatively easy to significantly change the outcome of the simulation by relatively small changes in the input conditions. This may be correct, indicating that minor changes in biofilm environmental parameters could greatly influence biofilm architecture, or it could mean that we have not yet honed in on the most important parameters to use in the models.

This brings us back to the original question: *What is the most important factor determining biofilm architecture and hence transport processes within?* At the present time, we cannot answer this question, based on modeling efforts. Models can be used to depict the impacts of changes in important parameters, but the models were originally developed based on a given set of parameters as being important. This does not allow true validation. We need to independently determine what the major parameters are, and then use them in the development of biofilm models. Modeling techniques are becoming sufficiently sophisticated that, given the proper boundary conditions, useful models could be developed fairly quickly.

Section 2 of this book addresses the question of whether physicochemical and microbial population analyses can be used to explain and predict biofilm system performance. The findings from this section should be used in the future to construct more comprehensive models.

# PART TWO

## ARCHITECTURE, POPULATION STRUCTURE AND FUNCTION

# Architecture, population structure and function: Introduction

*Stefan Wuertz*

Once it becomes obvious that biofilms are not homogeneous, the next question is about the consequences for their functioning. Researchers have long wondered about the microscopic details of biofilms and whether their physical structure (architecture) has any effect on nutrient gradients, mass transport, metabolic conversions, and stability of biofilms. This section addresses not only the architecture but also the community structure of the biofilm as a determinant of function. Experimental techniques and examples of modeling approaches are introduced throughout and integrated into the various chapters.

*Can information obtained by physicochemical analytical techniques and microbial population studies be used to explain and predict the performance of biofilm systems?* Similar to the previous question in part 1, we must consider spatial scales such as microscopic versus reactor. State-of-the-art measurements of chemical gradients like those made possible by miniaturized sensors, microscopic detection of specific microorganisms leading to enumeration of key bacterial players like nitrifiers, and analysis of the phylogenetic diversity of microbial populations by gene cloning have been done on model laboratory-scale systems. The following chapters address many of the techniques involved and provide insights into how such detailed knowledge can be used to understand the way biofilms work.

Chapter 6 gives an overview of things to come by summarizing biofilm structural components and giving examples of experimental investigations intended to introduce basic concepts including extracellular polymeric substances, detachment, and microbiological processes. The effect of biofilm architecture on mass transport processes and the incorporation of data obtained

---

© 2003 IWA Publishing. *Biofilms in wastewater treatment.* Edited by S. Wuertz, P.L. Bishop and P.A. Wilderer. ISBN: 1 84339 007 8.

from microelectrode and confocal laser scanning microscopy studies into mathematical models is discussed in Chapter 7.

Extracellular polymeric substances (EPS) used to be referred to as 'slime' by engineers and were assumed to consist primarily of polysaccharides by microbiologists. Their role in microbial aggregation and settling characteristics of activated sludge flocs were investigated in the 1970s and 1980s, yet the fundamental role of EPS in determining biofilm properties and the varied functions of EPS have only recently begun to be discovered, as described in Chapter 8. Adhesion is a fundamental process in biofilm formation, and in cases where biofilm growth is undesirable, such as in drinking water distribution systems, both engineering and treatment concepts are designed to disrupt the initial colonization of surfaces. In wastewater treatment, biofilms are encouraged to form and Chapter 9 deals with the theoretical aspects of adhesion. This is followed by considerations of surface properties of growth supports relevant for adhesion, underscored by numerous examples of laboratory studies.

Structure to a microbiologist first of all means population structure; that is, a general or detailed knowledge of the types of microorganism present in a biofilm or biofilm reactor. Second, the spatial distribution of the key microorganisms inside a biofilm is of interest as well as their numerical abundance. Until recently, most of this information was not available. Does it matter in environmental engineering? Only if such knowledge can be related to function. For example, it is important to know which type of nitrifying or phosphate-removing bacteria are present in a reactor system to be able to respond to environmental changes that lead to a loss of these key microorganisms. Chapter 10 discusses examples where the population structure of biofilm reactors – including higher microorganisms that are not bacteria – has been correlated with performance. Biofilms are dynamic systems experiencing continuous growth of new biomass as well as loss of biomass. The mechanisms involved in detachment and the influence of detachment dynamics on nutrient removal processes are the subject of Chapter 11.

# 6
# The effect of biofilm heterogeneity on metabolic processes

*Paul L. Bishop*

## 6.1 INTRODUCTION

In the past decade, great strides have been made towards a better understanding of the structure and function of biofilms used in water and wastewater treatment. We now have a great deal of knowledge about the physical, chemical and biological makeup of biofilms; microbial attachment, detachment and growth mechanisms in biofilms; and factors controlling mass transport in biofilms. However, two questions are commonly asked: 'what effects do these biofilm properties have on metabolic properties?' and 'can information obtained by physicochemical analytical techniques and microbial population analysis be used to explain and predict the performance of biofilm systems?'

The following is a discussion of what is currently known about biofilm properties and how this information can be obtained. The effect of these properties on metabolism is also provided.

© 2003 IWA Publishing. *Biofilms in wastewater treatment.* Edited by S. Wuertz, P.L. Bishop and P.A. Wilderer. ISBN: 1 84339 007 8.

## 6.2 BIOFILM PROPERTIES

Recently, conventional concepts about the internal structure of wastewater biofilms have been brought into question. It was previously assumed that biofilms could be modeled in one dimension (perpendicular to the substratum). It is now known that this is not normally the case. Biofilms are highly heterogeneous in structure. This has a significant impact on how the biofilm grows and on mass transport of materials into and out of the biofilm. It also is one of the factors governing detachment of biomass from the biofilm, a process that keeps the remaining biofilm in a more active state, but which also results in suspended solids in the effluent. These solids usually must be removed before final discharge of the wastewater to a receiving stream.

### 6.2.1 Heterogeneity

It has been known for about two decades that biofilms are usually spatially and temporally non-uniform (Bishop and Rittmann 1995). Wanner and Gujer (1986) and others proposed mechanistic models that could describe the development and metabolic functioning of biofilms in one dimension, perpendicular to the substratum. The validity of these models was later demonstrated by several researchers using microelectrodes and cryo-sectioning techniques to establish actual constituent profiles in the biofilm (Revsbech 1994; Zhang and Bishop 1994b,c,d; Zhang et al. 1994, 1995a,b; Bishop et al. 1995; Bishop and Rittmann 1995; Kudlich et al. 1996; Schramm et al. 1996; Bishop 1997; de Beer et al. 1997; Zhou and Bishop 1997; Santegoeds et al. 1998; Yu and Bishop 1998, 1999; Bishop and Yu 1999; de Beer and Schramm 1999). Figure 6.1 shows actual constituent profiles in a biofilm undergoing both heterotrophic biodegradation and nitrification, measured using microelectrodes. As can be seen, dissolved oxygen concentrations decrease rapidly from that present in the bulk solution, ammonia concentrations decrease and nitrate increases because of nitrification, and pH decreases, also due to nitrification. The local oxidation–reduction potential changes as the mode of metabolism changes from aerobic oxidation of organics to nitrification. One-dimensional models can readily be used to model these effects, as will be described later. More recently, such techniques as confocal laser scanning microscopy (CLSM) (de Beer et al. 1994a,b; de Beer and Stoodley 1995; DeLeo et al. 1997; de Beer and Schramm 1999; Ebihara and Bishop 1999; Lawrence and Nieu 1999; Kloep et al. 2000; Yang et al. 2000) and nuclear magnetic resonance (NMR) (Costerton et al. 1995; Lens et al. 1999) have been used to demonstrate the heterogeneity of biofilms.

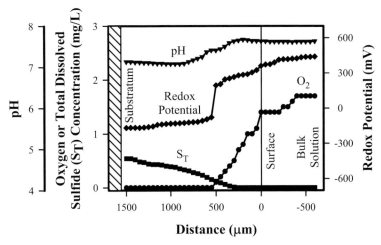

**Figure 6.1.** Variations in constituent profiles in a biofilm with depth (Yu and Bishop 1998).

Recent research has shown, however, that many biofilms have non-uniformities in physical properties (density, porosity, diffusivity, etc.) with depth, as well as in constituent concentrations (Zhang and Bishop 1994a,b,d; Zhang et al. 1994, 1995a,b; Bishop 1997; Stewart 1993, 1998; Beyenal et al. 1998; Lens et al. 1999; Beyenal and Lewandowski 2000). As can be seen from Figure 6.2, biofilm density increases with depth into the biofilm, whereas porosity, permeability, and effective diffusivity decrease. Biofilms may also be non-uniform laterally, leading to advective transport, as well as diffusive transport, in the biofilm (de Beer et al. 1994b; de Beer and Stoodley 1995; Massol-Deya et al. 1995). Figure 6.3 depicts several possible biofilm growth modes. The cluster-shaped and columnar biofilms may allow for advective transport deep into the biofilm. This can lead to variabilities in microbial diversity and reaction processes in different locations within the biofilm (Bishop and Kinner 1986). Finally, the surfaces of biofilms are not usually flat; the rough contours of the surface impact the hydrodynamics of flow past the biofilm, and consequently mass transport from the bulk liquid into the biofilm, as well as promoting detachment of biomass from the biofilm through scour (Wanner 1993; Zahid and Ganczarczyk 1994a,b; Gibbs and Bishop 1995; Hermanowicz et al. 1995; Bishop et al. 1997; Picioreanu et al. 1998, 1999; Morgenroth and Wilderer 2000). Figure 6.4 is a depiction of how the conformation of the biofilm surface can influence mass transport into the biofilm (Picioreanu et al. 2000).

128    Biofilms in wastewater treatment

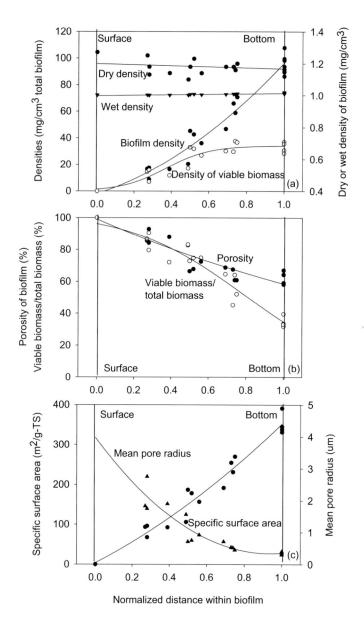

**Figure 6.2.** Spatial distributions of biofilm properties: (a) density distributions; (b) distributions of porosity and ratio of viable biomass to total biomass; (c) distributions of specific surface area and mean pore radius (Zhang and Bishop 1994a).

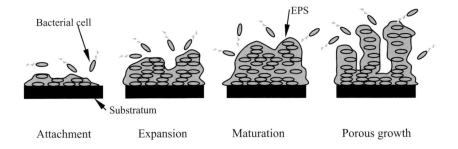

**Figure 6.3.** Modes of biofilm growth.

Smooth biofilms transport substrate into the biofilm in approximately equal flux rates over the entire biofilm surface, whereas in very rough surfaced biofilms the microorganisms below the troughs see very little substrate.

Many have questioned whether these microscale variabilities have any impact when the scale is extrapolated several orders of magnitude to a full-scale biofilm system. There is still no definitive answer to this question because useful scale-up models have not been fully developed yet. However, knowledge of the inner workings of a biofilm can provide useful insight into critical processes that govern biofilm behavior under various environmental and operational conditions. This should be of benefit in designing, controlling and troubleshooting full-scale biofilm systems (refer to Chapter 5).

## 6.2.2 Extracellular polymeric substances

Biofilms are not composed solely of microorganisms; in fact, microorganisms generally only make up about 2–5% of the total biomass (Flemming 1995; Lazarova and Manem 1995; Zhang *et al.* 1997; Jahn and Nielsen 1998; Zhang *et al.* 1998a,b, 1999). The greatest amount of the biofilm is composed of extracellular polymeric substances (EPS) produced by the microbes or resulting from cell lysis (see Chapter 8). Most of this material is made up of carbohydrates and proteins, but there are also other constituents such as DNA and humic substances (Zhang *et al.* 1997, 1998a,b, 1999). Jahn and Nielsen (1998) reported a typical composition of the EPS of biofilms present in a gravity sewer system (see Figure 6.5). Exopolymers are important to the biofilm as they have been associated with the initial adhesion of bacteria to a surface, protecting the microbes from dehydration and toxic substances, and providing ion exchange properties due to negatively charged surface functional groups which allow them to bind cationic species such as heavy metals (Sutherland 1977).

EPS production is not constant in biofilms, and it varies from one biofilm to another. EPS materials are also spatially variable. Figure 6.6 depicts the spatial distribution with depth for carbohydrates and proteins in biofilms of different thicknesses. It is evident that EPS yields decrease with depth into a biofilm in these samples.

We are beginning to understand the role that environmental conditions play in governing the EPS production rate and composition. It may be that in the future, by controlling operating conditions, we will be able to induce the biofilm microorganisms to produce beneficial EPS materials that will protect the microbes from toxicity and enhance contaminant mass transport and biodegradation.

## 6.2.3 Diffusivity

Solutes are transported in microbial biofilms by a combination of advection and diffusion. The heterogeneous structure of a biofilm may allow some advective transport through channels in the biofilm, but the most important mechanism is generally molecular diffusion through the pores and small channels located in the EPS material (Fu *et al.* 1994; Zhang and Bishop 1994a,b,c,d; Zhang *et al.* 1994; Beyenal *et al.* 1998; Stewart 1998; Brito and Melo 1999; Beyenal and Lewandowski 2000). The largest component of biofilm is water, but the diffusivity of a molecule in the biofilm is generally less than that in water because of the tortuosity of the pores and the minimal permeability of the biofilm. The effective diffusion coefficient must be used. Stewart (1998) presents an excellent compilation of measured effective diffusion coefficients in biofilms.

It has been experimentally demonstrated that the effective diffusion coefficient $D_e$ for a given solute can vary with depth into the biofilm (Fu *et al.* 1994; Zhang and Bishop 1994a,b,c,d; Zhang *et al.* 1994; Beyenal *et al.* 1998; Stewart 1998; Beyenal and Lewandowski 2000). Figure 6.7 depicts a typical variation in effective diffusivity for oxygen with depth in a biofilm. $D_e$ decreases with depth because of the increasing density and decreasing porosity and permeability of the biofilm with depth. Flow velocity past the biofilm is a major controlling factor determining biofilm density (Zhang and Bishop 1994a,b,c,d; Beyenal and Lewandowski 2000). Figure 6.8 depicts variations in effective diffusivities with depth in a biofilm exposed to various bulk liquid flow velocities and substrate concentrations (Zhang and Bishop 1994b,c; Zhang *et al.* 1994; Beyenal and Lewandowski 2000). As can be seen, varying the flow rate past the biofilm in a reactor can influence the effective diffusivity of solutes in the biofilm, and consequently affect the rate of mass transfer and contaminant biodegradation. Picioreanu *et al.* (2001) showed that the flow régime, expressed as the Reynolds number, has a significant impact on the structure of a biofilm and on the resulting substrate concentration isoconcentration lines in the biofilm

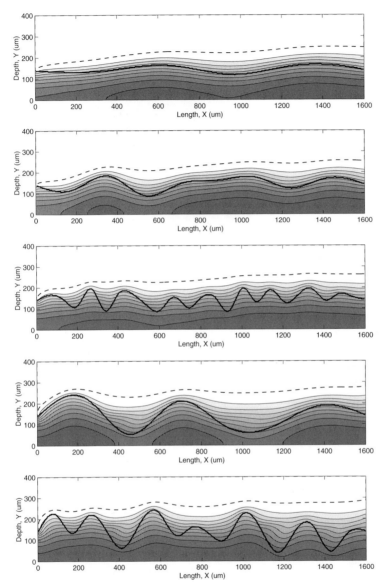

**Figure 6.4.** Substrate isoconcentration lines at Reynolds number, $Re$, = 4.1 for biofilms with various roughnesses. Dashed lines indicate the boundary layer (Picioreanu *et al.* 2000).

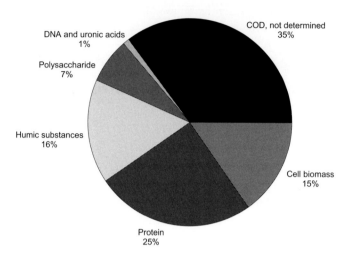

**Figure 6.5.** Typical composition of a gravity sewer biofilm (Jahn and Nielsen 1998).

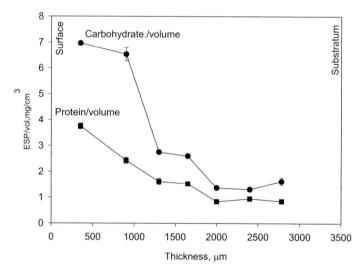

**Figure 6.6.** Spatial distribution of EPS carbohydrates and proteins, normalized to volume, with depth into a biofilm (Zhang *et al.* 1998b).

(Figure 6.9). At an *Re* of 13.3, the biofilm colonies grew closer together and quickly became compact with few channels, owing to substrate mass transfer limitations; at a lower *Re* of 6.7, plenty of nutrients are available, growth occurs in all directions, and columnar and mushroom-shaped biofilm colonies developed.

Brito and Melo (1999) found that periodic changes in the upflow bulk fluid velocity in a fluidized bed reactor could be used as a tool to increase the transport of soluble substrates inside biofilms. The biofilm structure was also found to be important for transport of inert particulates into biofilms. Little research has been done on particulate removal and degradation by biofilms. However, Okabe et al. (1998) found that macropores (20–200 µm diameter) in biofilms facilitated particulate transport into the biofilm.

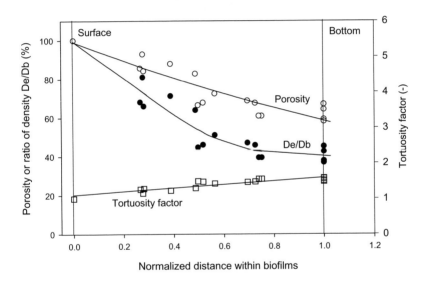

**Figure 6.7.** Spatial distributions of biofilm properties: distributions of porosity, ratio of $D_e/D_b$ (effective diffusivity / diffusivity in water), and tortuosity factor distributions (adapted from Zhang and Bishop 1994b).

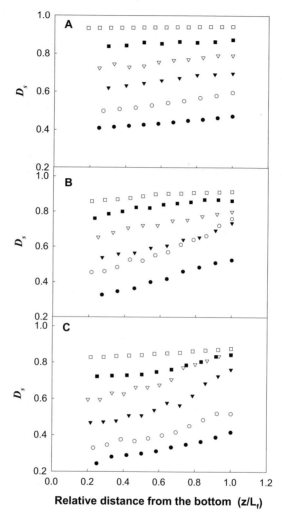

**Figure 6.8.** Variations of $D_s$ with relative distances from the bottom of biofilms (Beyekal and Lewandowski 2000).

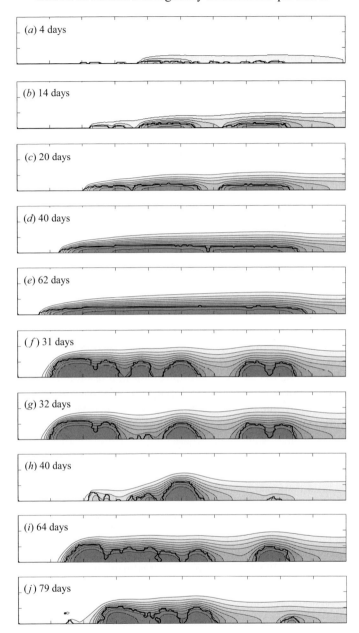

**Figure 6.9.** Simulated biofilm evolution in time for (*a*–*e*) $Re = 13.3$ and (*f*–*j*) $Re = 6.7$. Arrows represent the vector velocity. Substrate isoconcentration lines are also shown (Picioreanu *et al.* 2001).

## 6.2.4 Detachment

Biomass does not remain attached to a surface forever. Stewart (1993) described four processes by which biomass can be detached from a biofilm: (1) abrasion, (2) erosion, (3) sloughing, and (4) predator grazing. This biofilm detachment also plays a significant role in the heterogeneous structure of biofilms (see Chapter 11). Morgenroth and Wilderer (2000) found that overall reactor performance and biofilm structure are significantly affected by detachment. Different detachment patterns have a significant influence on the organism distribution within the biofilm and on overall process performance. Faster-growing heterotrophs are favored over autotrophic biomass in situations with long intervals between detachment, with resulting high variations in biofilm thickness. Within a biofilm, different bacterial populations have to find their optimal niche (Bishop and Kinner 1986). Faster-growing heterotrophs are dominant at the surface where they can take advantage of high substrate concentrations (Zhang *et al.* 1994). Detached organisms can be quickly replaced. Slower growing organisms, such as autotrophs, find a niche deeper in the biofilm where they can balance their need for nutrients with their slower replacement due to possible detachment, which would be greater nearer the surface. This can be readily seen in Figure 6.10, which shows microbial population distributions with depth into the biofilm in several types of biofilms (Bishop *et al.* 1995).

It may be possible to operate a biofilm reactor in a fashion that encourages a desired biofilm growth pattern. If high populations of fast-growing heterotrophs are desired, the flow rate past the biofilm should be increased to allow shear to minimize the thickness of the biofilm and to allow only faster-growing organisms to become established. Alternatively, if growth of a slow-growing organism is desired, such as those that degrade recalcitrant azo dyes, then development of a thicker biofilm should be the goal. This method of population control has been demonstrated for degradation of azo dyes (Harmer and Bishop 1992; Jiang and Bishop 1994; Seshadri *et al.* 1994; Willis and Bishop 1995; Zhang *et al.* 1995a; Kudlich *et al.* 1996; Coughlin *et al.* 1997, 1999).

Detachment processes may also be important for controlling biofilm predators. An example is the damage done to a biofilm through grazing by snails and worms. The nitrification towers at a treatment plant in Dayton, Ohio, experienced numerous upsets due to loss of biofilm. During these periods, ammonia oxidation and nitrate production decreased substantially, resulting in unacceptable ammonia concentrations in the plant effluent. The cause of this was found to be large infestations of the snail *Physa gyrina* in the nitrifying biofilm. The snails grazed across the biofilm, devouring much of it and causing other parts of the biofilm to detach (Palsdottir and Bishop 1997).

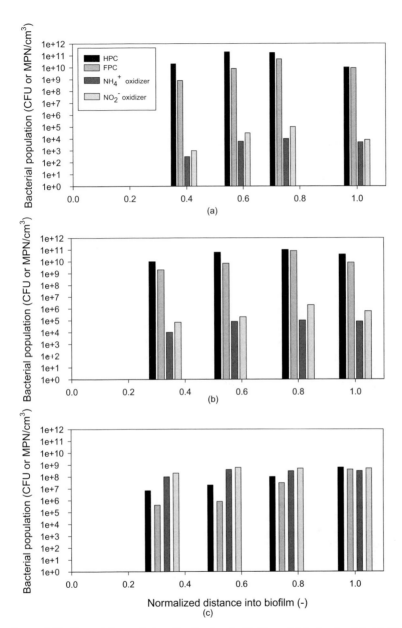

**Figure 6.10.** Bacterial population distribution in biofilms: (a) heterotrophic biofilm; (b) heterotrophic/autotrophic biofilm; (c) autotrophic biofilm (Zhang 1994).

The biofilm's microbial community structure plays a major role in determining the heterogenic properties of a biofilm. This goes both ways, however, the physical structure of the biofilm also influences the types and populations of microorganisms that can become established there. Three related topics will be discussed here: population dynamics in the biofilm, the role of cell-to-cell communication in establishing the biofilm architecture, and the use of molecular probes to describe the microbial community.

## 6.3 CONFOCAL LASER SCANNING MICROSCOPY

After colonizing a surface, microbial cells undergo phenotypic changes and produce exopolysaccharides, as described above. The cells subdivide, forming microcolonies. Some of these microcolonies form cone-shaped structures (Wimpenny and Colasanti 1997). Eventually, these cluster or mushroom-shaped colonies may fuse together, in many cases leaving water channels that penetrate almost to the base of the film. The size of these channels may range from sub-micrometer to tens or hundreds of micrometers in diameter. In other cases, the microcolonies can form stacks attached to the substratum at the base, but generally separated from their neighbors (Walker *et al.* 1995). Often, there is a thin continuous biofilm at the base between the stacks. A third general biofilm architecture is that of a dense biofilm, in which the microcolonies grow together leaving only small diameter channels throughout the structure.

The validity of these predicted structural forms has been confirmed by analytical techniques. The use of microscopic techniques to observe the structure of a biofilm, such as scanning electron microscopy or transmission electron microscopy, was limited until recently, because the procedures required to prepare the biofilm specimen for observation often resulted in distortion and artifacts that led to misleading results. This has changed with the introduction of the confocal laser scanning microscope (CLSM). CLSM can be used to examine a biofilm in its natural hydrated state, avoiding the problems of fixation and dehydration, and can provide a computer-generated three-dimensional depiction of the biofilm's structure (Wimpenny and Colasanti 1997). The CLSM has been used to verify that the biofilm architecture is quite heterogeneous and that it is filled with multi-sized water channels. It has also been used to show that some of these channels are continuous and large enough to allow convective transport of particulates through the biofilm (Stoodley *et al.* 1994). In a few cases, microelectrodes have shown that concentration gradients in the water channels are different than in the adjacent biofilm (de Beer *et al.* 1993, 1994a,b) indicating that convective transport is occurring, although this evidence is missing in most microelectrode studies.

## 6.4 CELL-TO-CELL COMMUNICATION

Recently, it has been shown that some bacteria, in the appropriate environment, possess the ability to communicate with other bacteria. It is postulated that they can use this ability to control the construction of their biofilm in such a way as to maximize mass transport of nutrients to microorganisms within the biofilm and thus enhance growth (Davies *et al.* 1998; Kolter and Losick 1998; Møller *et al.* 1998; De Kievit and Iglewski 1999). They can control both biofilm formation and detachment. The process is controlled by the same acylated homoserine lactone (acyl-HSL) signal molecules that control quorum sensing in several Gram-negative bacteria, including *Pseudomonas aeruginosa*. The acyl-HSLs are secreted by bacteria and act as extracellular signaling molecules that activate genes to control many phenomena, such as bioluminescence, exoenzyme synthesis, and virulence factor production (Fuqua *et al.* 1994). The acyl-HSLs are secreted in proportion to the total number of cells present and provide an index of population densities. It is postulated that the bacteria may use these signaling molecules to prevent overcrowding and create channels for nutrient transport (Kolter and Losick 1998). They may signal both where growth should occur and where detachment is needed for the overall health of the biofilm community.

## 6.5 MOLECULAR PROBES

A major development in biofilm research has occurred in recent years because of the application of molecular probes for estimating specific biofilm microbial populations, their growth kinetics, and their spatial distributions. These techniques are based on the detection and quantification of target molecules that are unique for specific microbial populations, such as lipids and membrane components, proteins, and nucleic acids. The most widely used techniques rely on the identification of signature sequences present in the DNA of a population (e.g., ribosomal RNA (rRNA) genes or metabolic genes) (Oerther *et al.* 1999). Short-chain oligonucleotide probes can be designed to locate and attach themselves to these specific rRNA sequences. These probes are usually designed so that they fluoresce, allowing the locations of these hybridized probes to be easily determined using confocal or epifluorescence microscopy.

There are two general methods for using oligonucleotide probes. In the first method, the microbial populations' nucleic acids are extracted and then probed using the polymerase chain reaction (PCR) procedure and determination of nucleic acid sequences. The result is a measure of the total amount of rRNA in a sample. Knowing the amount of rRNA present in an individual cell, these results can be used to estimate populations of specific microorganisms in a sample. The second technique assays the nucleic acids present in whole cells using a process called fluorescent *in situ* hybridization (FISH). Using this technique, individual cells of a specific type of microorganism can be detected. They can not only be counted to determine populations, but their spatial locations within a biofilm can be determined by using CLSM (Manz 1999) or cryosectioning into thin slices and using fluorescence microscope visualization of the target organisms in each slice (Lisle *et al.* 1999). Fry *et al.* (1997) provide an excellent overview of available technologies.

A few studies have demonstrated that the signal intensity of whole-cell hybridizations can be used to estimate cellular growth rate, for example, using flow cytometry (Amann *et al.* 1990). These procedures assume a relationship between rRNA content and cell growth rate. These methods, and interpretation of the results, must be used with care, however. It is essential to document that the correlation between growth rate and ribosomal number per organism is valid (Sternberg *et al.* 1999). Second, the stability of ribosomes under a variety of conditions may be different in different bacteria, and third, the correlation between growth rate and ribosome concentration may be relevant only for bacteria able to grow fast under optimal conditions.

Møller *et al.* (1998) investigated toluene degradation during mixed-culture microbial growth in a biofilm. They found that the two dominant organisms responsible for toluene degradation occupied different niches in the biofilm, with *Acinetobacter* sp. strain C6 primarily attaching to the substratum and *Pseudomonas putida* R1 being the dominant organism in the outermost layers of the biofilm. They suggested that close physical association between the two organisms was not essential for biofilm proliferation.

Nitrification and the microbial community structure responsible for the nitrification in a fluidized bed reactor were studied using FISH (Kloep *et al.* 2000). Probes for several types of bacteria were used. It was found that several of the ammonia-oxidizing bacteria grew in dense, spherical- to oval-shaped cell clusters. These represented a large proportion of the bacteria present. Using different probes, they found a population shift within the ammonia-oxidizers over time, suggesting the importance of temporal characterization of microbial communities.

Sternberg *et al.* (1999) investigated the influence of adding a new carbon source to a biofilm community on gene expression and microbial growth. They found that when bacterial colonies reached a certain critical size, microbial activity at the center of the microcolony decreased, compared with that of cells in the periphery of the colony, owing to a signaling reporter system. However, they found that the shift to a better carbon source for the bacteria present in the biofilm activated the cells in the microcolonies and enhanced their growth rates.

The activation penetrated all the way through the microcolonies to the inactive organisms in the center of the colony. The experiments revealed a modulated growth activity in a biofilm. In the initial phases, all cells are highly active, with ribosomal promoter activity corresponding to that of rapidly growing cells. After primary colonization and formation of small colonies, activity decreased in the center of the colony and gradually radiated out to the colony surface.

Fry et al. (1997) used two sets of oligonucleotide hybridization probes, designed to monitor sulfate-reducing bacteria (SRB) and methanogens, to follow microbial structural changes in anaerobic biofilms in response to perturbation of the influent medium composition. They found that methanogens and SRB co-existed, regardless of sulfate availability as electron acceptors, which is contrary to the generally held view that SRB have a competitive advantage over methanogens. Several groups of SRB were present at significant levels. Finally, they found that *Methanosarcina* and *Methanosaeta* (formerly *Methanothrix*) species co-existed under all conditions tested, indicating that bulk liquid acetate concentrations were probably significantly different from acetate concentrations in the biofilm.

## 6.6 CONCLUDING REMARKS

It is now well established that wastewater biofilms are highly heterogeneous in nature. Not only do they typically contain a wide variety of microorganisms, the locations of those microbes are also varied. It appears that the diversity and populations of microorganisms can have an impact on the architecture of the biofilm. Cell-to-cell communication through the excretion of signaling molecules may lead to a structured biofilm that allows efficient growth of the microbial colony as a whole at the expense of individual cells. These signaling molecules may turn metabolic systems on and off in cells to promote a desired biofilm structure, or may lead to sloughing of undesired biomass. The importance of this phenomenon in a mixed culture, where organisms that do not respond to the signaling molecule may colonize the channels created by other microorganisms (substrate concentrations will be highest there and most conducive to colonization), is not yet known.

Microbial populations present in a biofilm are not the only determinants of the heterogeneity of the biofilm structure. Hydrodynamics also plays a significant role. Higher liquid flow velocities past the biofilm can lead to better mass transport of substrate to the biofilm. Hydrodynamics, owing to erosion of surface organisms and control of substrate transport, also partly controls the shape of the biofilm. Smooth biofilms transport substrate into the biofilm equally across the surface of the biofilm, whereas transport is unequal in rough surfaced biofilms, where microorganisms in the troughs or at the bottom of channels may see little substrate. This can significantly affect the spatial distribution of metabolic activity in the biofilm.

Finally, the structural properties of the biofilm vary with depth into the biofilm, whatever the conformation of the biofilm. In general, biofilm densities increase with depth, while porosity and pore diameter decrease with depth. This leads to a decrease in effective diffusivity of substrates and nutrients as they move into the biofilm, which in turn decreases metabolic activity in the deeper reaches of the biofilm. The EPS also varies considerably with depth into the biofilm. This is important as the EPS plays a significant role in adsorption and sequestering of diffusing materials, and in obstructing transport of metabolites.

The microbiology of a biofilm and the resulting physicochemical reactions occurring there are very complex. Much more research needs to be done to develop the interrelationships taking place there. Mechanistic modeling of biofilm growth is still in its infancy, but is rapidly improving. This modeling will be needed to effectively and accurately describe what is actually occurring during biofilm development and waste degradation. Experimental results can be used to verify model findings, but because of the complexity of biofilm growth, models will be needed to make sense of the highly varied biofilm architectures that can be observed in nature.

## 6.7 REFERENCES

Amann, R., Ludwig, W. and Schleifer, K. (1990). Phylogenetic identification and in situ detection of individual microbial cells without cultivation. *Microbiol. Rev.* **59**, 143–169.
Beyenal, H. and Lewandowski, Z. (2000) Combined effect of substrate concentration and flow velocity on effective diffusivity in biofilms. *Wat. Res.* **34**(2), 528–538.
Beyenal, H., Tanyolac, A. and Lewandowski, Z. (1998) Measurement of local effective diffusivity in heterogeneous biofilms. *Wat. Sci. Tech.* **38**(8/9), 171–178.
Bishop, P. (1997) Biofilm structure and kinetics. *Wat. Sci. Tech.* **36**(1), 287–294.
Bishop, P. and Kinner, N. (1986) Aerobic fixed-film processes. *Biotechnology*. H. Rehm and G. Reed. Weinheim, Germany, VCH Verlagsgellschaft. **8**, 113–176.
Bishop, P. and Yu, T. (1999) A microelectrode study of redox potential change in biofilms. *Wat. Sci. Tech.* **39**(7), 179–185.
Bishop, P., Zhang, T. and Fu, F.-Y. (1995) Effect of biofilm structure, microbial distributions and mass transport on biodegradation processes. *Wat. Sci. Tech.* **31**(1), 143–152.
Bishop, P., Gibbs, J. and Cunningham, B. (1997) Relationship between concentration and hydrodynamic boundary layers over biofilms. *Environ. Tech.* **18**, 375–386.
Bishop, P. and Rittmann, B. (1995) Modelling heterogeneity in biofilms: report of the discussion session. *Wat. Sci. Tech.* **32**(8), 263–265.
Brito, A. and Melo, L. (1999) Mass transfer coefficients within anaerobic biofilms: Effects of external liquid velocity. *Wat. Res.* **33**(17), 3673–3678.

Costerton, J., Lewandowski, Z., Caldwell, D., Korber, D. and Lappin-Scott, H. (1995) Microbial biofilms. *Annu. Rev. Microbiol.* **49**, 711–745.

Coughlin, M., Kinkle, B. and Bishop, P. (1999) Degradation of azo dyes derived from amino-2-naphthol by *Sphingomonas* strain 1CH. *J. Indust. Microbiol.* **21**, 341–346.

Coughlin, M., Tepper, A., Kinkle, B. and Bishop, P. (1997) Characterization of aerobic azo-dye degrading bacteria and their activity in a wastewater biofilm. *Wat. Sci. Tech.* **36**(1), 215–220.

Davies, D., Parsek, M., Pearson, J., Iglewski, B., Costerton, J. and Greenberger, E. (1998) The involvement of cell-to-cell signals in the development of a bacterial biofilm. *Science* **280**, 295–298.

de Beer, D. and Schramm, A. (1999) Microenvironments and mass transfer phenomena in biofilms studied with microsensors. *Wat. Sci. Tech.* **39**(7), 173–178.

de Beer, D., Schramm, A., Santegoeds, C. and Michael, K. (1997) A nitrite microsensor for profiling environmental biofilms. *Appl. Environ. Microbiol.* **63**, 973–977.

de Beer, D. and Stoodley, P. (1995) A relation between the structure of an aerobic biofilm and transport phenomena. *Wat. Sci. Tech.* **32**(8), 11–18.

de Beer, D., Stoodley, P. and Lewandowski, Z. (1994a) Liquid flow in heterogeneous biofilms. *Biotech. Bioengng* **44**, 636.

de Beer, D., Stoodley, P., Roe, F. and Lewandowski, Z. (1994b) Effects of biofilm structures on oxygen distribution and mass transfer. *Biotech. Bioengng* **43**, 1131–1138.

de Beer, D., Van den Heuvel, J. and Ottengraf, S. (1993) Microelectrode measurement of the activity distribution in nitrifying bacterial aggregates. *Appl. Environ. Microbiol.* **59**, 573–579.

De Kievit, T. and Iglewski, B. (1999) Quorum sensing, gene expression and *Pseudomonas* biofilms. *Methods Enzymol.* **310**, 117.

DeLeo, P., Baveye, P. and Ghiorse, W. (1997) Use of confocal laser scanning microscopy on soil thin-sections for improved characterization of microbial growth in unconsolidated soils and aquifer materials. *J. Microbiol. Methods* **30**, 193–203.

Ebihara, T. and Bishop, P. (1999) Biofilm structural forms utilized in bioremediation of organic compounds. *Wat. Sci. Tech.* **39**(7), 203–210.

Flemming, H.-C. (1995) Sorption sites in biofilms. *Wat. Sci. Tech.* **32**(8), 27–34.

Fry, N., Raskin, L., Sharp, R., Alm, E., Mobarry, B. and Stahl, D, (1997) In situ analyses of microbial populations with molecular probes. In *Bacteria as Multicellular Organisms*. (ed. J. Shapiro and M. Dworkin), pp. 292–336. New York, Oxford University Press.

Fu, Y.-C., Zhang, T. and Bishop, P. (1994) Determination of effective oxygen diffusivity in biofilms in a completely mixed biodrum reactor. *Wat. Sci. Tech.* **29**(10/11), 455–462.

Fuqua, W., Winans, C. and Greenberg, E. (1994) Quorum sensing in bacteria: the LuxR-LuxI family of cell density-responsive transcriptional regulators. *J. Bacteriol.* **176**(2), 269–276.

Gibbs, J. and Bishop, P. (1995) A method for describing biofilm surface roughness using geostatistical techniques. *Wat. Sci. Tech.* **32**(8), 91–98.

Harmer, C. and Bishop, P. (1992) Transformation of azo dye AO-7 by wastewater biofilms. *Wat. Sci. Tech.* **26**(3/4), 627–636.

Hermanowicz, S., Schindler, U. and Wilderer, P. (1995) Fractal structure of biofilms: New tools for investigation of morphology. *Wat. Sci. Tech.* **32**(8), 99–106.

Jahn, A. and Nielsen, P. (1998) Cell biomass and exopolymer composition in sewer biofilms. *Wat. Sci. Tech.* **37**(1), 17–24.

Jiang, H. and Bishop, P. (1994) Aerobic biodegradation of azo dyes in biofilms. *Wat. Sci. Tech.* **29**(10/11), 525–530.

Kloep, F., Roske, I. and Niel, T. (2000) Performance and microbial structure of a nitrifying fluidized-bed reactor. *Wat. Res.* **24**, 311–319.

Kolter, R. and Losick, R. (1998) One for all and all for one. *Science* **280**(10), 226.

Kudlich, M., Bishop, P., Knackmuss, H.-J. and Stoltz, A. (1996) Synchronous anaerobic and aerobic degradation of the sulfonated azo dye Mordant Yellow 3 by immobilized cells from a naphthalenesulfonate-degrading mixed culture. *Appl. Microbiol. Biotech.* **46**, 597–603.

Lawrence, J. and Nieu, T. (1999) Confocal laser scanning microscopy for analysis of biofilms. *Methods Enzymol.* **310**, 131–144.

Lazarova, V. and Manem, J. (1995) Biofilm characterization and activity analysis in water and wastewater treatment. *Wat. Res.* **29**, 2227–2245.

Lens, P., Vergeldt, F., Lettinga, G. and van As, H. (1999) $^1$H NMR characterization of the diffusional properties of methanogenic granular sludge. *Wat. Sci. Tech.* **39**(7), 187–194.

Lisle, J., Stewart, P. and McFeters, G. (1999) Fluorescent probes applied to physiological characterization of bacterial biofilms. *Methods Enzymol.* **310**, 166–178.

Manz, W. (1999) In situ analysis of microbial biofilms by rRNA-targeted oligonucleotide probing. *Methods Enzymol.* **310**, 79–91.

Massol-Deya, A., Whallon, J., Hickey, R. and Tiedje, J. (1995) Channel structures in aerobic biofilms of fixed-film reactors treating contaminated groundwater. *Appl. Environ. Microbiol.* **61**, 769–777.

Møller, S., Sternberg, C., Andersen, J., Christensen, B., Ramos, J., Givskov M. and Molin, S. (1998) In situ gene expression in mixed-culture biofilms: Evidence of metabolic interactions between community members. *Appl. Environ. Microbiol.* **64**(2), 721–732.

Morgenroth, E. and Wilderer, P. (2000) Influence of detachment mechanisms on competition in biofilms. *Wat. Res.* **34**, 417–426.

Oerther, D., de los Reyes,F. and Raskin, L. (1999) Interfacing phylogenetic oligonucleotide probe hybridizations with representations of microbial populations and specific growth rates in mathematical models of activated sludge processes. *Wat. Sci. Tech.* **39**(1), 11–20.

Okabe, S., Kuroda, H. and Watanabe, Y. (1998) Significance of biofilm structure on transport of inert particulates into biofilms. *Wat. Sci. Tech.* **38**(8/9), 163–170.

Palsdottir G, and Bishop P. L. (1997) Nitrifying biotower upsets due to snails and their control. *Wat. Sci. Tech.* **36**(1), 247–254.

Picioreanu, C., van Loosdrecht, M. and Heijnen, J. (1998) Mathematical modeling of biofilm structure with a hybrid differential-discrete cellular automaton approach. *Biotech. Bioengng* **58**(1), 101–116.

Picioreanu, C., van Loosdrecht, M. and Heijnen, J. (1999) Discrete-differential modeling of biofilm structure. *Wat. Sci. Tech.* **39**(7), 115–122.

Picioreanu, C., van Loosdrecht, M.C.M., Heijnen, J.J. (2000) A theoretical study on the effect of surface roughness on mass transport and transformation in biofilms, *Biotech. Bioengng* **68**, 354–369.

Picioreanu, C., van Loosdrecht M.C.M., Heijnen J.J. (2001) Two-Dimensional Model of Biofilm Detachment Caused by Internal Stress from Liquid Flow. *Biotech. Bioengng* **72**, 205–218.

Revsbech, N. (1994). Analysis of microbial mats by use of electrochemical microsensors: Recent advances. In *Microbial Mats* (ed. L. Stal and P. Caumette), pp. 135–147. Berlin, Springer-Verlag. G 35, NATO ASI Series.

Santegoeds, C., Schramm, A. and de Beer, D. (1998) Microsensors as a tool to determine chemical microgradients and bacterial activity in wastewater biofilms and flocs. *Biodegradation* **9**, 159–167.

Schramm, A., Larsen, L., Revsbech, N., Ramsing, N., Amann, R. and Schleifer, K.-H. (1996) Structure and function of a nitrifying biofilm as determined by in situ hybridization and the use of microelectrodes. *Appl. Environ. Microbiol.* **62**, 4641–4647.

Seshadri, S., Bishop, P. and Mourad Agha, A. (1994) Anaerobic/aerobic treatment of selected azo dyes in wastewater. *Waste Mgmt* **14**, 127–137.

Sternberg, C., Christensen, B., Johansen, T., Nielsen, A., Andersen, J., Givskov, M. and Molin, S. (1999) Distribution of bacterial growth activity in flow-chamber biofilms. *Appl. Environ. Microbiol.* **65**(9), 4108–4117.

Stewart, P. (1993). A model for biofilm detachment. *Biotech. Bioengng* **41**, 111–117.

Stewart, P. S. (1998) A review of experimental measurements of effective diffusive permeabilities and effective diffusion coefficients in biofilms. *Biotech. Bioeng.* **59**(3), 261–272.

Stoodley, P., DeBeer, D. and Lewandowski, Z. (1994) Liquid flow in biofilm systems. *Appl. Environ. Microbiol.* **60**, 2711–2716.

Sutherland, I. (1977) Bacterial exopolysaccharides – Their nature and production. In *Surface Carbohydrates of the Prokaryotic Cell.* (ed. I. Sutherland), pp. 27–95. Academic Press, New York.

Walker, J., Mackerness, C. and Keevil, C. (1995) Heterogeneous mosaic – A haven for waterborne pathogens. In *Microbial Biofilms,* (ed. H. Lappin-Scott and J. Costerton), pp. 196–204. Cambridge University Press, Cambridge, U.K.

Wanner, O. (1993) Modeling of mixed-population biofilm accumulation. In *Biofouling and Biocorrosion in Industrial Water Systems.* (ed. G. Geesey, Z. Lewandowski and H.-C. Flemming), pp. 37–62. Lewis Publishers, Boca Raton, FL.

Wanner, O. and Gujer, W. (1986) A multispecies biofilm model. *Biotech. Bioengng* **28**, 314–328.

Willis, H. and Bishop, P. (1995) Solids retention time and biofilm detachment in fixed film biological reactors. *Buckeye Bulletin* **69**(2), 24–29.

Wimpenny, J. and Colasanti, R. (1997) A unifying hypothesis for the structure of microbial biofilms based on cellular automaton models. *FEMS Microbiol. Ecol.* **22**, 1–16.

Yang, X., Beyenal, H., Harkin, G. and Lewandowski, Z. (2000) Quantifying biofilm structure using image analysis. *J. Microbiol. Methods* **39**, 109–119.

Yu, T. and Bishop, P. (1998) Stratification of microbial metabolic processes and redox potential change in sulfate-reducing biofilms studied using oxygen, sulfide, pH and redox potential microelectrodes. *Wat. Sci. Tech.* **37**(4/5), 195–198.

Yu, T. and Bishop, P. (1999) A microelectrode study of redox potential change in biofilms. *Wat. Sci. Tech.* **39**(7), 179–185.

Zahid, W. and Ganczarczyk, J. (1994a) A technique for a characterization of RBC biofilm surface. *Wat. Res.* **28**, 2229–2231.

Zahid, W. and Ganczarczyk, J. (1994b) Structure of RBC biofilms. *Wat. Environ. Res.* **66**, 100.

Zhang, T. (1994) Influence of Biofilm Structure on Transport and Transformation Processes in Biofilms. Ph.D. dissertation, University of Cincinnati, Cincinnati, OH.

Zhang, T. and Bishop, P. (1994a) Density, porosity and pore structure of biofilms. *Wat. Res.* **28**, 2267–2277.

Zhang, T. and Bishop, P. (1994b) Evaluation of tortuosity factors and effective diffusivities in biofilms. *Wat. Res.* **28**, 2279–2287.

Zhang, T. and Bishop, P. (1994c) Experimental determination of the dissolved oxygen boundary layer and mass transfer resistance near the fluid-biofilm interface. *Wat. Sci. Tech.* **30**(11), 47–58.

Zhang, T. and Bishop, P. (1994d) Structure, activity and composition of biofilms. *Wat. Sci. Tech.* **29**(7), 335–344.

Zhang, X., Bishop, P.L. and Kupferle, M. (1997) Measurement of polysaccharides and protein in biofilm extracellular polymers. In *Proceedings of the 2nd IAWQ International Conference on Microorganisms in Activated Sludge and Biofilm Processes, Berkeley, CA*, pp. 551–554.

Zhang, T., Fu, Y. and Bishop, P. (1994) Competition in biofilms. *Wat. Sci. Tech.* **29**(10/11), 263–270.

Zhang, T., Fu, Y., Bishop, P., Kupferle, M., Fitzgerald, S., Jiang, S. and Harmer, C. (1995a) Transport and biodegradation of toxic organics in biofilms. *J. Haz. Materials* **41**, 267–285.

Zhang, T., Fu, Y. and Bishop, P. (1995b) Competition for substrate and space in biofilms. *Wat. Environ. Res.* **67**, 992–1003.

Zhang, X., Bishop, P. and Kinkle, B. (1998a) Extraction of extracellular polymers from biofilms using five different methods. In *Proc., 71st Annual Water Environment Federation Conference, Orlando, FL.*, vol. 1, pp. 13–28. WEF, Alexandria, VA.

Zhang, X., Bishop, P. and Kupferle, M. (1998b) Measurement of polysaccharides and protein in biofilm extracellular polymers. *Wat. Sci. Tech.* **37**(4/5), 345–348.

Zhang, X., Bishop, P. and Kinkle, B. (1999) Comparison of extraction methods for quantifying extracellular polymers in biofilms. *Wat. Sci. Tech.* **39**(7), 211–218.

Zhou, Q. and Bishop, P. (1997) Determination of oxygen profiles and diffusivity in encapsulated biomass k-carrageenan gel beads. *Wat. Sci. Tech.* **36**(1), 271–277.

# 7
# Mass transport in heterogeneous biofilms

*Zbigniew Lewandowski and Haluk Beyenal*

## 7.1 INTRODUCTION

Bennett and Mayers (1974), in their book on momentum, heat, and mass transport, concisely defined diffusional mass transport as 'the tendency of a component in a mixture to travel from a region of high concentration to one of low concentration'. From this definition it can be inferred that diffusional mass transport is an expression of the second law of thermodynamics: systems spontaneously move toward the state of highest probability, uniform distribution of components, and equilibrium. It is, therefore, expected that in the presence of any gradient within a system – concentration, density, temperature, or pressure – the system will spontaneously move toward the state of highest probability, i.e. toward uniform distribution of its components. To reach this state, the components of the system are transported down the respective gradients at a rate equal to the product of the gradient's magnitude and a proportionality constant, which reflects the physical mechanisms responsible for the transport of the component.

© 2003 IWA Publishing. *Biofilms in wastewater treatment.* Edited by S. Wuertz, P.L. Bishop and P.A. Wilderer. ISBN: 1 84339 007 8.

Microorganisms, when suspended in a solution of nutrients, disturb the uniform distribution of the dissolved nutrients by locally depleting them and generating nutrient concentration gradients. The rate at which the nutrients are replenished, by mass transport, is of vital interest for the microorganisms because it determines the availability of the vital resources. For a single bacterium in suspension, the rate of nutrient supply by diffusion is adequate because the nutrients diffuse through the relatively thin stationary liquid layer surrounding a single cell. However, microorganisms tend to aggregate, and for the aggregated colonies, the rate of nutrient supply by diffusion alone is not adequate. It seems natural that the aggregated microorganisms should develop strategies to overcome this deficient nutrient delivery. The propensity exhibited by many microorganisms to form biofilms may be seen as an implication of such a strategy. By attaching to surfaces, microorganisms increase the relative flow velocity between the immobile microbial aggregates and the bulk liquid, and benefit from the convective transport of nutrients. Hypothetically, these benefits explain why microorganisms prefer to remain attached to surfaces.

Once the microorganisms attach to a surface, they change their mode of growth, and accumulate in layers called biofilms. To quantify the rate of nutrient consumption and mass transport dynamics in biofilms, the space occupied by the aggregated microorganisms is conceptually separated from the space occupied by the bulk solution. The terms 'internal mass transport' and 'external mass transport' are used in reference to these two zones. The conceptual image of the space occupied by the aggregated microbes in biofilms has recently evolved from displaying the microorganisms as uniformly and randomly distributed in the matrix of extracellular biopolymers to structurally heterogeneous biofilms, displaying microorganisms aggregated in microcolonies (Costerton et al. 1995; Keevil and Walker 1992; Wolfaardt et al. 1994; Massol-Deya 1994; Bishop and Rittmann 1995). Although conceptual models of heterogeneous biofilms produced by various research groups disagree in details, like the shape and size of the microbial aggregates, all models depict microcolonies (dense aggregates of microorganisms) as building blocks of biofilms. According to these models, biofilms are porous, the nutrients penetrate the pores and reach the deep layers of the biofilm. Quantifying mass transport dynamics in structurally heterogeneous biofilms is challenging because it has to take into account the rates of nutrient transport from bulk solution to the biofilm, nutrient transport within the pores, nutrient transport within the microcolonies, and the rate of nutrient consumption. By comparison, quantifying transport and nutrient consumption, in uniform biofilms is less complicated because the model of uniform biofilms does not allow for nutrient transport within the pores.

An important goal of biofilm engineering is to relate microscale microbial activity in biofilms to the macroscale performance of biofilm reactors. Major tasks leading to that goal are: (1) to construct mathematical models of heterogeneous biofilms using parameters that can be characterized both at the micro- and macroscales of observation; and (2) to verify that the solutions of these models predict the behavior of real biofilms. No doubt, the relation between the mass transport and microbial activity must play a vital role in such models. Before these models are constructed, the relations between local biofilm activity and the extent of biofilm heterogeneity need to be understood and quantified. To do so, biofilm heterogeneity must be quantified, and used as a variable that can be correlated with other measurable variables in biofilms, like the rates of local mass transport and local flow velocity.

Attempts have been made to quantify biofilm heterogeneity with respect to biofilm density, porosity, pore structures, tortuosity (Zhang and Bishop 1994a,b) and fractal dimension (Hermanowicz et al. 1995) as well as to introduce some of these parameters into the mathematical modeling of biofilm activity (Wimpenny and Colasanti 1997; Picioreanu et al. 1998a,b). In our laboratory we have constructed a computer program to extract quantitative parameters from microscope images of biofilms: areal porosity, diffusion distance, fractal dimension, and textural entropy (Lewandowski et al. 1999; Yang et al. 2000, 2001). The trend toward quantifying biofilm structure is becoming evident (Heydorn et al. 2000; Xavier et al. 2001) and it is expected that understanding the effects of biofilm heterogeneity on mass transport dynamics and on microbial activity in biofilms will lead to improved mathematical descriptions of biofilm processes and biofilm reactors.

This chapter discusses mass transport in heterogeneous biofilms. To develop the concepts presented here, we define:

*biofilm*: as aggregates of microorganisms attached to a surface;

*biofilm processes*: as microbial attachment to surfaces, metabolic activity of microorganisms on surfaces, and microbial detachment from surfaces;

*nutrients*: as substances dissolved in water (no particulate matter), which are used by biofilm organisms in metabolic reactions;

*products*: as dissolved substances (no particulate matter) generated by biofilm microorganisms;

*substratum*: as any surface supporting microbial growth;

*biofilm system*: as an assemblage of substratum, covered with biofilm, and immersed in the solution of nutrients.

## 7.2 BIOFILM HETEROGENEITY AND BIOFILM MODELS

There are currently two conceptual models of biofilms in use: the model of homogeneous biofilms, and the model of heterogeneous biofilms. Initially, the model of homogeneous biofilms, typically represented by a uniform layer of microorganisms attached to a surface, was used exclusively. The model of heterogeneous biofilms, typically represented by microbial aggregates attached to a surface, was introduced much later and is now used to interpret experimental results that are difficult to explain using the model of homogeneous biofilms. The following three results exemplify such difficulties. (1) Drury (1992) introduced small fluorescent beads into a biofilm and studied their fate. After the experiment was terminated, he dissected the biofilm and recovered the beads. Many of the beads, to everyone's surprise, were found at the bottom of the biofilm, a result that is difficult to explain using the conceptual model of homogeneous biofilms. (2) Lewandowski *et al.* (1993) used nuclear magnetic resonance imaging to study hydrodynamics in biofilm systems, and detected water movement within the space occupied by the biofilm, which was, again, difficult to explain using the concept of homogeneous biofilms. (3) de Beer *et al.* (1994) showed that oxygen concentration profiles in biofilms varied significantly from one location to another. Attempts to use statistics to find a 'representative profile' for a given biofilm failed because the differences between individual profiles were much bigger than those expected in homogeneous biofilms.

Many research groups have expressed dissatisfaction with the conceptual model of homogeneous biofilms (Keevil and Walker 1992; de Beer *et al.* 1994; Wolfaardt *et al.* 1994; Massol-Deya 1994; Hermanowicz *et al.* 1995). Having the benefit of hindsight, it seems that the introduction of confocal laser microscopy (CLM), and the first confocal images of biofilms, precipitated the idea that the conceptual model of homogeneous biofilms is inadequate in some instances. These images showed that microorganisms in biofilms were aggregated in microcolonies, and not uniformly distributed as previously thought (Lawrence *et al.* 1991). As a result, the model of homogeneous biofilms has been appended by the model of heterogeneous biofilms, displaying microorganisms aggregated in microcolonies instead of being uniformly distributed throughout the matrix. At present, the two conceptual models, of homogeneous and heterogeneous biofilms, are functioning side by side. The conceptual model of heterogeneous biofilms is preferred by life scientists because it can easily accommodate many recent concepts, like cell–cell communication. The model of homogeneous biofilms is favored by the engineering community because it simplifies mathematical modeling of biofilm processes. There are signs that these two models are converging, however, as the growing popularity of the heterogeneous biofilm model among life scientists stimulates the engineering community to construct mathematical models that take into account biofilm heterogeneity. Part of this chapter is also devoted to bridging these two conceptual models.

Mathematical models of homogeneous biofilms were generated as an extension of pre-existing models of diffusion with reaction systems explored mostly by chemical engineers. The task of constructing mathematical models of heterogeneous biofilms is more challenging because heterogeneous biofilms need to be modeled in 3-D, which complicates both the models and their solutions. New approaches to mathematical modeling of heterogeneous biofilms use cellular automata (Picioreanu *et al.* 1998a,b, 2000, 2001; Hermanowicz, 2001) to simulate biofilm structure.

Figure 7.1 exemplifies heterogeneous biofilms: the bottom is covered with a discontinuous layer of cells. Above this layer grow mushroom-shaped microcolonies separated by voids. Liquid moves in the voids, allowing for convective transport of nutrients *within voids, in the biofilm.*

It has been well documented that biofilm heterogeneity influences rates of nutrient transport and consumption near and within biofilms (Beyenal *et al.* 1998; Bishop and Rittmann 1995; Lewandowski *et al.* 1993; Zhang and Bishop 1994a,b; Yang and Lewandowski 1995; de Beer *et al.* 1994). To quantify the effects of biofilm heterogeneity on mass transport rates, the heterogeneity itself needs to be quantified, and used as a parameter in the appropriate mathematical

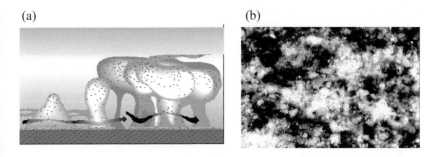

**Figure 7.1.** Heterogeneous biofilms. (a) Diagrammatic representation of the structure of a hypothetical bacterial biofilm drawn from confocal laser microscopy examination of a large number of mixed-species biofilms. The discrete microcolonies are surrounded by a network of interstitial voids filled with water. The arrows are used to indicate the possibility of convective flow within the water channels. (b) Light microscope image of a heterogeneous biofilm (Beyenal and Lewandowski 2002).

models, in the same manner as other parameters affecting mass transport in biofilms. Tools to quantify biofilm structure have been developed: density, porosity, specific surface area, mean pore radius (Zhang and Bishop 1994a,b), and fractal dimension (Hermanowicz et al. 1995) make the task of correlating biofilm structure with mass transport dynamics technically possible.

We routinely quantify parameters characterizing biofilm heterogeneity from microscope images of biofilms, such as areal porosity, diffusion distance, fractal dimension, and textural entropy (Lewandowski et al. 1999; Yang et al. 2000, 2001). The structural elements of the biofilm that we choose to measure are called *features*. Biomass cluster size and cluster shape are examples of features. There are many structural features that we study, and an objective strategy is required to select the most useful ones because some characteristics of biofilm structure, such as size, shape, color and texture, are not easily converted to quantitative values. One way to evaluate the relevance of a feature is by how well it correlates with changes in the underlying processes that created the biofilm. There are some obvious starting points. We conjecture that there exist a finite number of features and associated parameters which describe the structure of a biofilm and contain enough information either to reflect variations in the growth dynamics or to predict the functional characteristics of the biofilm. As a starting point for finding such features, we rely on the fact that biofilms achieve steady, or at least pseudo-steady state, in which their physical structure is dynamic at the molecular level but static at the scale corresponding to our microscopic field of view. If a parameter (measured feature) appears to approach a stable level, it is behaving in the expected manner and it has passed our initial screening test. If it does not, we provisionally reject the parameter, assuming it is not a useful descriptor of biofilm structure. It appears from our results that areal porosity, diffusion distance, fractal dimension and textural entropy all tend to reach stable levels in some biofilms (Lewandowski et al. 1999).

Mathematical models of biofilm activity that assume homogeneous biofilm structure require a prior knowledge of nutrient diffusivity and use it as a control parameter. Some authors use such models to fit experimental data and to calculate the effective diffusivity (Beyenal et al. 1997; Livingston and Chase 1989), whereas others assume nutrient diffusivity and calculate biofilm activity and/or accumulation rate (Wood and Whitaker 1998; Elmaleh 1990; Grady 1983; Rittmann and McCarty 1980; Atkinson et al. 1967). Mathematical models of biofilm activity that accept heterogeneous biofilm structure, such as AQUASIM (Reichert 1994), require porosity, density, effective diffusivity, and reactor geometry as input parameters (Horn and Hempel 1997; Wanner and Reichert 1996). Only recently a new approach to biofilm modeling, cellular automata, was introduced (Kreft and Wimpenny 2001; Kreft et al. 2001;

Hermanowicz 2001; Picioreanu *et al.* 1998a,b, 2000, 2001; Eberl *et al.* 2000; Noguera *et al.* 1999; Wimpenny and Colasanti 1997). This allows biofilm density, porosity and the shape of the microcolonies to be predicted from first principles, although the link between modeling and experiment is weak.

## 7.3 QUANTIFYING NUTRIENT UPTAKE KINETICS FROM THE NUTRIENT CONCENTRATION PROFILES

Biofilm engineering requires a quantitative description of biofilm heterogeneity and its effect on mass transport rates. A step in this direction is to quantify nutrient uptake kinetics from nutrient concentration profiles. The following conceptual image of mass transport and microbial activity serves as a base for developing such tools. At a distance from the biofilm surface, within the bulk solution, nutrients are uniformly distributed and the solution is well mixed. It is assumed that there is no nutrient consumption in the bulk solution, and that nutrients are consumed only within the space occupied by the biofilm; the nutrients are delivered to the biofilm by mass transport. Figure 7.2 shows an oxygen profile in a biofilm; oxygen concentration decreases from 6 mg $L^{-1}$ in bulk to 0.6 mg $L^{-1}$ near the biofilm surface, indicating that mass transport near the biofilm surface is impeded. This is a typical situation; the thickness of the mass transport boundary layer can be estimated from such data by using simple geometrical procedures (Revsbech and Jorgensen 1986).

To quantify the shape of a nutrient profile, it is important to find the position of the biofilm surface; mathematical models quantify nutrient concentration as a function of distance, which can be measured from the bottom or from the biofilm surface (Lewandowski 1993). No matter which level, biofilm surface or bottom, is selected as the reference for measuring the distance, determining the position of the biofilm surface is important. Exact positioning of the biofilm surface is not trivial. Conceptually, the biofilm surface is at the inflection point of the nutrient concentration profile. However, as can be seen in Figure 7.2, the position of the inflection point on such a profile can be rather ill-defined; it is not easy to decide where the inflection point is and it is even harder for noisy signals, in which the data points are dispersed. To find a rule for how to position the biofilm surface based on the nutrient concentration profiles, we glued together an oxygen microsensor and a fiber optic microsensor so that their tips were at the same level (Lewandowski *et al.* 1991). This combined sensor simultaneously measured two profiles, oxygen concentration and optical density; superimposing these two profiles enabled exact positioning of the biofilm surface (Figure 7.3). The optical density profile in Figure 7.3 shows the biofilm surface was at the end of the curved part of the profile within the biofilm.

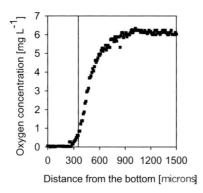

**Figure 7.2.** Oxygen concentration profile measured with a microelectrode. Perpendicular line designates approximate position of biofilm surface (Rasmussen and Lewandowski 1998).

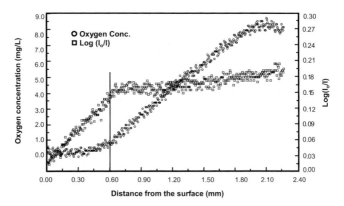

**Figure 7.3.** Profiles of oxygen concentration and optical density in a biofilm. A combined microsensor, an oxygen microelectrode and an optical density microprobe, permitted positioning the biofilm surface 0.60 mm from the bottom. This distance, when marked on the oxygen concentration profile, indicates that the biofilm surface is at the beginning of the linear part of the oxygen profile within the mass transport boundary layer (Lewandowski *et al.* 1991); I is the local light intensity, and $I_o$ is the maximum light density.

It is quite tedious to construct combined microsensors and they are not used very often. Based on the results in Figure 7.3, we have accepted as a rule that the biofilm surface should be positioned at the end of the curved part of the profile measured within the biofilm. This rule has been corroborated by several other measurements. We used this rule to position the biofilm surface on the nutrient concentration profiles in Figure 7.2.

Once the position of the biofilm surface has been determined, the distribution of nutrient concentration can be conveniently described in a new system of coordinates, and parameters of interest, such as flux at the biofilm surface and thickness of the boundary layer, can be estimated (Lewandowski 1993). The origin of the new coordinate system is fixed at the biofilm surface ($\aleph_s, C_s$), where $\aleph$ is the distance measured from the biofilm surface, $C$ is the concentration, and the subscript s stands for surface. The part of the profile above the biofilm surface is then adequately described in this new coordinate system by the following empirical equation:

$$\frac{C - C_s}{C_b - C_s} = 1 - \exp[-B(\aleph - \aleph_s)] \tag{7.1}$$

where $C$ is the local nutrient concentration, $C_s$ is nutrient concentration at the biofilm surface, $C_b$ is the bulk liquid nutrient concentration (considered constant), and $B$ is an experimental coefficient. Two parameters most often calculated from nutrient concentration profiles are: (1) the slope of the profile at the biofilm surface and (2) the thickness of the mass transport boundary layer. The slope of the profile is used to calculate the nutrient flux; the procedure calls for multiplying the slope by the molecular diffusion coefficient of the diffusing substance in the liquid, according to Fick's first law.

To calculate the coefficient B, Equation 7.1 is linearized (Equation 7.2) and coefficient $B$ is the slope of the line when the data points are arranged in coordinates $[\aleph - \aleph_s]$ versus $\ln[1 - [(C - C_s)/(C_b - C_s)]]$:

$$\ln\left(1 - \frac{C - C_s}{C_b - C_s}\right) = -B(\aleph - \aleph_s) \tag{7.2}$$

Once parameter $B$ has been evaluated, the slope of the profile at the biofilm surface is calculated as:

$$\left(\frac{dC}{d\aleph}\right)_{(\aleph - \aleph_s) = 0} = B(C_b - C_s) \tag{7.3}$$

The thickness of the mass transport boundary layer, defined here as the distance from the biofilm surface to the point where the nutrient concentration is 95% of the nutrient concentration in the bulk solution, can be calculated from Equation 7.1 by substituting $C = 0.95C_b$ and solving for $\aleph$.

Nutrient concentration profiles are measured at a pseudo-steady state, which means that the shape of the concentration profile does not change for reasonable lengths of time. Each data point in Figure 7.2 represents a unique equilibrium between the rate of nutrient delivery to a point in space and the rate of nutrient uptake from that point in space. To be exact, the nutrients are metabolically transformed only within the space occupied by the biofilm: there is no nutrient consumption outside the biofilm, and the gradient of nutrient concentration in that zone is due to nutrient transport toward the biofilm surface at a rate determined by the microbial metabolic activity in the biofilm and by hydrodynamics. Because each point of the concentration profile reflects equilibrium between nutrient delivery and nutrient uptake, the shape of the profile is affected by all factors influencing the rate of microbial metabolism and all factors influencing the rate of mass transport. Because of this complex arrangement of factors influencing the profile, judging local biofilm activity or local mass transport dynamics from nutrient concentration profiles alone is difficult. Changes in nutrient uptake rate, caused by microbial metabolism, may have the same effect on the shape of the profile as changes in the flow rate.

## 7.4 QUANTIFYING MASS TRANSPORT MECHANISMS FROM FLOW VELOCITY PROFILES IN BIOFILMS

The next step toward quantifying the effect of biofilm heterogeneity on mass transport dynamics is to understand the mechanism of mass transport in such biofilms. The key to the mass transport mechanism is hydrodynamics, which determines the overall rate of nutrient delivery from the bulk solution to the biofilm and the distribution of nutrients within the biofilm. Flow velocity near a biofilm changes from a maximum in the bulk solution to zero near the surface or bottom of the biofilm, forming a velocity profile similar to that formed by the nutrient concentration. Because of biofilm porosity, water moves in the space occupied by the biofilm, and the flow velocity reaches zero near the bottom (Lewandowski *et al.* 1993). The rate of convection decreases near the biofilm surface because flow velocity decreases there, and it is expected that molecular diffusion becomes the principal mechanism of mass transport in that zone.

To determine the mechanisms of mass transport near the biofilm surface and within the biofilm, we use a set of equations quantifying the rates of convection and diffusion, as reported by Schwarzenbach *et al.* (1993).

The time needed for a molecule to travel a distance $L$ by diffusion is

$$t = L^2/2D \tag{7.4}$$

where $D$ is diffusivity (in cm$^2$/s). The time needed for a molecule to travel a distance $L$ by convection is

$$t = L/v \tag{7.5}$$

where $v$ is flow velocity (in cm/s). Equating the right sides of these equations and solving the resulting equation for the distance, $L$, we arrive at the critical distance, $L_{crit}$, which a molecule may travel during time $t$ by either convection or diffusion; the mechanisms of transport are different but the distance is the same:

$$L_{crit} = 2D/v. \tag{7.6}$$

Thus, by measuring the distance traveled by a particle within the biofilm, and knowing the diffusivity, we may distinguish the contributions from convection and diffusion in the overall mass transport. If the distance traveled by the molecule is longer than the critical distance, then the prevailing mass transport mechanism is convection. Alternatively, we can simplify this calculation by dividing Equation 7.4 by Equation 7.5, and conclude that when $Lv/2D > 1$, then convection exceeds diffusion.

In such analyses it is important to select the characteristic distance the molecules have to travel in a way that is meaningful for the system under study. In our system, we think that the thickness of the mass transport boundary layer is a good choice for the characteristic length, and it can be calculated from Equation 7.1, for a given nutrient concentration profile. We assume that the molecular diffusivity of the dissolved nutrients (e.g. oxygen), $D$, is $2 \times 10^{-5}$ cm$^2$/s (Beyenal et al. 1997) and that the characteristic distance, $L$, is the thickness of the boundary layer, say 0.5 mm, which is equal to the thickness of the boundary layer in Figure 7.2. Equation 7.4 shows that a molecule needs 62.5 s to diffuse 0.5 mm, which is equivalent to a linear velocity of 8 µm/s. Thus, 8 µm/s is the characteristic flow velocity in our system that separates mass transport by diffusion from mass transport by convection. At locations where flow velocity exceeds the characteristic flow velocity, mass transport is dominated by convection. Having estimated the characteristic flow velocity, we can look up the available experimental data reporting flow velocities near and within biofilms and estimate where in the biofilm molecular diffusion exceeds convection. We use the results of flow velocity measurements reported by Lewandowski et al. (1993), shown in Figure 7.4.

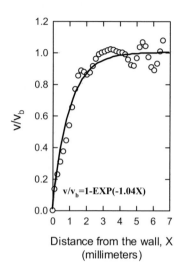

**Figure 7.4.** Flow velocity profile measured by nuclear magnetic resonance imaging (Lewandowski *et al.* 1993).

The shape of the flow velocity profile near the biofilm is described by an empirical exponential equation (Lewandowski *et al.* 1993):

$$\frac{v}{v_b} = 1 - \exp(-A\,X) \tag{7.7}$$

where $X$ is the distance from the bottom, $v$ is the local flow velocity and $v_b$ is the average bulk flow velocity in the conduit. The results in Figure 7.4 were collected at the average flow velocity in the conduit, $v_b = 4.6$ cm/s; parameter $A$ was estimated as $1.04 \times 10^{-3}$ $\mu m^{-1}$. Note that $A$ is expressed per micron, not per millimeter as in Figure 7.4. Now, solving Equation 7.7 for $X$ yields that the flow velocity exceeded 8 µm/s at distances exceeding 1.7 µm from the bottom, which is so close to the bottom that it is safe to assume that the mass transport in biofilm voids is entirely due to convection. This conclusion satisfies those researchers who adopted the model of heterogeneous biofilms, because it reflects their expectations.

Equations 7.1 and 7.7 have the same form, with one exception. In Equation 7.7 we positioned the origin of the coordinate system at the bottom, not at the biofilm surface as in Equation 7.1. That is how the data were originally published, and the difference does not affect the calculations; positioning the origin of coordinate system is a matter of computational convenience. If, for any reason, positioning the origin of the coordinate system at the biofilm surface is preferable, then the entire coordinate system can be shifted to the biofilm surface using the same notation as that used in Equation 7.1. The flow velocity calculated in the previous paragraph is relevant to the interstitial spaces in the biofilm, but not to the inner space of the microcolonies. To test the mechanism of mass transport in the microcolonies, we use the same approach as above, but assume that $L$, the characteristic length in the system, equals the size of a single microorganism in a microcolony, say 1 micrometer, and that $D$ is the diffusivity in the matrix of extracellular polymers, say $2 \times 10^{-5}$ cm$^2$/s (Beyenal *et al.* 1997). The same computational procedure as the one used in the preceding paragraph shows that for convection to become the dominant mass transport mechanism the flow velocity within the microcolony would have to exceed 40 μm/s, an unlikely event as shown by Xia *et al.* (1998). Therefore, the dominant mechanism of mass transport within microcolonies is molecular diffusion, as expected.

In summary, we have demonstrated that molecular diffusion is the dominant mass transport mechanism in microcolonies, and that convection is dominant within the interstitial voids. These conclusions, although not unexpected, justify the popular notion that biofilm porosity increases mass transport rates *within the pores* but does not affect the mass transport *within the microcolonies*, at least not directly. An increase in mass transport outside of the microcolonies may affect the mass transport mechanisms within microcolonies indirectly, by affecting the nutrient concentration surrounding the microcolonies in the pores deep in the biofilm. Nevertheless, the predominant mass transport mechanism within microcolonies remains diffusion.

## 7.5 LOCAL MASS TRANSPORT RATES IN HETEROGENEOUS BIOFILMS

The most striking differences between the models of homogeneous and heterogeneous biofilms are related to the mechanisms of mass transport in the two systems. Heterogeneous biofilms are porous, water moves within the space occupied by the biofilm, and mass transport rates vary from one place to another. Large differences exist between the mass transport rates measured in the microcolonies and those measured in the voids because mass transport is controlled by diffusion in the microcolonies and by convection in the voids. These differences make using a single mass transport coefficient for the entire biofilm questionable. Whether using a single mass transport coefficient for the entire biofilm is justified or not depends on how much it varies from location to location. To evaluate the variations in mass transport rates between locations in biofilms we introduced the concept of a local mass transport coefficient: the mass transport coefficient at a single point within the biofilm (Yang and Lewandowski 1995).

The definition of the local mass transport coefficient has been derived from the measurement procedure: 'mass transport coefficient of electroactive species to the tip of an electrically polarized microelectrode.' Because the tips of microelectrodes are small, a few micrometers in diameter, they form highly localized sinks for electroactive species in the biofilm. Practically, the limiting current generated by the local reduction of ferricyanide to ferrocyanide is measured by cathodically polarized microelectrodes, a well-known procedure used for large and stationary electrodes (Burgman and Sides 1988; Cammarata *et al.* 1990; Wooster *et al.* 1991). Our version of the measurement differs from those described in the literature by two factors: (1) our electrodes have very small tips; and (2) our electrodes are mobile. To measure the local mass transport coefficient, we replace the nutrient solution with a solution of ferricyanide in a supporting electrolyte (Yang and Lewandowski 1995). During the measurement, the biofilm remains inactive but its structure is preserved (Lewandowski and Beyenal 2001).

When the local nutrient concentrations measured across a biofilm are plotted versus distance, they form a nutrient concentration profile. It is expected that the shape of the nutrient concentration profile will follow the shape of the local mass transport coefficient profile when both are measured at the same location. It is also expected that at locations where the local mass transport coefficient is high the local nutrient concentration will be high as well, at least higher than at a location where the local mass transport coefficient is low. Figure 7.5 shows profiles of oxygen concentration and local mass transport coefficient measured at the same location in a biofilm (Rasmussen and Lewandowski 1998).

The data in Figure 7.5 show that the changes in mass transport coefficient do not correlate well with the changes in oxygen concentration. Approaching the biofilm surface, from the bulk solution, the oxygen concentration decreases rapidly and reaches quite low levels at the biofilm surface, whereas the local mass transport coefficient remains quite high at that location. This observation is unexpected and difficult to explain, because there is no oxygen consumption in the bulk, the oxygen concentration profile should follow the shape of the mass transport coefficient profile much closer than it does in Figure 7.5. However, although these two profiles do not match, each of them is consistent with our knowledge of the system's behavior. We expect to measure a low concentration of oxygen at the biofilm surface; such a result fits the concept of mass transport boundary layer of high mass transport resistance above the biofilm surface. To measure a high value of mass transport coefficient near the biofilm surface is also not surprising because, as we have estimated, convection is the predominant mass transport mechanism in that zone. These two facts cannot coexist: high mass transport resistance and convection. It is difficult to explain why the oxygen concentration decreases rapidly above the biofilm surface while the local mass transport coefficient remains almost constant.

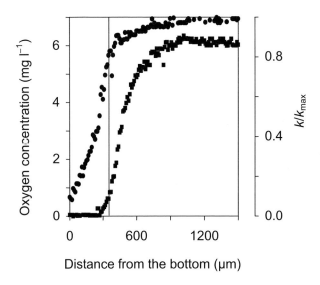

**Figure 7.5.** Superimposed oxygen concentration (■) and local mass transport coefficient (•) profiles. The local concentration of oxygen is about 6 mg $l^{-1}$ in the bulk and about 0.6 mg $l^{-1}$ near the surface, which accounts for a 90% decrease. However, the local mass transport coefficient decreases by only about 10% between these two locations (Rasmussen and Lewandowski 1998).

To explain this apparent discrepancy we need to examine the procedure for measuring flow velocity in biofilms. All available flow velocity measurements in biofilms report only one component of the flow velocity vector, parallel to the bottom. Based on these results, we estimated that mass transport is controlled by convection near biofilms. However, the convective mass transport rate equals the nutrient concentration times the flow velocity component *normal to the reactive surface*. The component of the flow velocity *parallel to the surface* has nothing to do with the convective mass transport toward that surface. Consequently, the estimate of the mass transport mechanism based on flow velocity holds only in the direction in which the flow velocity was measured. Indeed, when the flow near a surface is laminar, the laminas of liquid slide parallel to the surface and there is little or no convection across these layers; the mass transport parallel to the surface is convective, while the mass transport perpendicular to the surface remains diffusive (Figure 7.6).

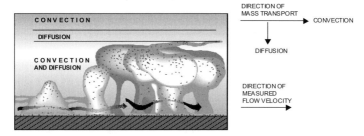

**Figure 7.6.** Alternating zones of convective and diffusive mass transport in heterogeneous biofilms. This hypothetical model of mass transport is consistent with the results in Figure 7.5. Mass transport in the space occupied by the biofilm is convective, but the amount of nutrient delivered to this space is limited by the diffusive mass transport just above the biofilm surface.

## 7.6 THE CONCEPT OF BIOFILMS COMPOSED OF DISCRETE LAYERS

It is becoming clear that mass transport and microbial activity in heterogeneous biofilms are too complex to be modeled exactly. Therefore, simplifying concepts are needed to bridge the concepts of uniform biofilms and heterogeneous biofilms. We hope that the concept of biofilms composed of discrete layers may fulfil the requirements. In a sense, all heterogeneous biofilms are discrete because they are composed of individual microcolonies. However, using individual microcolonies as units of structure in biofilm modeling is not very productive because the number and sizes of microcolonies vary from one biofilm to another and the biofilm structure for all practical purposes remains outside of our control. Instead, we see advantages in subdividing (heterogeneous) biofilms into a finite number of layers parallel to the bottom and use these layers as building blocks of heterogeneous biofilms. This gives the modeler more control of the conceptual assemblage being modeled by manipulating the numbers and thickness of the uniform layers that represent the biofilm; a procedure that is well known in analysis of sediments (Berg and Petersen 1998). Another advantage of this approach is in having a system of microelectrode measurements in place that may be used to validate the mathematical models (Figures 7.7 and 7.8). By using the proposed approach, we can subdivide biofilms into a finite number of discrete, uniform and continuous layers; each of these layers has the average properties of heterogeneous biofilms at the selected distance from the bottom. Such a biofilm can then be modeled as an assemblage of layers having uniform properties. The non-uniformity of the heterogeneous biofilm is reflected by the changes in the average properties of each biofilm layer, and by the resulting gradients in these properties across the biofilm. Because the average properties are determined within certain limits, set by the standard deviations from the average, the results of modeling may, if necessary, be expressed within confidence limits as well. To describe the activity of the entire biofilm, each layer is modeled individually, and then the results are integrated over the thickness of the biofilm.

No matter which mathematical model of heterogeneous biofilms we eventually select, it has to accurately reflect the mechanism of mass transport in such biofilms. The best model would be one that considers the activity of each microcolony separately, and then integrates the results over the entire biofilm, but such an approach is not practical. The next best model is to subdivide the biofilms into a finite number of uniform layers. Figure 7.7 demonstrates the utility of subdividing heterogeneous biofilms into a finite number of discrete layers, and its consistency with previously introduced concepts. To evaluate variations in mass transport rates between locations in heterogeneous biofilms we introduced the concept of a local mass transport coefficient, the mass transport coefficient at a single point within the biofilm (Yang and Lewandowski 1995). Later we expanded this technique and evaluated the local diffusivities in heterogeneous biofilms (Beyenal *et al.* 1998; Beyenal and Lewandowski 2000). Local diffusivities vary and the average diffusivity decreases near the bottom. Figure 7.8 shows a typical distribution of diffusivity in biofilms, measured and interpreted using the system in Figure 7.7C.

Local diffusivity fluctuates among locations within the biofilm, as expected. We often evaluate the local diffusivity at different distances from the bottom and present the results as a sequence of maps quantifying the variations above and within the biofilm (Figure 7.8). To interpret such results it is convenient to calculate average diffusivities at different levels (distances from the bottom) and to calculate the standard deviations of the individual diffusivities, representing the variability and reflecting biofilm heterogeneity. Such a system can then describe the spatial distribution of the measured parameter and its dependence on biofilm heterogeneity. Each layer of such a biofilm can then be modeled as uniform, having average properties calculated from individual measurements within the layer.

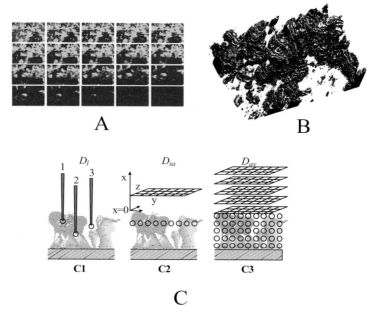

**Figure 7.7.** A discrete biofilm: The concept and implementation. (A) A series of confocal images is taken at different heights from the bottom to the top of the biofilm, upper left corner to lower right corner. (B) A three-dimensional image of the biofilm is reconstructed from the series of confocal images (Lewandowski 2000a,b). (C) The system of diffusivity measurements in heterogeneous biofilms. C1, local relative effective diffusivity ($D_l$) measured by microelectrodes at arbitrarily selected locations at different distances from the bottom. C2, the $D_l$ are measured at grid points equally distant from the bottom. The measured $D_l$ are then averaged, which gives the surface-averaged relative effective diffusivity, $D_{av} = \sum_{n=1}^{k} D_{ln}/k$. C3, the average relative effective diffusivity, $D_{av} = \sum_{n=1}^{p} D_{fxn}/p$, is an average of all local measurements for the entire biofilm (Beyenal and Lewandowski 2000).

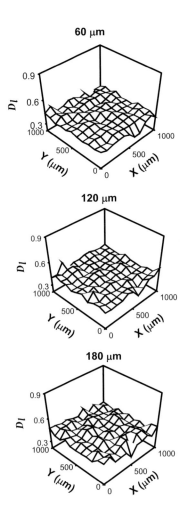

**Figure 7.8.** Horizontal distributions of local relative effective diffusivities ($D_l$) for biofilms grown at a flow velocity of 3.2 cm/s and a glucose concentration of 50 mg/L at distances 60 µm, 120 µm, and 180 µm from the bottom. The surface-averaged relative effective diffusivities were 0.409, 0.426, and 0.449, and the standard deviations were 0.0179, 0.0195, and 0.0284, respectively (Beyenal and Lewandowski 2000).

## 7.7 MODELING MASS TRANSPORT AND ACTIVITY IN BIOFILMS COMPOSED OF DISCRETE LAYERS

At the macroscale, microbial activity in biofilms is controlled by the rates of nutrient transfer to the biofilm and consumption within the homogeneous biofilm. Equation 7.8 equates the biofilm activity with the internal mass transport, assuming constant effective diffusivity and constant biofilm density (uniform biofilm or biofilm layer).

$$D_f \frac{d^2 C}{dx^2} = \frac{\mu_{max} C X_f}{Y_{X/S}(K_S + C)} \qquad (7.8)$$

where
$D_f$ = averaged effective diffusivity of growth-limiting nutrient (m²/s);
$X_f$ = averaged biofilm density (kg/m³);
$Y_{x/s}$ = yield coefficient (kg microorganisms / kg nutrient);
$\mu_{max}$ = maximum specific growth rate (s⁻¹);
$K_s$ = Monod half-rate constant (kg/m³);
$C$ = growth limiting substrate concentration (kg/m³).

Following the approach used to model uniform biofilms, we assume that the mass transport is one-dimensional, and that the nutrients are transferred only toward the bottom of the biofilm and consumed by the microorganisms located in the biofilm. With these assumptions the mass balances for nutrients around a differential element shown in Figure 7.9 are described below.

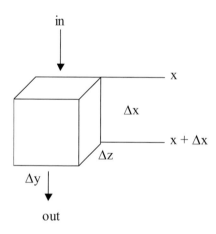

**Figure 7.9.** A differential element of a biofilm.

As described in previous sections we attempted to calculate the average effective diffusivity over a representative surface area of the biofilm $\Delta x$ thick. By accepting the average properties of the biofilms over the differential volume ($\Delta y \times \Delta x \times \Delta z$) we ignore the variations in the $z$ and $y$ directions (they are lumped as an average into our discrete layers), but not in the $x$ direction. Using this assumption we model the activity of discrete biofilms and test if it is meaningfully different from the activity of uniform biofilms.

If $N$ is the flux of the nutrient, the rate of the nutrient supply into the differential element along $x$, is:

Nutrient (in) = $N\Delta y\Delta z$ (7.9)

Similarly, the rate of nutrient flow from the differential element at $x+\Delta x$ is:

Nutrient (out) = $N\Delta y\Delta z + dN/dx(N\Delta y\Delta z\Delta x)$ (7.10)

The nutrient consumption rate in the differential element is:

$$\frac{dN}{dx}\Delta y\Delta z\Delta x = \frac{\mu_{max} C X_{fl}}{Y_{X/S}(K_S+C)}\Delta y\Delta z\Delta x \qquad (7.11)$$

where $X_{fl}$ is the average cell density in the differential element.
In summary, the difference between the rates of nutrient in and out from the differential element equals to the nutrient consumption rate within the volume of the differential element, and the mass balance around the differential element is:

$$N\Delta y\Delta z - (N\Delta y\Delta z + \frac{dN}{dx}\Delta y\Delta z\Delta x) = \frac{\mu_{max} C X_{fl}}{Y_{X/S}(K_S+C)}\Delta y\Delta z\Delta x \qquad (7.12)$$

Because we have assumed one-dimensional mass transport, the diffusive flux is given by Fick's first law (Bird et al. 1960):

$$N = -D_f \frac{dC}{dx} \qquad (7.13)$$

where $D_f$ is effective diffusivity. However, in discrete biofilms diffusivity changes in the $x$ direction, and should be represented by the variable effective diffusivity which is averaged over $\Delta y \times \Delta z$:

$$D_f = D_f(x) = D_{fx} \qquad (7.14)$$

$D_{fx}$ represents the average effective diffusivity in the volume of the differential element ($\Delta y\Delta x\Delta z$). In similar fashion, biofilm density also changes in the $x$ direction: biofilms are denser near the bottom than near the surface. The relation between effective diffusivity and biofilm density can be approximated from the following equation (Fan et al. 1990):

$$D_{fx} = 1 - \frac{0.43 \, X_{fl}^{0.92}}{11.19 + 0.27 \, X_{fl}^{0.99}} \qquad (7.15)$$

Inserting Equations 7.13 and 7.14 into Equation 7.12 gives the following:

$$\frac{dN}{dx} = \frac{d}{dx}\left(D_{fx}\frac{dC}{dx}\right) = D_{fx}\frac{d^2C}{dx^2} + \frac{dD_{fx}}{dx}\frac{dC}{dx} = \frac{\mu_{max} \, C X_{fl}}{Y_{X/S}(K_S + C)} \qquad (7.16)$$

Because experimentally we found that the variation of effective diffusivity across the biofilm is constant (see Figure 7.10), we define the diffusivity gradient ($\zeta$) across the biofilm as:

$$dD_{fx}/dx = \zeta \qquad (7.17)$$

Combining Equation 7.17 with Equation 7.16 results in a mass transport continuity equation, which may be used to compute nutrient concentration profiles in discrete biofilms:

$$D_{fx}\frac{d^2C}{dx^2} + \zeta \frac{dC}{dx} = \frac{\mu_{max} \, C X_{fl}}{Y_{X/S}(K_S + C)} . \qquad (7.18)$$

## 7.8 EXPERIMENTAL VALIDATION OF THE MODEL OF BIOFILMS COMPOSED OF DISCRETE LAYERS

Mathematical models of heterogeneous biofilms should be validated experimentally. We see this requirement as an obstacle to using biofilm models based on cellular automata with confidence. For mathematical models of biofilms constructed based on the cellular automata, the link between mathematical modeling and experimentation is weak. This is not surprising, considering that the cellular automata model is essentially a matrix that may accommodate any number of rules, and that the resulting structure will reflect these rules. For biofilms, these rules have not been formulated, and therefore the models based on cellular automata are difficult to validate experimentally. The model of biofilms composed of discrete layers, on the other hand, can be validated experimentally. We have a system of measurements in place that can validate results of mathematical modeling of such discrete layers. For example, Equation 7.18 introduces variable diffusivity and a diffusivity gradient. We have demonstrated experimentally that effective diffusivity profiles exist, are linear, and can be quantified in biofilms: Figure 7.10 shows changes in diffusivity across a heterogeneous biofilm. These results are average diffusivities, and their standard deviations are determined from the results in Figure 7.8 (Beyenal and Lewandowski 2000).

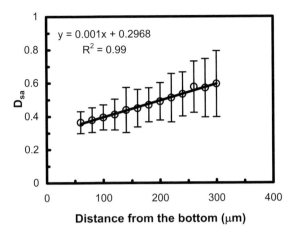

**Figure 7.10.** $D_{sa}$ versus the distance from the bottom for biofilms grown at 7.5 cm/s flow velocity; $D_{sa}$ (= $D_{fx}/D_w$) is relative effective diffusivity, where $D_w$ is molecular diffusivity of the nutrient. Parameter $\zeta$ can be calculated as the slope × $D_w$. The standard deviations increase toward the biofilm surface, which reflects the fact that biofilms are more heterogeneous near the surface than near the bottom. It is also worth noticing that the effective diffusivity gradient ($\zeta$) has the same dimensions as the mass transport coefficient (length per unit time); we termed $\zeta$ a secondary mass transport coefficient (Beyenal and Lewandowski 2000).

## 7.9 CAN DISCRETIZING BIOFILMS REFLECT THE EFFECT OF BIOFILM HETEROGENEITY?

The question of how important it is to discretize biofilms can be answered by comparing fluxes of nutrients to (1) a non-discretized uniform biofilm and (2) a biofilm that has been discretized.

Equation 7.18 is a nonlinear differential equation and can be solved using numerical techniques. It is a second-order equation, and we have two boundary conditions:

@      $x = 0$      $dC/dx = 0$      (7.19)

@      $x = L_f$      $C = C_s$      (7.20)

The first boundary condition (7.19) states that there is no nutrient flux across the boundary at the bottom. The other boundary condition (7.20) states that the biofilm defines the position of the second boundary, the biofilm surface, and gives the concentration of nutrient at this location. These two cases, a non-discretized and a discretized biofilm, are represented by the following adjustments in Equation 7.18:

1. For non-discretized biofilms (no diffusivity profile):

$$\zeta = 0 \quad \text{and} \quad D_{sa} = D_{av} \tag{7.21}$$

2. For discretized biofilms (diffusivity forms a profile):

$$D_{sa} = (a + bx) \tag{7.22}$$

Equation 7.22 represents the linear correlation between the surface-averaged relative effective diffusivity and the distance, and the numerical values are taken from the data in Figure 7.10. The effective diffusivity for each layer was calculated from Equation 7.22:

$$D_{fx} = D_{sa} \times D_w \tag{7.23}$$

To quantify the differences between discrete and uniform biofilms we calculated fluxes of glucose using the biokinetic parameters given by Bailey and Ollis (1986) as $K_s = 22$ mg/L and $\mu_{max}/Y_{x/s} = 6.37 \times 10^{-5}$ s$^{-1}$. Assuming that the biofilm is 1000 μm thick, the concentration of glucose at the biofilm–liquid interface is 500 mg/L, and $a = 0.2$ in Equation 7.22. Figure 7.11 shows that the predicted glucose consumption rates in discretized biofilms are four to five times higher than they are in uniform biofilms, which reflects the magnitude to which biofilm heterogeneity affects biofilm activity in the model biofilm. Note that the value of $\zeta$ effects the glucose consumption by the biofilms in a complicated way; when $\zeta$ increases, the glucose consumption rate decreases. This effect is counterintuitive, because the glucose consumption should increase as biomass concentration increases. However, the increase in biomass density decreases diffusivity, and counterbalances the expected effect of increasing nutrient consumption rate. Biofilm density in Equation 7.18 was calculated from Equation 7.15. To solve Equation 7.18 we used Matlab's boundary value solver function (bvp4c). The bvp4c is a finite difference code that implements the three-stage Lobatto IIIa formula. Mesh selection and error control are based on the residual of the continuous solution. We used Matlab's defaults to control precision of the solution.

Results in Figure 7.11 show that discretized biofilms are metabolically more active than uniform biofilms, which is, qualitatively, consistent with the expected effect of biofilm heterogeneity.

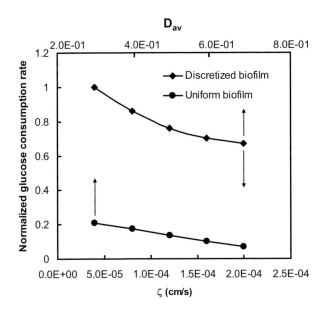

**Figure 7.11.** The predicted normalized fluxes (= nutrient consumption rate) to the uniform and discretized biofilms. The fluxes have been normalized with respect to the maximum glucose flux calculated for the discretized biofilm. The predicted fluxes of glucose to the discretized biofilm are consistently four to five times higher than the fluxes of glucose to the uniform biofilm. Both biofilms have th same $D_{av}$. However, the uniform biofilm has $\zeta = 0$, as required by Equation 7.21. Both biofilms have the same diffusivity and density but show different activity.

## 7.10  BIOFILMS GROWN AT HIGH FLOW VELOCITIES

Most information about mass transport in biofilms has been generated by studying biofilms at low flow velocities, not exceeding a few centimeters per second. There are, however, many situations where biofilms grow at high flow velocities, a few meters per second. These biofilms are much less accessible for *in situ* inspection because they usually accumulate in closed conduits. We have demonstrated that biofilms grown under such conditions have a different morphology than that shown in Figure 7.1: microcolonies are tapered, with long filamentous streamers (Lewandowski and Stoodley 1995). We hypothesized that the reason for the microcolonies to assume such a shape is the viscoelastic nature of extracellular biopolymers holding the microcolonies together. When a single microcolony attached to the bottom of a reactor (as in Figure 7.1) it is subjected to high shear stress and high flow velocity, the boundary layer separates downstream of the microcolony, causing formation of a low-pressure zone. Because the microcolonies are held together by viscoelastic polymers, the material flows slowly, assuming elongated shapes, or streamers. In effect, the biofilm surface looks like the one in 7.12.

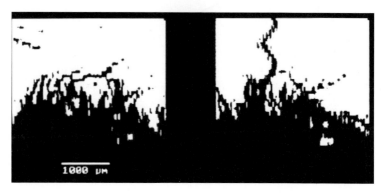

**Figure 7.12.** Surface of a biofilm grown at high flow velocity (Lewandowski and Stoodley 1995).

There are several features that distinguish biofilms grown at high flow velocities from those grown at moderate and low flow velocities. From the image in Figure 7.12, we see that it is difficult to determine the position of the biofilm surface. Consequently, the concepts of mass transport and hydrodynamic boundary layers become vague. For mass transport dynamics, the most significant difference is the whipping action of streamers introducing turbulence to that zone.

Little is known about biofilms grown at high flow velocities: nutrient concentration profiles, mass transport coefficient profiles, and effective diffusivity profiles are all inaccessible because microelectrodes cannot withstand such conditions and flow velocity profiles cannot be measured by the known techniques. Most of the knowledge of such biofilms relies on visual observations and speculations.

## 7.11   CONCLUDING REMARKS

The morphology of biofilms grown at low flow velocities often resembles the model in Figure 7.1, microcolonies separated by interstitial voids. There is enough experimental data to evaluate the nature of mass transport in such systems, at least qualitatively. For the purpose of discussing mass transport dynamics, we subdivide the biofilm system into two zones: the external zone, between the biofilm surface and the bulk solution; and the internal zone, between the substratum and the biofilm surface. The external zone can be further subdivided into two parts: (1) away from the biofilm surface in the bulk solution; and (2) in proximity to the biofilm surface. In the bulk solution, away from the biofilm surface, mass transport is controlled by convection, if the water is flowing. Just above the biofilm surface, water flow is laminar. It is important to notice that the water flow is parallel to the biofilm surface and the mass transport is normal to that surface. Therefore, nutrients are transported along the surface with the flowing water and toward the surface by diffusion. This duality of the mass transport mechanism near the biofilm surface produces confusing experimental results because most tools we use to study biofilms at microscale do not discriminate between directions. Within the biofilm, the mass transport mechanism changes: the same component of flow velocity that was parallel to the bottom above the biofilm surface becomes perpendicular to some microcolonies below the biofilm surface. Therefore, the mass transport in biofilm channels is due to convection and, perhaps, dispersion, as the biofilm may act as a porous medium. Convection within biofilm pores makes mass transport quite efficient and may allow the nutrients to penetrate deep into the biofilm.

From all available information, mass transport within microcolonies is due to diffusion. This is a somewhat uncertain assertion because of the microchannels, which exist in some microcolonies (Yang and Lewandowski 1995). These microchannels have dimensions of a few micrometers and it is not very likely that they change the overall conclusion that within microcolonies mass transport is controlled by diffusion.

In biofilms grown at low and moderate flow velocities, mass transport above the biofilm surface and below the biofilm surface is both convective and diffusive. Above the biofilm surface, nutrients are transported along the biofilm surface by convection and toward the biofilm surface by diffusion. Below the biofilm surface, nutrients are transported in biofilm pores by convection and in microcolonies by diffusion. Consequently, microorganisms within microcolonies are supplied with nutrients from the bulk as the result of a chain of alternating convection and diffusion processes.

Mass transport in biofilms is affected by biofilm structure and by hydrodynamics. The fact that hydrodynamics also affects biofilm structure complicates the picture. To simplify matters, it is convenient to discuss biofilms grown at low flow velocities, a few cm/s, and biofilms grown at high flow velocities, a few m/s, separately. Biofilms grown at low flow velocities and biofilms grown at high flow velocities have distinctly different morphologies (as exemplified by Figures 7.1 and 7.12). For biofilms grown at low flow velocities, the influence of biofilm structure on mass transport can be separated from the influence of hydrodynamics because the effect of hydrodynamics on biofilm structure can be neglected. For biofilms grown at high flow velocities, this simplification is not warranted: at high flow velocities, hydrodynamics strongly affects biofilm structure. Available data on nutrient concentration gradients and flow velocity gradients are limited to biofilms grown at low flow velocities. Therefore, most information about the nature of mass transport in biofilms comes from measurements in biofilms grown at low flow velocities; very little is known about mass transport in biofilms grown at high flow velocities.

In summary:

(1) To describe the dynamics of mass transport in biofilms, it is convenient to inspect the distribution of mass transport resistance throughout the system. In slowly flowing water, nutrient supply from the bulk to the biofilm is limited by diffusion across the mass transport boundary layer above the biofilm surface, a zone of high mass transport resistance. As a result, convection below the biofilm surface helps distributing nutrients within the biofilm, but the amount of nutrients available within the volume occupied by the biofilm remains determined by the flux across the zone of high mass transport resistance above the biofilm.

(2) It is important to develop biofilm models that use biofilm heterogeneity as a parameter. However, considering microcolonies as building blocks of biofilms may produce models that are difficult to validate experimentally. After all, it is not practical, or even possible, to measure activity of every single microcolony in a biofilm. Therefore, we are attempting to simplify modeling of heterogeneous biofilms by using discretized biofilms. Models of discretized biofilms can be validated because we have experimental tools, and a compatible conceptual approach, to make their validating possible. Discretized biofilms are composed of several uniform layers; the exact number of these layers can be adjusted, and each of these layers is modeled as a uniform biofilm. This approach allows using biofilm porosity as a parameter influencing the biomass density and nutrient diffusivity within each layer, and using existing experimental protocols to verify the predictions of that model.

## 7.12 REFERENCES

Atkinson B., Busch B., Swilley A.W. and Williams, D.A. (1967) Mass transfer and organism growth in a biological film reactor. *Trans. Inst. Chem. Eng.* **45**, 57–264.

Bailey J.E. and Ollis, D.F. (1986) *Biochemical Engineering Fundamentals*, 2nd edition. McGraw-Hill, NY.

Berg P. and Petersen, N.R. (1998) Interpretation of measured concentration profiles in sediment pore water. *Limnol. Oceanogr.* **43**, 1500–1510.

Bennett C.O. and Mayers J.E. (1974) *Momentum, Heat, and Mass Transfer*. McGraw-Hill.

Beyenal, H. and Lewandowski, Z. (2000) Combined effect of substrate concentration and flow velocity on effective diffusivity in biofilms. *Wat. Res.* **34**, 528–538.

Beyenal, H. and Lewandowski, Z. (2002) Internal and external mass transfer in biofilms grown at various flow velocities. *Biotech. Prog.* **18**, 55–61.

Beyenal, H., Şeker, Ş., Salih, B. and Tanyolaç, A. (1997) Diffusion coefficients of phenol and oxygen in a biofilm of *Pseudomonas putida*. *Amer. Inst. Chem. Eng. J.* **43**, 243–250.

Beyenal, H., Tanyolac, A. and Lewandowski, Z. (1998) Measurement of local effective diffusivity in heterogeneous biofilms. *Wat. Sci. Tech.* **38**(8/9), 171–178.

Bird, R.B., Steward, W.E. and Lightfoot, E.N. (1960) *Transport Phenomena*. John Wiley, Singapore.

Bishop, P.L. and Rittmann, B.E. (1995) Modeling heterogeneity in biofilms: report of the discussion session. *Wat. Sci. Tech.* **32**(8), 263–265.

Burgman, J.W. and Sides P.J. (1988) Electrochemical measurements of effective diffusivity in Hall-Heroult electrolyte. *Amer. Inst. Chem. Eng. J.* **34**, 1649–1655.

Cammarata, V., Talham R.T., Crooks, R.M. and Wrighton, M. S. (1990) Use of microelectrode arrays to directly measure diffusion of ions in solid electrolytes: Physical diffusion of $Ag^+$ in a solid polymer electrolyte. *J. Phys. Chem.* **94**, 2680–2684.

Costerton J.W., Lewandowski, Z., Caldwell, D.E., Korber, D.R. and Lappin-Scott, H.M. (1995) Microbial biofilms. *Annu. Rev. Microbiol.* **49**, 711–745.

de Beer, D., Stoodley, P., Roe, F. and Lewandowski, Z. (1994) Effects of biofilm structures on oxygen distribution and mass transport. *Biotech. Bioengng* **43**, 1131–1138.

Drury, W.J. (1992) Interactions of 1 micron latex microbeads with biofilm. Ph.D. thesis, Center for Biofilm Engineering, Montana State University.

Eberl, H.J., Picioreanu, C., Heijnen, J.J. and van Loosdrecht, M.C.M. (2000) A three-dimensional numerical study on the correlation of spatial structure, hydrodynamic conditions, and mass transfer and conversion in biofilms. *Chem. Engng Sci.* **24**, 6209–6222.

Elmaleh, S. (1990) Rule of thumb modeling of biofilm reactors. *Wat. Sci. Tech.* **22**(1/2), 405–418.

Fan, L.S., Ramos, R.L., Wisecarver, K.D. and Zehner, B.J. (1990) Diffusion of phenol through a biofilm grown on activated carbon particles in a draft-tube three-phase fluidized bed bioreactor. *Biotech. Bioengng* **35**, 279–286.

Grady, C.P. Jr (1983) Modeling of biological fixed films–A state of the art review, In *Fixed Film Biological Process for Wastewater Treatment* (ed. Y. C. Wu and E. D. Smith), pp. 75–134. Noyes Data Corporation, New Jersey.

Hermanowicz, S.W. (2001) A simple 2D biofilm model yields a variety of morphological features. *Math. Biosci.* **169**, 1–14.

Hermanowicz, S.W., Schindler U. and Wilderer P. (1995). Fractal structure of biofilms: new tools for investigation of morphology. *Wat. Sci. Tech.* **32**(8), 99–105.

Heydorn, A, Nielsen., A.T., Hentzer, M., Sternberg, C., Givskov, M., Ersboll, B. K. and Molin S. (2000) Quantification of biofilm structures by the novel computer program COMSTAT. *Microbiology –UK.* **146**, 2395–2407.

Horn, H. and Hempel, D.C. (1997) Substrate utilization and mass transfer in an autotrophic biofilm system: Experimental results and simulation. *Biotech. Bioengng* **53**, 363–371.

Keevil, C.W., Walker, J.T. (1992) A Normarski DIC microscopy and image analysis of biofilms. *Binary.* **4**, 93–95.

Kreft, J.U., Picioreanu, C., Wimpenny, J.W.T. and van Loosdrecht, M.C.M. (2001) Individual-based modeling of biofilms. *Microbiol–SGM.* **147**, 2897–2912.

Kreft, J.U. and Wimpenny, J.W.T. (2001) Effect of EPS on biofilm structure and function as revealed by an individual-based model of biofilm growth. *Wat. Sci. Tech.* **43**(6), 135–141.

Lawrence, J.R., Korber, D.R., Hoyle, B.D., Costerton, J.W. and Caldwell, D.E. (1991) Optical sectioning of microbial biofilms. *J. Bacteriol.* **173**, 6558–6567.

Lewandowski, Z. (2000a) Rapid Communication: Notes on Biofilm Porosity. *Wat. Res.* **34**, 2620–2624.

Lewandowski, Z. (2000b) Structure and Function of Biofilms. In *Biofilms: Advances in their Study and Control* (ed. L.V. Evans), pp. 1–17. Harwood Academic Publishers.

Lewandowski, Z. (1993) Dissolved oxygen gradients near microbially colonized surfaces. In *Biofouling and Biocorrosion in Industrial Water Systems* (ed. G. Geesey, Z. Lewandowski, H.-C. Flemming), pp. 175–189. Lewis Publishers, Boca Raton.

Lewandowski, Z., Altobelli, S. A. and Fukushima, E. (1993) NMR and microelectrode studies of hydrodynamics and kinetics in biofilms. *Biotech. Progress.* **9**, 40–45.

Lewandowski, Z. and Stoodley, P. (1995) Flow induced vibrations, drag force, and pressure drop in conduits covered with biofilm. *Wat. Sci. Tech.* **32**(8), 19–26.

Lewandowski, Z., Walser, G., Characklis, W.G. (1991) Reaction kinetics in biofilms. *Biotech. Bioengng* **38**, 877–882.

Lewandowski, Z., Webb, D., Hamilton, M. and Harkin, G. (1999) Quantifying biofilm structure. *Wat. Sci. Tech.* **39**(7), 71–76.

Lewandowski, Z. and Beyenal, H. (2001) Limiting-Current-Type Microelectrodes for Quantifying Mass Transport Dynamics in Biofilms. In *Microbial Growth in Biofilms*: Part B: *Methods Enzymol.* (ed R.J. Doyle), vol. 337, pp. 339–359.

Livingston, A. G., Chase, H. (1989) Modeling of phenol biodegradation in a fluidized-bed bioreactor. *Amer. Inst. Chem. Eng. J.* **35**, 1980–1992.

Massol-Deya, A.A., Whallon, J., Hickey, R.F., Tiedje, J.M. (1994) Channel structures in aerobic biofilms of fixed-film reactors treating contaminated groundwater. *Appl. Environ. Microbiol.,* **61**, 769–777.

Noguera, D.R., Okabe, S. and Picioreanu, C. (1999) Biofilm modeling: Present status and future directions. *Wat. Sci. Tech.* **39**(7), 273–278.

Picioreanu, C., van Loosdrecht, M.C.M. and Heijnen, J.J. (2001) Two-dimensional model of biofilm detachment caused by internal stress from liquid flow. *Biotech. Bioengng* **72**, 205–218.

Picioreanu, C., van Loosdrecht, M.C.M. and Heijnen, J.J. (2000) Effect of diffusive and convective substrate transport on biofilm structure formation: A two-dimensional modeling study. *Biotech. Bioengng* **69**, 504–515.

Picioreanu, C., van Loosdrecht, M.C.M. and Heijnen, J.J. (1998a) A new combined differential-discrete cellular automaton approach for biofilm modeling: Application for growth in gel beads. *Biotech. Bioengng* **57**, 718–731.

Picioreanu, C., van Loosdrecht, M.C.M. and Heijnen J.J. (1998b) Mathematical modeling of biofilms structure with a hybrid differential – discrete cellular automaton approach. *Biotech. Bioengng* **58**, 101–116.

Rasmussen, K. and Lewandowski, Z. (1998) Microelectrode measurements of local mass transport rates in heterogeneous biofilms. *Biotech. Bioengng* **59**, 302–309.

Revsbech, N.P. and Jorgensen, B.B. (1986) Microelectrodes: Their Use in Microbial Ecology. In *Advances in Microbial Ecology* (ed. K.C. Marshall), pp. 293–352. Plenum Publishing Corporation.

Reichert, P. (1994) AQUASIM – a tool for simulation and data analysis of aquatic systems. *Wat. Sci. Tech.* **30**(2), 21–30.

Rittmann, B.E. and McCarty, P.L. (1980) Model of steady-state biofilm kinetics. *Biotech. Bioengng* **22**, 2343–2357.

Schwarzenbach, R. P., Gschwend, P. M. and Imboden, D.M. (1993) *Environmental Organic Chemistry*, p. 202. J. Wiley, New York.

Wanner, O. and Reichert, P. (1996) Mathematical modeling of mixed culture biofilms. *Biotech. Bioengng* **49**, 172–184.

Wimpenny, J.W. and Colasanti, R. (1997) A unifying hypothesis for the structure of microbial biofilms based on cellular automaton models. *FEMS Microbiol. Ecol.* **22**, 1–16.

Wolfaardt, G.M., Lawrence, J.R., Robarts, R.D. and Caldwell, D.E. (1994) A multicellular organization in a degradative biofilm community. *Appl. Environ. Microbiol.* **60**, 434–446.

Wood, B. D. and Whitaker, S. (1998) Diffusion and reaction in biofilms. *Chem. Engng Sci.* **53**, 397–425.

Wooster, T.T., Longmire, M.L., Watanable, M. and Murray, R.W. (1991) Diffusion and heterogeneous electron-transfer rates in acetonitrile and in polyether polymer melts by alternating current voltammetry at microdisk electrodes. *J. Phys. Chem.* **95**, 5315–5321.

Xavier, J.B., Schnell, A., Wuertz, S., Palmer, R., White, D.C., and Almeida, J.S. (2001). Objective threshold selection procedure (OTS) for segmentation of scanning laser confocal microscope images, *J. Microbiol. Methods* **47**, 169–180.

Xia, F., Beyenal, H., Lewandowski, Z. (1998). En electrochemical technique to measure local flow velocity in biofilms. *Wat. Res.* **32**, 3631–3636.

Yang, S. and Lewandowski, Z. (1995) Measurement of local mass transfer coefficient in biofilms. *Biotech. Bioengng* **48**, 737–744.

Yang, X., Beyenal, H., Harkin, G. and Lewandowski, Z. (2000) Quantifying biofilm structure using image analysis. *J. Microbiol. Methods* **39**, 109–119.

Yang, X., Beyenal, H., Harkin, G. and Lewandowski, Z. (2001) Evaluation of biofilm image thresholding methods. *Wat. Res.* **35**, 1149–1158.

Zhang, T. C. and Bishop, P.L. (1994a) Evaluation of tortuosity factors and effective diffusivities in biofilms. *Wat. Res.* **28**, 2279–2287.

Zhang, T. C. and Bishop, P. L. (1994b) Density, porosity, and pore structure of biofilms. *Wat. Res.* **28**, 2267–2277.

# 8
# The crucial role of extracellular polymeric substances in biofilms

*Hans-Curt Flemming and Jost Wingender*

## 8.1 INTRODUCTION

Most microorganisms live and grow in aggregated forms such as biofilms, flocs ('planktonic biofilms') and sludges (Costerton *et al.* 1995; Wimpenny 2000). These forms of growth are frequently lumped in the unifying expression *biofilm*. The common feature is that the microorganisms exist in close associations at high cell densities and are embedded in a matrix of extracellular polymeric substances (EPS), which are responsible for the morphology, architecture, coherence, physicochemical properties and biochemical activity of these aggregates (Wingender *et al.* 1999a; Flemming and Wingender 2001a). Biofilms are ubiquitously distributed in aquatic environments, on tissues of plants, animals and man, as well as on surfaces of technical systems such as filters and other porous materials, reservoirs, pipelines, ship hulls, heat exchangers,

© 2003 IWA Publishing. *Biofilms in wastewater treatment*. Edited by S. Wuertz, P.L. Bishop and P.A. Wilderer. ISBN: 1 84339 007 8.

separation membranes, etc. (Costerton *et al.* 1987; Flemming and Schaule 1996); biofilms may also develop on medical devices, thus initiating persistent infections in humans (see Costerton *et al.* 1987, 1999). Biofilms develop at phase boundaries; they can be frequently found adherent to a solid surface (substratum) at solid-water interfaces, but they also develop at water-air and at solid-air interfaces (Wimpenny 2000). Biofilms and flocs are accumulations of microorganisms (prokaryotic and eukaryotic unicellular organisms), EPS, multivalent cations, inorganic particles, and biogenic material (detritus), as well as colloidal and dissolved compounds. EPS are considered as the key components that determine the structural and functional integrity of microbial aggregates. EPS form a three-dimensional, gel-like, highly hydrated and locally charged biofilm matrix, in which the microorganisms are more or less immobilized. EPS create a microenvironment for sessile cells, which is conditioned by the nature of the EPS matrix. In general, the proportion of EPS in biofilms can vary between roughly 50 and 90% of the total organic matter (Christensen and Characklis 1990; Nielsen *et al.* 1997). In activated sludge and sewer biofilms, 85–90% and 70–98%, respectively, of total organic carbon were found to be extracellular, indicating that cell biomass may constitute only a minor fraction of the organic matter of microbial aggregates in wastewater environments (Frølund *et al.* 1996; Jahn and Nielsen 1998). EPS play an essential role in wastewater treatment processes, because they are involved in the formation of activated sludge flocs (bioflocculation) and the development of fixed biofilms, e.g. in trickling filters, rotating biological contactors, fluidised-bed reactors or submerged fixed-bed reactors (Bryers and Characklis 1990; Bitton 1994); in addition, EPS influence the dewaterability of wastewater sludges (Poxon and Darby 1997).

## 8.2 DEFINITION OF EPS

EPS production is a general microbial property that seems to be expressed in most environments. The ability to form EPS is widespread among prokaryotic organisms (Bacteria, Archaea), but has also been shown to occur in eukaryotic microorganisms including microalgae such as diatoms (see Cooksey 1992; Khandeparker and Bhosle 2001), and fungi such as yeasts and molds (McCourtie and Douglas 1985; Sutherland 1996). Microbial EPS are biosynthetic polymers (Table 8.1). Geesey (1982) defined EPS as 'extracellular polymeric substances of biological origin that participate in the formation of microbial aggregates'. Another definition was given in a glossary to the report of the Dahlem Workshop on Structure and Function of Biofilms in Berlin 1988 (Characklis and Wilderer 1989); here, EPS were defined as 'organic polymers of microbial origin which in biofilm systems are frequently responsible for binding cells and other particulate materials together (cohesion) and to the substratum (adhesion)'. In a critical evaluation of EPS isolation techniques, Gehr and Henry (1983) described extracellular material as 'that material which can be removed from microorganisms (and in particular, bacteria) without disrupting the cell, and without which the microorganism is still viable'. This definition alludes to the observation that EPS are not essential structures of bacteria, as loss of EPS does not impair growth and viability of cells in laboratory cultures. Under natural conditions, however, EPS production seems to be an important feature of survival, as most environmental bacteria occur in microbial aggregates such as flocs and biofilms, whose structural and functional integrity are based essentially on the presence of an EPS matrix.

**Table 8.1.** General composition of some bacterial EPS; humic substances are included in the table, because they are sometimes considered as part of the EPS matrix.

| EPS | Principal components (subunits, precursors) | Main type of linkage between subunits | Structure of polymer backbone | Substituents (examples) |
| --- | --- | --- | --- | --- |
| Polysaccharides | Monosaccharides Uronic acids Amino sugars | Glycosidic bonds | Linear Branched Side-chains | Organic: O-acetyl, N-acetyl, succinyl, pyruvyl |
|  |  |  |  | Inorganic: sulfate, phosphate |
| Proteins (polypeptides) | Amino acids | Peptide bonds | Linear | Oligosaccharides (glycoproteins) Fatty acids (lipoproteins) |
| Nucleic acids | Nucleotides | Phosphodiester bonds | Linear |  |
| (Phospho)lipids | Fatty acids Glycerol Phosphate Ethanolamine Serine Choline Sugars | Ester bonds | Side-chains |  |
| Humic substances | Phenolic compounds Simple sugars Amino acids | Ether bonds C-C bonds Peptide bonds | Cross-linked |  |

The abbreviation *EPS* has been used specifically for *extracellular polysaccharides* and *exopolysaccharides* or as a collective term for *exopolymeric substances, exopolymers, extracellular polymeric secretions* and *extracellular polymeric substances*. In early biofilm research, extracellular polysaccharides have often been assumed to be the dominant components of microbial aggregates (see, for example, Costerton *et al.* 1981), and have been confirmed to be abundant EPS components in flocs and biofilms (see Sutherland 1999a, 2001). That may be the reason why the term *EPS* has frequently been used as an abbreviation for *extracellular polysaccharides* or *exopolysaccharides*. However, other extracellular macromolecules such as proteins can also appear in significant amounts or even predominate (see below). In the following text, the abbreviation *EPS* is used for *extracellular polymeric substances* as a more general and comprehensive term for different classes of organic macromolecule such as polysaccharides, proteins, nucleic acids, lipids/phospholipids or humic substances, which have been identified as components in the EPS of microbial aggregates (Table 8.1).

The structure of EPS varies obviously quite markedly. Figure 8.1a,b shows scanning electron micrographs of biofouling layers on various reverse osmosis membranes. The supporting membrane materials and the preparation procedures were identical; however, both membranes were exposed to river water of different origin. In Figure 8.1a, the cells are embedded in a thick slime matrix, whereas in Figure 8.1b fibrillar structures are dominant. Although it is acknowledged that the dewatering procedure required for scanning electron microscopy (SEM) imaging produces drying artifacts, morphological differences are still obvious and are attributed to the nature of the different EPS.

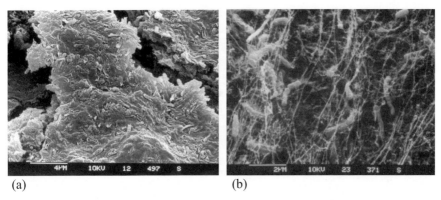

(a)                         (b)

**Figure 8.1.** Biofilms irreversibly blocking reverse osmosis membranes. (a), microorganisms embedded in a thick EPS layer; (b), microorganisms in fibrillar structures which presumably are drying artifacts (courtesy of G. Schaule).

## 8.3 COMPOSITION AND PROPERTIES OF EPS

EPS consist of varying proportions of carbohydrates, proteins, nucleic acids, lipids/phospholipids and humic acids. Analysis of EPS is usually done after separation of these macromolecules from microbial cells. Different physical and chemical methods, including centrifugation, filtration, heating, blending, sonication as well as treatment with sodium hydroxide, complexing agents and ion-exchange resins, have been described for the extraction of EPS from microbial aggregates (for reviews, see Nielsen and Jahn 1999; Spaeth and Wuertz 2000). Examples of suitable methods for the isolation of EPS from many biofilms and flocs without causing significant cell lysis, are the use of a cation-exchange resin (Dowex) (Jahn and Nielsen 1995; Nielsen and Jahn 1999) or a crown ether (Wuertz *et al.* 2001), combined with stirring under defined conditions. These methods are based on the removal of calcium ions, destabilizing the EPS structure and facilitating the separation of EPS from the cells. However, as the extraction efficiency for EPS varies depending on the type and origin of the microbial aggregate under study as well as on the fractionation technique used, the distinction between cell-bound and extracellular polymers must be considered as strictly operational. In Table 8.2, examples of literature data for the chemical composition of EPS from wastewater biofilms and activated sludge obtained by ion-exchange extraction methods are shown.

In earlier studies, isolation and purification procedures were often focused on the carbohydrate fraction of the EPS. However, when more extensive analyses of EPS were done, proteins were frequently shown to be abundant in the EPS, independent of the extraction method used. This observation has been reported for pure cultures of Gram-negative and Gram-positive bacteria (see, for example, Brown and Lester 1980; Platt *et al.* 1985; Arvaniti *et al.* 1994; Jahn and Nielsen 1995; Pereira and Vieira 2001) as well as for mixed-population biofilms and flocs (see, for example, Pavoni *et al.* 1972; Brown and Lester 1980; Jahn and Nielsen 1995; Nielsen *et al.* 1997; see Table 8.2). Several publications have demonstrated that proteins were even predominant over polysaccharides and represented the largest fraction in the EPS of biofilms and activated sludge from wastewater systems (e.g., Rudd *et al.* 1983; Nielsen *et al.* 1997; Bura *et al.* 1998; Dignac *et al.* 1998; Jorand *et al.* 1998; Liao *et al.* 2001; Martin-Cereceda *et al.* 2001; see Table 8.2). Among the nucleic acids, DNA has regularly been found in the EPS from wastewater biofilms and flocs (Pavoni *et al.* 1972; Brown and Lester 1980; Urbain *et al.* 1993; Nielsen *et al.* 1997; Bura *et al.* 1998; Liao *et al.* 2001; Martin-Cereceda *et al.* 2001; see Table 8.2), but also in the extracellular material from pure cultures (Platt *et al.* 1985; Arvaniti *et al.* 1994; Jahn and Nielsen 1995; Watanabe *et al.* 1998). Sometimes, the content of nucleic acids may even exceed that of proteins and polysaccharides, as has been shown for the EPS from the self-flocculating photosynthetic bacterium *Rhodovulum* sp. (Watanabe *et al.* 1998, 1999); here, RNA was the major constituent, whereas DNA only appeared in minor quantities. Other macromolecular components of EPS may be lipids and phospholipids (Goodwin and Forster 1985; Gehrke *et al.* 1998, 2001). In addition, the accumulation of humic substances in the EPS matrix of wastewater biofilms and activated sludges seems to be common (Nielsen *et al.* 1997; Jahn and Nielsen 1998; Martin-Cereceda *et al.* 2001; see Table 8.2).

**Table 8.2.** Chemical composition of EPS extracted from wastewater biofilms and activated sludge by cation-exchange resin treatment (examples). (RBC, rotating biological contactor; n.d., not determined; TOC, total organic carbon; VS, volatile solids; VSS, volatile suspended solids.)

| Sample | Protein | Carbo-hydrate | Uronic acids | DNA | Humic compounds | Unit | References |
|---|---|---|---|---|---|---|---|
| Sewer biofilm | 154 | 12 | 6 | 12 | 293 | mg/g TOC | Jahn and Nielsen 1995 |
| Sewer biofilm | 351 | 46 | 11 | 26 | 221 | mg/g TOC | Jahn and Nielsen 1995 |
| Activated sludge | 212 | 40 | 12.1 | 16 | 101 | mg/g VS | Nielsen *et al.* 1996 |
| Activated sludge | 243 | 48 | 6.1 | n.d. | 126 | mg/g VS | Frølund *et al.* 1996 |
| Activated sludge | 162 | 12.7 | 4.5 | 11.2 | n.d. | mg/g VSS | Bura *et al.* 1996 |
| Activated sludge | 51 | 21 | 3.8 | 25 | 18 | mg/g VS | Martin-Cereceda *et al.* 2001 |
| Biofilm in RBC | | | | | | | Martin-Cereceda *et al.* 2001 |
| Unit 1 | 200 | 40 | 2.5 | 22 | 24 | mg/g VS | |
| Unit 2 | 161 | 38 | 1.8 | 17 | 45 | mg/g VS | |
| Unit 3 | 159 | 53 | 3.8 | 28 | 50 | mg/g VS | |

EPS may consist of neutral molecules such as the polysaccharides dextran and levan; more frequently, EPS contain ionic groups which confer net negative or positive charges on the polymers at near neutral pH values. Extracellular polysaccharides owe their negative charge either to carboxyl groups of uronic acids or to non-carbohydrate substituents such as phosphate, sulfate, glycerate, pyruvate or succinate (Sutherland 2001). Extracellular polysaccharides often contain variable proportions of hexuronic acids such as glucuronic acid, galacturonic acid or mannuronic acid (Fazio et al. 1982; Uhlinger and White 1983; Sutherland 2001). Exceptions are polysaccharides, which completely consist of uronic acid residues; an example is the well-studied extracellular polysaccharide alginate of *Pseudomonas* and *Azotobacter* species; alginates are composed of mannuronate and guluronate residues, with mannuronate being partly O-acetylated. In wastewater biofilms and activated sludge flocs, uronic acid-containing polysaccharides seem to be common; however, uronic acid fractions in the EPS have been shown to be relatively low (Table 8.2). Sometimes, sulfate groups have been described to occur in sugar molecules such as in extracellular polysaccharides from the slime of *Staphylococcus epidermidis* strains (see Arvaniti et al. 1994). Cationic groups in polysaccharides may be due to the presence of amino sugars (Hejzlar and Chudoba 1986, Veiga et al. 1997).

Proteins can also contribute to the anionic properties of EPS. The negative charge of proteins is due to the presence of diacid amino acids such as aspartic and glutamic acid, which have been found to be quantitatively important constituents of extracellular proteins extracted from activated sludge (Higgins and Novak 1997; Dignac et al. 1998). Nucleic acids are polyanionic owing to the phosphate residues in the nucleotide moiety of the polymer molecule. Uronic acids, acidic amino acids and phosphate-containing nucleotides as negatively charged components of EPS can be crucial structural elements of flocs and biofilms, since they are expected to be involved in electrostatic interactions with multivalent cations (e.g., $Ca^{2+}$, $Mg^{2+}$, $Fe^{3+}$), thus mediating the formation and/or the stabilization of the network of the EPS matrix (see below).

EPS may contain non-polymeric substituents of low molecular weight. Extracellular polysaccharides often carry organic substituents such as acetyl, succinyl or pyruvyl groups or inorganic substituents such as sulfate or phosphate (Sutherland 2001). These substituents greatly alter the structure and physicochemical properties of the polysaccharides. For example, acetyl groups in alginates decrease the capacity and selectivity of divalent cation binding (Geddie and Sutherland 1994), increase solution viscosity of alginate (Lee et al. 1996), enhance the water-holding ability of the polysaccharide (Skjåk-Bræk et al. 1989), protect alginate from enzymatic degradation by alginate lyases (Lange et al. 1989) and affect biofilm formation (Nivens et al. 2001). Proteins can be substituted with fatty acids to form lipoproteins or can be glycosylated with oligosaccharides to form glycoproteins, which, for example, have been detected in the EPS from activated sludge (Horan and Eccles 1986).

Extracellular proteins contribute to hydrophobic properties of EPS due to their high proportions of the hydrophobic amino acids alanine, leucine and glycine (Higgins and Novak 1997; Dignac et al. 1998). The combination of acidic and hydrophobic amino acids in extracellular proteins was supposed to contribute to the electrostatic and hydrophobic interactions observed in flocs of activated sludges (Urbain *et al.* 1993; Dignac *et al.* 1998). Hydrophobic fractions of EPS extracted from activated sludge were made up of proteins but not carbohydrates, indicating that polysaccharides were not involved in hydrophobic interactions (Jorand *et al.* 1998). In another study on mixed-culture biofilms, *in situ* characterization of biofilm EPS demonstrated the presence of positively and negatively charged microzones as well as hydrophobic regions that were spatially in close association and were distributed non-uniformly in the biofilm matrix (Wolfaardt *et al.* 1998). In conclusion, the hydrophilic/hydrophobic regions within EPS molecules and their extent and distribution in the EPS matrix is expected to largely determine the ion-exchange potential and the sorption properties of biofilms and flocs.

Yield, composition and properties of EPS can vary spatially and temporally in response to the availability of nutrients and other environmental conditions. In aerobic heterotrophic biofilms grown in a wastewater-seeded rotating drum reactor, the study of the spatial distribution of extracellular polysaccharides and proteins indicated a stratified biofilm structure with non-uniform EPS production, which was highest in the upper layers of the biofilms (Zhang *et al.* 1998; Zhang and Bishop 2001). The heterogeneous structure of the biofilm was not only represented by vertical gradients of EPS yields, but also by the development of aerobic/anoxic zones, increased density, decreased porosity, lowered nutrient availability and decreased viable biomass along the biofilm depth (Zhang and Bishop 2001). de Beer *et al.* (1996) described a different spatial distribution of exopolysaccharides in dense granules and loose flocs from an upflow anaerobic sludge blanket reactor. In granules, approximately 50% of the total amount of exopolysaccharides were located in a 40-μm-thick zone on the surface, while the remainder was dispersed in the rest of the aggregate; in flocs, the highest concentration of exopolysaccharides were present in the centre. The hydrophilic polysaccharide coating of the granules was supposed to prevent the attachment of gas bubbles, explaining the lower susceptibility to flotation of granules as compared to flocs (de Beer *et al.* 1996). Thus, the spatial distribution of EPS may be related to functions of the EPS and control or reflect a certain degree of organization and differentiation within the community of biofilm organisms.

Production, composition and amount of EPS in flocs and biofilms are subject to temporal variations. Attachment of bacteria to surfaces as a first step in biofilm formation has been shown to stimulate the formation of extracellular polysaccharides (Vandevivere and Kirchman 1993). Temporal variation of EPS production has been reported in batch cultures of *Pseudomonas atlantica*; the carbohydrate composition of the EPS changed during the growth cycle, with a marked increase in the proportions and absolute amounts of uronic acids on prolonged incubation (Uhlinger and White 1983). Changes in the composition of EPS have been observed in activated sludge during anaerobic storage

(Nielsen et al. 1996); significant degradation of extracellular proteins and total carbohydrates took place within a few days, whereas the content of DNA and uronic acids showed only minor changes. Storage of activated sludge at a low temperature of 4 °C resulted in a considerable loss of EPS constituents; the most labile components were acidic polysaccharides and DNA (Bura et al. 1998). Thus, the extracellular matrix of flocs and biofilms seems to be a dynamic structure, where a certain degree of EPS turnover as a result of differential production and degradation processes can be expected.

EPS production is affected by the availability and composition of nutrients. In sequencing batch reactors with sludge growing on synthetic wastewater, the COD (chemical oxygen demand):N:P ratio was found to influence the hydrophobicity, surface charge, and EPS composition of microbial flocs (see Bura et al. 1998). For example, under P-depleted conditions, there was an increase in uronic acids and DNA of the EPS, but a decrease of the surface charge of the microbial flocs. In batch cultures of a methanogenic bacterium, polymer production was enhanced under low-phosphate and low-nitrogen conditions (Veiga et al. 1997); it was suggested that carbon utilization shifted toward EPS production when the C:N and/or C:P ratio was enhanced. The composition of EPS from biofilms of a rotating biological contactor (RBC) system has been reported to be different in successive RBC wastewater treatment stages (Martin-Cereceda et al. 2001). Along the treatment train, extracellular protein content was highest in biofilms from the first RBC unit, whereas the amounts of polysaccharides and humic substances in the EPS were highest in the third RBC unit (Table 8.2); in addition, the hydrophobic character of EPS increased along the RBC treatment train. The observed differences in EPS composition and hydrophobicity were supposed to be due to variations of nutritional conditions in the system, with cells in the last RBC unit presumably more starved than those of the first unit (Martin-Cereceda et al. 2001). In the same study, the comparison of EPS from RBC biofilms and activated sludge flocs showed that the biofilms contained significantly higher quantities of EPS; the amount of proteins was 3.5 times higher and the amounts of polysaccharides and humic substances were twice as high for biofilms than for sludge flocs. These observations could be attributed to different operational and environmental conditions, different wastewater composition in both wastewater treatment plants or differences in microbial growth patterns (suspended aggregates in sludge flocs versus attached microorganisms in RBC biofilms (Martin-Cereceda et al. 2001)).

The extracellular localization of EPS and their composition may be the result of different processes: active secretion, shedding of cellular material, cell lysis and adsorption from the environment. EPS can be actively secreted by living cells. Various specific pathways of biosynthesis and discrete export machineries have been described for the transport of proteins and polysaccharides to the microbial cell surface or into the surrounding environment (examples of reviews are Koster *et al.* (2000) and Whitfield and Valvano (1993)). Extracellular DNA and RNA can also be produced by living bacteria without autolysis of the cells (e.g., Watanabe *et al.* 1998); however, it is not known if nucleic acids are actively secreted or passively released due to an increase in cell envelope permeability. Another mechanism of secretion of polymers is the spontaneous release through formation of outer membrane-derived vesicles, which has been described as a common secretion mechanism in Gram-negative bacteria (Kadurugamuwa and Beveridge 1995; Beveridge *et al.* 1997). These membrane vesicles have been demonstrated in biofilms (Beveridge *et al.* 1997; Beveridge 1999). Surface blebbing occurs during normal growth and seems to represent a process, by which cellular macromolecules (nucleic acids, enzymes, lipopolysaccharide, phospholipids) are shed into the extracellular space in the form of membrane vesicles. Death and lysis of cells contribute to the release of cellular high-molecular-weight compounds into the EPS matrix where they are entrapped. Typically, intracellular organic polymers like poly-β-hydroxyalkanoates or glycogen as carbon and energy storage polymers or integral components of cell walls (e.g. peptidoglycan) and membranes (e.g. phospholipids, lipopolysaccharide) may thus become part of the EPS. Then, the biofilm represents a 'recycling yard' for intracellular components. finally, EPS, which is shed from microbial aggregates can be adsorbed in other places, so that the sites of synthesis, release and ultimate localization as an EPS component are not necessarily identical. This may especially be true for humic substances, which are primarily soil organic matter and are transported in aquatic environments where they can be adsorbed and accumulated in microbial flocs and biofilms.

## 8.4 MECHANICAL STABILITY MEDIATED BY EPS

The presence of EPS is considered as the basic prerequisite for the formation as well as for the maintenance of the integrity of biofilms or flocs. EPS are responsible for the mechanical stability of these microbial aggregates caused by intermolecular interactions between many different macromolecules; multivalent cations often promote and enforce the interactions between the EPS. At first glance, the EPS matrix looks simply random and without perceivable order. From a point of view of microbial ecology, however, it has an important function: to create the gel matrix in which the organisms can be fixed with a long retention time next to each other. This function is a prerequisite for the formation of stable microconsortia. It should be achieved with the lowest expense of energy and nutrients. The fact that only 1–2% of organic matter is required to bind 98–99% of water and to form a stable gel demonstrates how successful the strategy is.

What are the forces that keep this matrix together? Obviously, they are not covalent bonds. On the contrary, the cohesive forces are provided by weak interactions (Mayer et al. 1999). In principle, three types must be considered:

*London (dispersion) forces*—mainly in hydrophobic areas and not localized to functional groups; the binding energy is about 2.5 kJ mol$^{-1}$. These forces can be weakened by surface-active substances.

*Electrostatic* interactions—active between ions and between permanent and induced dipoles. The ionic interactions are relatively strong; it is mainly divalent cations, and in particular $Ca^{2+}$, which is responsible for a considerable proportion of the overall binding energy. The binding energy of non-ionic electrostatic bonds usually ranges from 12 to 29 kJ mol$^{-1}$. The binding force is strongly dependent upon the distance between the partners of the bond. These forces may also act repulsively. They can be influenced by ionic strength, complexing agents, acid and base.

*Hydrogen bonds*—mainly active between hydroxyl groups as is particularly frequent in polysaccharides, and water molecules. They also support the tertiary structure of proteins. The binding energy ranges from 10 to 30 kJ mol$^{-1}$. These forces can be influenced by chaotropic agents which disturb the water structure; examples are urea, tetramethyl urea and guanidine hydrochloride.

The individual binding force of any type of these interactions is relatively small compared with a covalent C–C bond (about 250 kJ mol$^{-1}$). However, the total binding energies of weak interactions between EPS molecules add up to bond values equivalent to those of covalent C–C bonds. As mentioned above, hydrophilic and hydrophobic properties of EPS have been demonstrated; on the basis of these results, all three types of binding force are expected to contribute to the overall stability of floc and biofilm matrices, probably to various extents. The result is the formation of a three-dimensional, gel-like network of EPS, whose composition, structure and properties may vary dynamically as the microorganisms respond to changes in environmental conditions.

## 8.5 ROLE OF EPS IN MICROBIAL AGGREGATION

EPS may contribute to the attachment of cells to surfaces (adhesion), but the extent of their involvement is still open to debate. There are indications that the production of certain EPS (extracellular polysaccharides) is triggered in response to adhesion (Allison and Sutherland 1987; Vandevivere and Kirchman 1993; Davies et al. 1993); however, after the initial events of adhesion, EPS are supposed to be always expressed and to be absolutely necessary for the development and maintenance of biofilms and flocs. Thus, network formation by the chemical forces mentioned above seems to be essential for the development and maintenance of microbial aggregates. On the basis of these observations, EPS are included as the key components in several models explaining the aggregation of microorganisms as well as the physicochemical properties of the extracellular matrix in flocs and biofilms (e.g., Pavoni *et al.* 1972; Harris and Mitchell 1973; Jorand *et al.* 1995; Nielsen *et al.* 1997; Higgins and Novak 1997).

In activated sludge systems, EPS have been implicated in determining the bioflocculation process, floc structure, floc charge, floc settleability and dewatering properties. In the polymer-bridging model, floc formation is considered as the result of the interaction of high molecular weight, long-chain EPS with microbial cells and other particles as well as with other EPS molecules, so that EPS bridge the cells into a three-dimensional matrix (Pavoni *et al.* 1972). Flocculation is associated with the formation of EPS. Cellular aggregation was found to depend on the physiological state of the microorganisms; flocculation of cultures of mixed populations from domestic wastewater did not occur until they entered into a restricted state of growth (Pavoni *et al.* 1972). There was a direct correlation between microbial aggregation and EPS accumulation; the ratio of EPS to microorganism mass rapidly increased during culture aggregation. The major EPS were polysaccharides, proteins and nucleic acids (RNA, DNA). Surface charge was not considered a necessary prerequisite for flocculation, because it remained constant throughout all growth phases regardless of the flocculability of the culture. Bacteria washed free of EPS formed stable dispersions, but re-addition of extracted EPS again resulted in flocculation.

In activated sludge cultivated on a carbon-limited medium, the release of EPS (carbohydrates, proteins, nucleic acids) by cell lysis resulted in an increase in flocculation (Vallom and McLoughlin 1984). Addition of DNA to activated sludge or to pure cultures of bacteria isolated from activated sludge promoted flocculation, indicating that DNA can act as a polyelectrolyte floc agent (Vallom and McLoughlin 1984). The importance of nucleic acids in microbial aggregation was also observed in the self-flocculating bacterium *Rhodovulum* sp., which produced EPS made up of carbohydrates, proteins and nucleic acids (Watanabe *et al.* 1998). Treatment of flocculated cells with nucleolytic enzymes (RNase, DNase) resulted in deflocculation of the cells, whereas polysaccharide- and protein-degrading enzymes had no effect; it was argued that extracellular nucleic acids were active in flocculation (Watanabe *et al.* 1998).

In batch cultures of the wastewater bacterium *Zoogloea*, it was shown that production of an extracellular polysaccharide was accompanied by flocculation of the bacterial cells (Unz and Farrah 1976). The polymer formation was initiated in mid-logarithmic growth phase, and the quantity produced appeared to be influenced by the level of carbon and nitrogen in the medium. Methanogenic bacteria were shown to produce EPS, which were supposed to help in the formation and maintenance of anaerobic granules (Veiga *et al.* 1997); high settleability of these granules is important for the performance of anaerobic sludge reactors. The detection of extracellular polymeric fibrils in natural and wastewater flocs by high-resolution transmission electron microscopy (TEM) confirmed the role of EPS as a structural support to the microbial aggregates (Liss et al. 1996).

The concept of EPS mediating microbial aggregation has been applied in field trials by using a slime-producing organism to enhance biomass settleability in aerobic wastewater treatment systems in the paper industry (Volpe et al. 1998); addition of EPS-producing bacteria (*Bacillus* sp.) to an aerated stabilization system and to an activated sludge unit resulted in an improvement of flocculation and effluent quality.

In addition to EPS, divalent cations are regarded as important constituents of microbial aggregates, as they bind to negatively charged groups present on bacterial surfaces, in EPS molecules and on inorganic particles entrapped in flocs and biofilms. It has been reported that extraction of $Ca^{2+}$ from flocs and biofilms by displacement with monovalent cations or by chelation with the general complexing agent EDTA or the more $Ca^{2+}$-specific chelant EGTA resulted in the destabilisation of flocs (Bruus et al. 1992; Higgins and Novak 1997) and biofilms (Turakhia et al. 1983). Practical implications are that weakening of activated sludge structure by removal of $Ca^{2+}$ leads to an increase in the number of small particles with subsequent decrease of filterability and dewaterability. Uniaxial compression measurements on biofilms of *Pseudomonas aeruginosa* demonstrated that above a critical calcium ion concentration, mechanical stability of the biofilms increased with increasing calcium content in the growth medium; this behavior was explained by crosslinking of negatively charged alginate molecules in the EPS by calcium ions (Körstgens et al. 2001). These observations suggest that divalent cations may be important for the maintenance of floc and biofilm structure by acting as bridging agents within the three-dimensional EPS matrix. Bruus et al. (1992) also integrated the role of divalent cations into their sludge floc model. The floc structure was proposed to be considered as a three-dimensional EPS matrix kept together by divalent cations with varying selectivity to the matrix ($Cu^{2+} > Ca^{2+} > Mg^{2+}$). It was argued that approximately half of the $Ca^{2+}$ pool was associated with EPS, forming a matrix that resembled gels of carboxylate-containing alginates. $Fe^{3+}$ ions may also be of importance in floc stabilisation. Specific removal of $Fe^{3+}$ from activated sludge flocs caused a weakening of floc strength, resulting in release of particles to bulk water, dissolution of EPS and partial floc disintegration (Nielsen and Keiding 1998). Watanabe et al. (1999) reported that EPS-mediated flocculation of *Rhodovulum* sp. was promoted by $Ca^{2+}$, $Mg^{2+}$, $Fe^{3+}$ and $Al^{3+}$, with trivalent cations having a stronger effect than divalent cations. The promotion of flocculation was explained by bridge formation between the EPS and the cations.

On the basis of investigations on laboratory-scale activated sludge reactors, Higgins and Novak (1997) emphasized the role of structural proteins in conjunction with divalent cations in flocculation. Increasing the concentrations of $Ca^{2+}$ or $Mg^{2+}$ resulted in an increase in bound protein, whereas there was little effect on bound polysaccharides. Addition of high concentrations of $Na^+$ led to a decrease of bound protein. It was supposed that the monovalent sodium ions displaced divalent cations from within the flocs. This displacement would reduce binding of protein within the floc and result in solubilization of protein. Further support for the involvement of extracellular protein in the aggregation of bacteria into flocs came from the observation that treatment of activated sludge flocs with a proteolytic enzyme (pronase) resulted in deflocculation, in a shift to smaller particles in the 5–40 µm range and in a release of polysaccharide. Gel electrophoretic analysis of extracted EPS from municipal, industrial and laboratory activated sludge revealed the presence of a single protein with a molecular mass of approximately 15,000 daltons (Da). Analysis of amino acid composition and sequence indicated that this protein displayed similarities to lectins (sugar-binding proteins); binding site inhibition studies demonstrated lectin-like activity of the 15,000 Da protein (Higgins and Novak 1997). On the basis of these results, a model of flocculation was proposed (Figure 8.2): lectin-like proteins bind polysaccharides that are cross-linked to adjacent proteins. Divalent cations bridge negatively charged functional groups on the EPS molecules. The cross-linking of EPS and cation bridges leads to the stabilization of the biopolymer network mediating the immobilization of microbial cells.

Urbain et al. (1993) concluded from their studies on 16 activated sludge samples from different origins that internal hydrophobic bonds were involved in flocculation mechanisms, and their balance with hydrophilic interactions determined the sludge settling properties. Hydrophobic areas in between the cells were considered as essential adhesives within the floc structure. As mentioned above, hydrophobic binding sites may be provided by the protein fraction of the EPS (Jorand et al. 1998). Cell-surface hydrophobicity was shown to be important for adhesion of bacteria to activated sludge flocs (Olofsson et al. 1998). Cells with high cell surface hydrophobicity attached in higher numbers to the flocs than bacteria with a more hydrophilic surface. The hydrophobic cells attached not only on the surface of the flocs, but also penetrated the flocs through channels and pores, whereas hydrophilic cells did not. It was assumed that adhesion of hydrophobic bacteria within flocs would increase the potential of the flocs to clear free-living cells from the water phase (Olofsson et al. 1998).

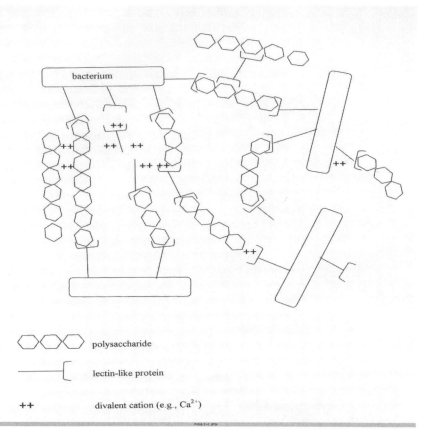

**Figure 8.2.** Role of polysaccharides, lectin proteins and divalent cations in the formation of microbial aggregates on a molecular scale (redrawn after Higgins and Novak 1997).

Generally, microcolonies and single cells within flocs and biofilms are separated by microscopically transparent voids. However, specific staining (e.g., Thiery's stain, ruthenium red) allowed microscopic detection of polysaccharidic material between microcolonies and single cells, indicating that the intercellular space was filled with EPS connecting the cells together (Jorand *et al.* 1995). This was confirmed when flocs from natural riverine systems and from wastewater were viewed by high-resolution TEM (Liss *et al.* 1996). Pores devoid of physical structures under optical microscopes were found to be filled with complex matrices of extracellular polymeric fibrils (4–6 nm diameter). These fibrils were found to represent the dominant bridging mechanism between organic and inorganic components of the flocs and contributed to the extensive surface area per volume unit of the flocs.

Size distribution studies demonstrated that activated sludge flocs varied in size between about 1 and 1,000 µm or more (Bitton 1994; Li and Ganczarczyk 1991; Jorand *et al.* 1995). Flocs smaller than 5 µm in diameter (dispersed microorganisms, aggregates of a few microorganisms) predominated by number, whereas most of the surface area, the volume and the biomass were made up of larger flocs (64–256 µm) (Li and Ganczarczyk 1991; Jorand *et al.* 1995). Activated sludge flocs could easily be dispersed into smaller flocs by short periods of ultrasound (30–60 s at 37 W), with minimal bacterial cell lysis (Jorand *et al.* 1995). Ultrasonic disruption resulted in the appearance of subunits with sizes of 2.5 µm and 13 µm from larger flocs of 125 µm with concomitant release of EPS (carbohydrates, proteins, DNA). The 13-µm aggregates proved particularly resistant to sonication. Similarly, bacterial microcolonies with a diameter of 10–20 µm embedded in EPS were shown to remain unaffected, when activated sludge flocs were disintegrated by sulfide treatment (Nielsen and Keiding 1998). On the basis of the microscopic observations and size distribution studies, a model of the aggregation mechanisms and structure of activated sludge flocs was suggested (Jorand *et al.* 1995; Figure 8.3).

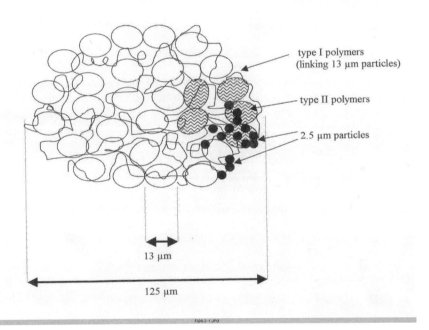

**Figure 8.3.** Role of EPS in the formation of activated sludge flocs on a microscopic scale (redrawn after Jorand *et al.* 1995).

Primary particles, 2.5 µm in size, form secondary particles (microcolonies), 13 µm in size. The secondary particles make up tertiary particles with a mean diameter of 125 µm. Some of the 2.5-µm particles may lie between the 13-µm aggregates. EPS form links between the 13-µm aggregates and between microorganisms within the 13-µm and 2.5-µm aggregates. Because only the EPS between the 13-µm flocs were extracted by sonication, it was assumed that these polymers ('type I polymers') must be different from those within the 13-µm and 2.5-µm flocs ('type II polymers'), which could not be removed by sonication.

## 8.6 FUNCTIONS OF EPS

After discovering the universal presence of EPS-containing structures, the formation of a gel-like network has been regarded as the most important general function allowing microorganisms to live in aggregated communities. This primary function of EPS seems to represent an important survival advantage for the immobilized microorganisms, which may be the basis of various other functions of EPS as summarized in Table 8.3 (for review, see Wolfaardt *et al.* 1999).

In general, one of the most important functions of extracellular polysaccharides is supposed to be their role as fundamental structural elements of the EPS matrix that determines the mechanical stability of biofilms, as has already been described in more detail above. More recent studies suggest that lectin-like proteins also contribute to the formation of the three-dimensional network of the biofilm matrix by cross-linking polysaccharides, directly or indirectly, through multivalent cations bridges (Higgins and Novak 1997; Figure 8.2).

However, the main function of extracellular proteins in biofilms is still seen in their role as enzymes which are located at the cell surface or become accumulated in the EPS matrix of micobial aggregates. Extracellular enzymes have been shown to be abundant in flocs and biofilms in natural and technical environments, including activated sludge and wastewater biofilms; these enzymes include polysaccharidases, proteases, lipases, esterases, peptidases, glycosidases, phosphatases and oxidoreductases (e.g., Frølund *et al.* 1995; Lemmer *et al.* 1994; Griebe *et al.* 1997; for review, see Wingender and Jaeger 2002). The primary function of extracellular enzymes is supposed to be the digestion of exogenous macromolecules from the environment, thus providing low molecular weight nutrients, which can readily be taken up and metabolized by the cells.

**Table 8.3.** Functions attributed to microbial EPS.

| Function | Relevance |
|---|---|
| Adhesion to surfaces | Colonization of inert and tissue surface, accumulation of bacteria on nutrient-rich surfaces in oligotrophic environments |
| Aggregation of bacterial cells, formation of flocs and biofilms | Bridging between cells and inorganic particles trapped from the environment, immobilization of mixed bacterial populations, basis for development of high cell densities, generation of a medium for communication processes, cause for biofouling and biocorrosion events |
| Structural elements of biofilms | Mediation of mechanical stability of biofilms (frequently in conjunction with multivalent cations), determination of the shape of EPS structure (capsule, slime, sheath) |
| Protective barrier | Resistance to non-specific and specific host immunity (complement-mediated killing, phagocytosis, antibody response, free radical generation), resistance to certain biocides including disinfectants and antibiotics |
| Retention of water | Prevention of desiccation under water-deficient conditions |
| Sorption of exogenous organic compounds | Scavenging and accumulation of nutrients from the environment, sorption of xenobiotics (detoxification) |
| Sorption of inorganic ions | Accumulation of toxic metal ions (detoxification), promotion of polysaccharide gel formation, mineral formation |
| Enzymatic activities | Digestion of exogenous macromolecules for nutrient acquisition, release of biofilm cells by degradation of structural EPS of the biofilm |
| Interaction of polysaccharides with enzymes | Accumulation/retention and stabilization of secreted enzymes |

As mentioned above, EPS in activated sludge have been shown to be subject to degradation during storage under anaerobic or low-temperature conditions (Nielsen *et al.* 1996; Bura *et al.* 1998); extracellular enzymes can be expected to be involved in these processes. Obayashi and Gaudy (1973) reported the aerobic digestion of exogenous EPS by wastewater microorganisms. The EPS used as growth substrates were isolated from slimes of five different bacterial species and consisted mainly of neutral or uronic-acid-containing heteropolysaccharides and low amounts of protein. These data indicate that EPS cannot be regarded as biologically inert, so that a limited degree of turnover mediated by endogenous or exogenous enzymes can be assumed to take place under certain conditions within the EPS matrix. However, as has been pointed out by Sutherland (1999b), in multi-species biofilms, the effects of enzymes involved in the degradation or alteration of EPS may be moderated if a mixture of EPS is present and removal of one polymer leaves others with similar physical properties intact. Thus, there are some indications that EPS may serve as a nutrient reserve, although the role of EPS in nutrition is not yet well understood (Wolfaardt *et al.* 1999).

The involvement in cell detachment and dispersal has been proposed as a novel function of extracellular enzymes in microbial aggregates. Enzymes within the biofilm matrix may degrade polysaccharidic EPS, causing the release of biofilm bacteria and the spreading of the organisms to new environments (Xun et al. 1990; Boyd and Chakrabarty 1994).

The role of other EPS components as structural elements of the biofilm matrix remains to be established. However, it is expected that EPS components such as nucleic acids, lipids or humic substances significantly influence the rheological properties and thus the stability of biofilms, as can be deduced from basic laboratory studies on the properties of polymer mixtures. The colonization of surfaces may also be determined by EPS other than polysaccharides or proteins. Extracellular lipids from *Serratia marcescens* with surface-active properties (serrawettins) have been proposed to help bacteria in surface environments to overcome the strong surface tension of surrounding water, thus facilitating growth on solid surfaces (Matsuyama and Nakagawa 1996).

A function frequently attributed to EPS is their general protective effect on biofilm organisms against adverse abiotic and biotic influences from the environment. For example, it has frequently been observed that biofilm cells can tolerate significantly higher concentrations of certain biocides, including disinfectants and antibiotics, than planktonic populations (LeChevallier et al. 1988). This is supposed to be due mainly to physiological changes of biofilm bacteria (adoption of a biofilm-specific phenotype) enhancing their intrinsic resistance to biocides, but also to a barrier function of EPS (Brown and Gilbert 1993; Morton et al. 1998). It is assumed that the EPS matrix delays or prevents biocides from reaching target microorganisms within the biofilm by diffusion limitation and/or chemical interaction with EPS molecules. In the case of chlorine, de Beer et al. (1994) have shown that the chlorine demand by polysaccharide may lead to biofilm areas with low concentrations of chlorine in which the cells can survive. In mucoid *P. aeruginosa* strains, the protective effect of slime against chlorine was supposed to be based on a chemical reaction of the biocide, with alginate as the major slime component, resulting in the neutralization of chlorine (Wingender et al. 1999c). However, in the case of hydrogen peroxide, no chemical reaction between the biocide and alginate was observed (Wingender et al. 1999c); here, penetration of hydrogen peroxide into biofilms was limited, because the biocide was degraded by catalase enzymes produced by the biofilm bacteria (Stewart et al. 2000). These results show that the contribution of EPS to the response of bacteria against biocides varies depending on the properties of the biocide applied. Other protective functions of EPS include the contribution to bacterial evasion from various host defense mechanisms (Smith and Simpson 1990), the protection from desiccation owing to enhanced water retention (Ophir and Gutnick 1994) or the prevention of oxygen-mediated inhibition of nitrogen fixation in cyanobacteria (De Philippis and Vincenzini 1998).

Extracellular polymeric substances in biofilms 197

The accumulation of organic nutrients is regarded as another function of EPS (Wolfaardt et al. 1999). In addition, the sorption of potentially toxic xenobiotic compounds by EPS has been described; even these substances could be utilized as nutrients by biofilm organisms, when provided as the sole carbon source (Wolfaardt et al. 1999). In this context, the accumulation and stabilization of extracellular enzymes by their interaction with polysaccharides of the floc or biofilm matrix (Wingender and Jaeger 2002) may be a mechanism, by which the enzymatic degradation of exogenous organic compounds can be supported.

## 8.7 TECHNICAL ASPECTS OF EPS

The EPS-mediated mechanical stability of biofilms and other microbial aggregates is of importance for the performance of technical biofilm processes as in wastewater treatment, but also in the control or removal of unwanted biofilms in other industrial systems (Flemming and Wingender 2001b). There are relatively few studies that are concerned with the performance of wastewater biofilms and activated sludge systems in correlation with physicochemical properties of EPS, based on the quantitative and qualitative analysis of biofilms, flocs and their EPS components. In biofilm reactors, the problem of 'sloughing off' of biomass can be a problem in stable process maintenance. It clearly depends on EPS-mediated biofilm cohesion and adhesion to the substratum. In the activated sludge process, EPS are considered to be of considerable importance in the flocculation process (EPS-mediated microbial aggregation), and they have been associated with variations in settleability, foam formation, dewaterability and sorption of organic substances and heavy metals. Also, the stability of flocs is of great importance, because floc size is crucial for effective settlement and separation of biomass from water as well as for dewatering of sludges (Bruus et al. 1992). On the other hand, if unwanted biofilms have to be removed, their mechanical stability has to be overcome. The mode of action of cleaners is to weaken the mechanical stability of the biofilm matrix by chemical means.

### 8.7.1 Settleability

Irrespective of the experimental conditions, a positive linear relationship between the sludge volume index (SVI) and the total amount of EPS in sludge flocs has been described in many studies (Urbain et al. 1993). In 16 activated sludge samples from seven different wastewater treatment plants, all of the analyzed EPS constituents (polysaccharides, proteins, DNA) positively correlated with the SVI, resulting in a worsening of sludge settleability (Urbain et al. 1993). Liao et al. (2001) studied the settling characteristics of sludge flocs cultivated in sequencing batch reactors at different sludge retention times. They demonstrated that a higher SVI, an indication of poor settleability or compression, was associated with a larger amount of total EPS (protein, carbohydrate, DNA); a significant correlation existed between the amount of total EPS, including protein, and the SVI, whereas no significant correlation existed between carbohydrate and the SVI (Liao *et al.* 2001). In contrast, Shin *et al.* (2001) reported that an increase in carbohydrate content at constant protein levels in the EPS inhibited floc formation and was supposed to be responsible for poor settling (SVI) of sludge samples from sequencing batch reactors. It was concluded that the ratio of carbohydrate to protein in the EPS, and not the total amount of EPS, determined the sludge settling characteristics. As mentioned above, flotation characteristics of flocs and granules can also be influenced by the spatial distribution of exopolysaccharides within the microbial aggregates (de Beer *et al.* 1996). Thus, the relationships between the amount of EPS and sludge settleability seem to vary in different activated sludge systems, but quantitative analysis of EPS may be suitable for the prediction of floc settleability in a well-characterized system.

### 8.7.2 Foam formation

In addition to poor settlement, foam formation is another unwanted phenomenon that adversely affects the separation of activated sludge flocs into the sludge fraction and clarified effluent. Foaming was usually thought to be caused primarily by filamentous bacterial species, but there are indications that acidic polysaccharides may also be involved in foam formation (Kerley and Forster 1995; Baxter-Plant *et al.* 1998). High hydrophobicity of sludge is usually associated with foam formation. In a study of activated sludge samples from foaming and non-foaming aeration tanks of different full-scale plants, an increase in the amount of uronic acid in the EPS has been found to correlate with an increase in the hydrophobicity of sludge and foam solids (Kerley and Forster 1995). Enhanced foam stability was supposed to be due to the partial neutralization of negative charges of the uronic acids by polyvalent metal ions, conferring a greater degree of hydrophobicity to surface structures of the flocs (Kerley and Forster 1995). As there is only limited information about the role of acidic polysaccharides in foaming, it remains to be seen if the analysis of uronic acids can be generally a suitable parameter for predicting the tendency of foam formation in the activated sludge process.

### 8.7.3 Dewaterability

The energy demand for dewatering of sewage sludge is a major economic factor. The polysaccharides and proteins of the EPS in biomass mainly bind this water. The molecular mechanisms of water binding are of crucial importance for a rational basis of the improvement of dewatering techniques. Schmitt and Flemming (1999) investigated water binding by measuring the exchange of $D_2O$ against $H_2O$ in a Fourier-transform infrared spectroscopy in attenuated total reflection mode (FTIR–ATR) study. First results showed that treatment with chlorine clearly facilitated the exchange, indicating that chlorine may have destroyed water binding structures, probably in the EPS.

Floc size as well as particle size distribution are considered important for the dewaterability of wastewater sludges. Smaller particle sizes seem to be associated with poor dewatering. Removal of polymer-bridging cations such as calcium (Bruus *et al.* 1992) or enzymatic degradation of proteins in the EPS (Higgins and Novak 1997) resulted in an increase in small particles, causing a deterioration in dewatering properties of activated sludge. The reduced dewaterability known to occur during anaerobic storage of activated sludge was explained by deflocculation due to the degradation of extracellular carbohydrates and proteins with subsequent weakening of the entire sludge floc (Nielsen *et al.* 1996). Poxon and Darby (1997) studied the influence of extracellular polyanions on dewaterability of digested sludges. They found no simple relationship between total extracellular polyanion concentration and sludge dewaterability. Because the relationship between dewaterability and the concentration of extracellular polyanions was dependent on digester feed composition, it was speculated the specific biochemical characteristics of EPS were more significant than total concentration as an influence on sludge dewaterability (Poxon and Darby 1997).

### 8.7.4 Sorption of organic and inorganic molecules

Heavy metals and organic pollutants can be sequestered from the water phase by sorption to biomass (Flemming and Leis 2002). Also, distribution, immobilization and remobilization of dissolved organic and inorganic substances, all depend on the sorption properties of biofilms, as most surfaces in natural and technical environments are covered by biofilms which participate in any sorption process (Flemming *et al.* 1996). To understand and to optimize these processes, as well as to design more accurate models for the fate of pollutants, it is important to know more about the actual sites, capacities and mechanisms of sorption; the EPS play an important role in this system. Some examples: Späth *et al.* (1998) showed that lipophilic substances such as benzene, toluene and xylene were significantly accumulated in the EPS matrix. In an FTIR–ATR study, Schmitt *et al.* (1995) investigated the response of a biofilm of *Pseudomonas putida* to toluene. At a concentration of 5 mg $L^{-1}$, an increase in EPS production was observed and at 15 mg $L^{-1}$, an increase in carboxylic groups occurred. Thus, the biofilm responds dynamically to sorbed substances by alteration of the EPS composition.

## 8.7.5 Diffusion

The performance of biofilm reactors is crucially dependent upon the diffusion processes which are, in many cases, rate limiting. The diffusion properties of biofilms are strongly influenced by the structure and properties of the EPS network and by the tortuosity imposed by impermeable particles as represented by bacteria.

## 8.7.6 Formation of a gel layer on surfaces

The biofilm can mask original surface properties; in particular, it will turn hydrophobic surfaces into hydrophilic ones. Also, it represents a hydrodynamic barrier on separation membranes (Ridgway and Flemming 1996) and increases friction resistance on ship hulls and in pipelines (Characklis 1990). In heat exchangers, biofilms form an insulation layer in which diffusive heat transport largely prevails over convective heat transport, decreasing the efficacy of the process (Characklis 1990).

## 8.8 ECOLOGICAL ASPECTS

From an ecological point of view, the EPS matrix provides the possibility that the microorganisms can form stable aggregates of mixed populations, leading to synergistic microconsortia. This facilitates the sequential degradation of substances that are not readily biodegradable. Many anthropogenic pollutants fall into that category. The spatial arrangement of the microorganisms gives rise to gradients in the concentration of oxygen and other electron acceptors as well as of substrates, products and pH value (Costerton *et al.* 1987; de Beer *et al.* 1993). Thus, aerobic and anaerobic habitats can arise in close proximity, and, as a consequence, the development of a large variability of species. The EPS matrix sequesters nutrients from the bulk water phase (Decho 1990), a particularly important mechanism in oligotrophic environments. Thus, biofilms seem to be a favorable form of life and part of a survival strategy of microorganisms (Marshall 1996). This matrix is involved in the sorption of dissolved and particulate substances, including biodegradable substances, xenobiotics and metal ions (Brown and Lester 1979; Spaeth *et al.* 1998; Wolfaardt *et al.* 1998).

An important modern concept is the role of EPS in allowing microorganisms to live continuously at high cell densities in stable mixed population biofilm communities. The EPS matrix constitutes a medium for communication processes between constituents cells of biofilms that is only made possible by the close proximity of the bacteria which are held together by the EPS.

Horizontal gene transfer may be facilitated because of the close contact of aggregated cells and the accumulation of DNA in the EPS matrix. Genetic material is conserved and more readily taken up than by planktonic cells. Gene transfer in biofilms is an important mechanism of information exchange, increasing metabolic capabilities and supporting adaptation of micobial communities to changing environmental conditions (for review, see Wuertz 2002). Several mechanisms of gene transfer are likely to occur in the natural environment: conjugation, transduction and natural transformation (Lorenz and Wackernagel 1994). Actually, a few studies have demonstrated gene exchange in pure culture biofilms (Angles *et al.* 1993, Lisle and Rose 1995) and plasmid transfer in river epilithon (Bale *et al.* 1988). These observations suggest that gene transfer can contribute to phenotypic changes of biofilm organisms.

Another form of information transfer resides in the phenomenon of quorum sensing, which is a signaling mechanism of Gram-negative bacteria, frequently in response to population density (Swift *et al.* 1996). This kind of cell-to-cell communication is mediated by low-molecular-weight diffusible signaling molecules. In Gram-negative bacteria, various types of signaling molecule have been detected, including N-acylhomoserine lactones (AHLs), 3-hydroxy-palmitic acid methyl ester, 2-heptyl-3-hydroxy-4-quinolone and diketopiperazines (cyclic dipeptides) (Withers *et al.* 2001). Among them, AHLs are the best-studied class of signaling molecules. Extracellular accumulation of AHLs above a critical threshold level results in transcriptional activation of a range of different genes with concomitant expression of new phenotypes. AHLs allow bacteria to sense when cell densities in their surroundings reach the minimal level for a coordinate population response to be initiated. As biofilms typically contain high cell concentrations, AHLs can be expected to function in biofilms. AHLs have actually been detected in naturally occurring biofilms (McLean *et al.* 1997) and in biofilms on corroded sewer pipes (Vincke *et al.* 2001). AHLs have been implicated in the shortened recovery process of nitrification; under nitrogen starvation conditions, *Nitrosomonas europaea* biofilm cells resumed their ammonium-oxidizing activity after resupply of ammonium without a lag phase due to accumulation of the signaling molecules within the biofilm to levels unobtainable in planktonic populations, which showed a significantly prolonged phase of recovery (Batchelor *et al.* 1997). With respect to biofilm formation, AHLs were considered as mediators of adhesion by switching cells to an attachment phenotype through expression of adhesive polymers, and they were supposed to facilitate induction of other genes essential for the maintenance of the biofilm mode of growth (Heys *et al.* 1997). Davies *et al.* (1998) described the involvement of an intercellular AHL in the differentiation of *P. aeruginosa* biofilms. Thus, cell-to-cell communication via AHLs seems to be of fundamental importance for biofilm bacteria to adapt dynamically in response to prevailing environmental conditions.

From a practical point of view, EPS as mediators of biofilm formation and stability have to be considered as target structures for remedial actions to remove, prevent or control undesirable biofilm formation (biofouling) and microbially influenced corrosion of materials (biocorrosion) in industry and medicine.

## 8.9 OUTLOOK

The considerations as compiled in this overview demonstrate that the phenomenon of EPS production is an important aspect of microbial aggregates and draws some attention to the space between the cells. Further investigation of the role of the extracellular proteins may lead to particularly interesting insights. A significant part of the extracellular proteins have been actually identified as enzymes. They are supposed to function in the extracellular degradation of macromolecules to low-molecular-weight products, which can be transported into the cells and are available for microbial metabolism. The degradation of particulate matter is performed by colonization of the material and the excretion of extracellular enzymes. In recent models of biofilm development, extracellular enzymes are included with the function of contributing to detachment processes of biofilm organisms, allowing the release of swarmer cells and the subsequent colonization of new environments (Costerton et al. 1999). However, many details of these important processes are still obscure. The EPS matrix prevents the loss of the enzymes. Specific interactions between extracellular enzymes and other EPS components have been observed. It has been shown (Wingender 1990; Wingender et al. 1999b) that the lipase of P. aeruginosa interacts functionally with the alginate formed by the same strain. It could be demonstrated that lipase of this strain displayed a higher stability in the presence of bacterial alginate; for example, the enzyme was less sensitive to heat inactivation when associated with the alginate. It was shown that this effect was specific to alginate because other polysaccharides such as dextran did not interact with the enzyme.

These observations suggest that the structure of the EPS matrix might not be purely random, but must be involved in the regulation of the activity of extracellular enzymes. Thus, the cell maintains a certain level of control over enzymes which otherwise are out of their reach. This can be considered as a strategy to form an organization in which the cells gain control about the space around them. Costerton et al. (1987) hypothesized that biofilms represent tissue-like structures. If this is the case, the space between the cells may become a new focus of attention, providing information about the cooperative effect of biofilm cells.

## 8.10 REFERENCES

Allison, D.G. and Sutherland, I.W. (1987) The role of exopolysaccharides in adhesion of freshwater bacteria. *J. Gen. Microbiol.* **133**, 1319–1327.

Angles, M.L., Marshall, K.C. and Goodman, A.E. (1993) Plasmid transfer between marine bacteria in the aqueous phase and biofilms in reactor microcosms. *Appl. Environ. Microbiol.* **59**, 843–850.

Arvaniti, A., Karamanos, N.K., Dimitracopoulos, G. and Anastassiou, E.D. (1994) Isolation and characterization of a novel 20-kDa sulfated polysaccharide from the extracellular slime layer of *Staphylococcus epidermidis*. *Arch. Biochem. Biophys.* **308**, 432–438.

Bale, M.J., Fry, J.C. and Day, M.J. (1988) Transfer and occurrence of large mercury resistance plasmids in river epilithon. *Appl. Environ. Microbiol.* **54**, 972–978.

Batchelor, S.E., Cooper, M., Chhabra, S.R., Glover, L.A., Stewart, G.S.A.B., Williams, P. and Prosser, J.I. (1997) Cell density-regulated recovery of starved biofilm populations of ammonia-oxidizing bacteria. *Appl. Environ. Microbiol.* **63**, 2281–2286.

Baxter-Plant, V.S., Hayes, E. and Forster, C.F. (1998) The examination of activated sludge from three plants in relation to the problem of stable foam formation. *Trans. Chem. Eng.* **76** B, 161–165.

Beveridge, T.J., Makin, S.A., Kadurugamuwa, J.L. and Li, Z. (1997) Interactions between biofilms and the environment. *FEMS Microbiol. Rev.* **20**, 291–303.

Beveridge, T.J. (1999) Structures of Gram-negative cell walls and their derived membrane vesicles. *J. Bacteriol.* **181**, 4725–4733.

Bitton, G. (1994) *Wastewater Microbiology*. Wiley-Liss, Inc., New York.

Boyd, A. and Chakrabarty, A.M. (1994) Role of alginate lyase in cell detachment of *Pseudomonas aeruginosa*. *Appl. Environ. Microbiol.* **60**, 2355–2359.

Brown, M.J. and Lester, J.N. (1979) Metal removal in activated sludge: the role of bacterial extracellular polymers. *Wat. Res.* **13**, 817–837.

Brown, M.J. and Lester, J.N. (1980) Comparison of bacterial extracellular polymer extraction methods. *Appl. Environ. Microbiol.* **40**, 179–185.

Brown, M.R.W. and Gilbert, P. (1993) Sensitivity of biofilms to antimicrobial agents. *J. Appl. Bacteriol. Symp. Suppl.* **74**, 87S–97S.

Bruus, J.H., Nielsen, P.H. and Keiding, K. (1992) On the stability of activated sludge flocs with implications to dewatering. *Wat. Res.* **26**, 1597–1604.

Bryers, J.D. and Characklis, W.G. (1990) Biofilms in water and wastewater treatment. In *Biofilms* (ed. W.G. Characklis and K.C. Marshall), pp. 671–696, John Wiley and Sons, New York.

Bura, R., Cheung, M., Liao, B., Finlayson, J., Lee, B. C., Droppo, I. G., Leppard, G. G. and Liss, S. N. (1998) Composition of extracellular polymeric substances in the activated sludge floc matrix. *Wat. Sci. Tech.* **37**(4–5), 325–333.

Characklis W.G. and Wilderer P.A. (1989) *Structure and Function of Biofilms*. John Wiley and Sons, Chichester.

Characklis, W.G. (1990) Microbial fouling. In *Biofilms* (ed. W.G. Characklis and K.C. Marshall), pp. 523–633, John Wiley and Sons, New York.

Christensen, B.E. and Characklis, W.G. (1990) Physical and chemical properties of biofilms. In *Biofilms* (ed. W.G. Characklis and K.C. Marshall), pp. 93–130. John Wiley and Sons, New York.

Cooksey, K.E. (1992) Extracellular polymers in biofilms. In *Biofilms – Science and Technology* (ed. L.F. Melo, T.R. Bott, M. Fletcher and B. Capdeville), pp. 137–147, Kluwer Academic Publishers, Dordrecht.

Costerton, J.W., Irvin, R.T. and Cheng, K.-J. (1981) The bacterial glycocalyx in nature and disease. *Annu. Rev. Microbiol.* **35**, 299–324.

Costerton, J.W., Cheng, K.-J., Geesey, G.G., Ladd, T.I., Nickel, J.C., Dasgupta, M. and Marrie, T.J. (1987) Bacterial biofilms in nature and disease. *Annu. Rev. Microbiol.* **41**, 435–464.

Costerton, J.W., Lewandowski, Z., Caldwell, D.E., Korber, D.R. and Lappin-Scott, H.M. (1995) Microbial biofilms. *Annu. Rev. Microbiol.* **49**, 711–745.

Costerton, J.W., Stewart, P.S. and Greenberg, E.P. (1999) Bacterial biofilms: a common cause of persistent infections. *Science* **284**, 1318–1322.

Davies, D.G., Chakrabarty, A.M. and Geesey, G.G. (1993) Exopolysaccharide production in biofilms: substratum activation of alginate gene expression by *Pseudomonas aeruginosa*. *Appl. Environ. Microbiol.* **59**, 1181–1186.

Davies, D.G., Parsek, M.R., Pearson, J.P., Iglewski, B.H., Costerton, J.W. and Greenberg, E.P. (1998) The involvement of cell-to-cell signals in the development of a bacterial biofilm. *Science* **280**, 295–298.

De Beer, D., van den Heuvel, J.C. and Ottengraf, S.P.P. (1993) Microelectrode measurements of the activity distribution in nitrifying bacterial aggregates. *Appl. Environ. Microbiol.* **59**, 573–579.

De Beer, D., Srinivasan, R. and Stewart, P.S. (1994) Direct measurement of chlorine penetration into biofilms during disinfection. *Appl. Environ. Microbiol.* **60**, 4339–4344.

De Beer, D., O'Flaharty, V., Thaveesri, J., Lens, P. and Verstrate, W. (1996) Distribution of extracellular polysaccharides and flotation of anaerobic sludge. *Appl. Microbiol. Biotech.* **46** 197–201.

Decho, A.W. (1990) Microbial exopolymer secretions in ocean environments: their role(s) in food webs and marine processes. *Oceanogr. Mar. Biol. Annu. Rev.* **28**, 73–153.

De Philippis, R. and Vincenzini, M. (1998) Exocellular polysaccharides from cyanobacteria and their possible applications. *FEMS Microbiol. Rev.* **22**, 151–175.

Dignac, M.-F., Urbain, V., Rybacki, D., Bruchet, A., Snidaro, D. and Scribe, P. (1998) Chemical description of extracellular polymers: implication on activated sludge floc structure. *Wat. Sci. Tech.* **38**(8–9), 45–53.

Fazio, S.A., Uhlinger, D.J., Parker, J.H. and White, D.C. (1982) Estimations of uronic acids as quantitative measures of extracellular and cell wall polysaccharide polymers from environmental samples. *Appl. Environ. Microbiol.* **43**, 1151–1159.

Flemming, H.-C. and Schaule, G. (1996) Measures against biofouling. In *Microbial Deterioration of Materials* (ed. E. Heitz, W. Sand and H.-C. Flemming), pp. 121–139. Springer, Heidelberg.

Flemming, H.-C., Schmitt, J. and Marshall, K.C. (1996) Sorption properties of biofilms. In *Environmental Behaviour of Biofilms* (ed. W. Calmano and U. Förstner), pp. 115–157. Lewis Publishers Chelsea, Michigan.

Flemming, H.-C. and Wingender, J. (2001a) Relevance of microbial extracellular polymeric substances (EPSs) – part I: structural and ecological aspects. *Wat. Sci. Tech.* **43**(6), 1–8.

Flemming, H.-C. and Wingender, J. (2001b) Relevance of microbial extracellular polymeric substances (EPSs) – part II: technical aspects. *Wat. Sci. Tech.* **43**(6), 9–16.

Flemming, H.-C. and Leis, A. (2002): Sorpion properties of biofilms. In *Encyclopedia of Environmental Microbiology* (ed. G. Bitton), vol. 5, pp. 2958–2967. John Wiley and Sons, New York.

Frølund, B., Griebe, T. and Nielsen, P.H. (1995) Enzymatic activity in the activated-sludge floc matrix. *Appl. Microbiol. Biotechnol.* **43**, 755–761.

Frølund, B., Palmgren, R., Keiding, K. and Nielsen, P.H. (1996) Extraction of extracellular polymers from activated sludge using a cation exchange resin. *Wat. Res.* **30**, 1749–1758.

Geddie, J.L. and Sutherland, I.W. (1994) The effect of acetylation on cation binding by algal and bacterial alginates. *Biotech. Appl. Biochem.* **20**, 117–129.

Geesey, G.G. (1982) Microbial exopolymers: ecological and economic considerations. *Am. Soc. Microbiol. News* **48**, 9–14.

Gehr, R. and Henry, J. G. (1983): Removal of extracellular material. Techniques and pitfalls. *Wat. Res.* **17**, 1743–1748.

Gehrke, T., Telegdi, J., Thierry, D. and Sand, W. (1998) Importance of extracellular polymeric substances from *Thiobacillus ferrooxidans* for bioleaching. *Appl. Environ. Microbiol.* **64**, 2743–2747.

Gehrke, T., Hallmann, R., Kinzler, K. and Sand, W. (2001) The EPS of *Acidithiobacillus ferrooxidans* – a model for structure–function relationships of attached bacteria and their physiology. *Wat. Sci. Technol.* **43**(6) 159–167.

Goodwin, J.A.S. and Forster, C.F. (1985) A further examination into the composition of activated sludge surfaces in relation to their settlement characteristics. *Wat. Res.* **19**, 527–533.

Griebe, T., Schaule, G. and Wuertz, S. (1997) Determination of microbial respiratory and redox activity in activated sludge. *J. Ind. Microbiol. Biotechnol.* **19**, 118–122.

Harris, R.H. and Mitchell, R. (1973) The role of polymers in microbial aggregation. *Annu. Rev. Microbiol.* **27**, 27–50.

Hejzlar, J. and Chudoba, J. (1986) Microbial polymers in the aquatic environment – I. Production by activated sludge microorganisms under different conditions. *Wat. Res.* **20**, 1209–1216.

Heys S.J.D., Gilbert P., Eberhard A. and Allison D.G. (1997) Homoserine lactones and bacterial biofilms. In *Biofilms: Community Interactions and Control* (ed. J. Wimpenny, P. Handley, P. Gilbert , H. Lappin-Scott and M. Jones), pp. 103–112. BioLine, Cardiff.

Higgins, M.J. and Novak, J.T. (1997) Characterization of exocellular protein and its role in bioflocculation. *J. Environ. Engng* **123**, 479–485.

Horan, N. and Eccles, C.R. (1986) Purification and characterization of extracellular polysaccharides from activated sludge. *Wat. Res.* **20**, 1427–1432.

Jahn, A. and Nielsen, P.H. (1995) Extraction of extracellular polymeric substances (EPS) from biofilms using a cation exchange resin. *Wat. Sci. Tech.* **32**(8), 157–164.

Jahn, A. and Nielsen, P.H. (1998) Cell biomass and exopolymer composition in sewer biofilms. *Wat. Sci. Tech.* **37**(1), 17–24.

Jorand, F., Zartarian, F., Thomas, F., Block, J.C., Bottero, J.Y., Villemin, G., Urbain, V. and Manem, J. (1995). Chemical and structural (2D) linkage between bacteria within activated sludge flocs. *Wat. Res.* **29**, 1639–1647.

Jorand, F., Boué-Bigne, F., Block, J.C. and Urbain, V. (1998) Hydrophobic/hydrophilic properties of activated sludge exopolymeric substances. *Wat. Sci. Tech.* **37**(4–5), 307–315.

Kadurugamuwa, J.L. and Beveridge, T.J. (1995) Virulence factors are released from *Pseudomonas aeruginosa* in association with membrane vesicles during normal growth and exposure to gentamicin: a novel mechanism of enzyme secretion. *J. Bacteriol.* **177**, 3998–4008.

Kerley, S. and Forster, C.F. (1995) Extracellular polymers in activated sludge and stable foams. *J. Chem. Tech. Biotech.* **62**, 401–404.

Khandeparker, R.D.S. and Bhosle, N.B. (2001) Extracellular polymeric substances of the marine fouling diatom *Amphora rostrata* Wm.Sm. *Biofouling* **17**, 117–127.

Körstgens, V., Flemming, H.-C., Wingender, J. and Borchard, W. (2001) Influence of calcium ions on the mechanical properties of a model biofilm of mucoid *Pseudomonas aeruginosa*. *Wat. Sci. Tech.* **43**(6), 49–57.

Koster, M., Bitter, W. and Tommassen, J. (2000) Protein secretion mechanisms in Gram-negative bacteria. *Int. J. Med. Microbiol.* **290**, 325–331.

Lange, B., Wingender J. and Winkler, U.K. (1989) Isolation and characterization of an alginate lyase from *Klebsiella aerogenes*. *Arch. Microbiol.* **152**, 302–308.

LeChevallier, M.W., Cawthon, C.D. and Lee, R.G. (1988) Inactivation of biofilm bacteria. *Appl. Environ. Microbiol.* **54**, 2492–2499.

Lee, J.W., Ashby, R.D. and Day, D.F. (1996) Role of acetylation on metal induced precipitation of alginates. *Carbohydr. Polym.* **29**, 337–345.

Lemmer, H., Roth, D. and Schade, M. (1994) Population density and enzyme activities of heterotrophic bacteria in sewer biofilms and activated sludge. *Wat. Res.* **28**, 1341–1346.

Li, D. and Ganczarczyk, J. (1991) Size distribution of activated sludge flocs. *J. Water Pollut. Control Fed.* **63**, 806–814.

Liao, B.Q., Allen, D.G., Droppo, I.G., Leppard, G.G. and Liss, S.N. (2001) Surface properties of sludge and their role in bioflocculation and settleability. *Wat. Res.* **35**, 339–350.

Lisle, J.T. and Rose, J.B. (1995) Gene exchange in drinking water and biofilms by natural transformation. *Wat. Sci. Tech.* **31**(5–6), 41–46.

Liss, S.N., Droppo, I.G., Flannigan, D.T. and Leppard, G.G. (1996) Floc architecture in wastewater and natural riverine systems. *Environ. Sci. Tech.* **30**, 680–686.

Lorenz, M.G. and Wackernagel, W. (1994) Bacterial gene transfer by natural genetic transformation in the environment. *Microbiol. Rev.* **58**, 563–602.

Marshall, K.C. (1996) Adhesion as a strategy for access to nutrients. In *Bacterial Adhesion: Molecular and Ecological Diversity* (ed. M. Fletcher), pp. 59–87. John Wiley and Sons, New York.

Martin-Cereceda, M., Jorand, F., Guinea, A. and Block, J.C. (2001) Characterization of extracellular polymeric substances in rotating biological contactors and activated sludge flocs. *Environ. Tech.* **22**, 951–959.

Matsuyama, T. and Nakagawa, Y. (1996) Surface-active exolipids: analysis of absolute chemical structures and biological functions. *J. Microbiol. Meth.* **25**, 165–175.

Mayer, C., Moritz, R., Kirschner, C., Borchard, W., Maibaum, R., Wingender, J. and Flemming, H.-C. (1999) The role of intermolecular interactions: studies on model systems for bacterial biofilms. *Int. J. Biol. Macromol.* **26**, 3–16.

McCourtie, J. and Douglas, L.J. (1985) Extracellular polymer of *Candida albicans*: isolation, analysis and role in adhesion. *J. Gen. Microbiol.* **131**, 495–503.

McLean, R.J.C., Whiteley, M., Stickler, D.J. and Fuqua, W.C. (1997) Evidence of autoinducer activity in naturally occurring biofilms. *FEMS Microbiol. Lett.* **154**, 259–263.

Morton, L.H.G., Greenway, D.L.A., Gaylarde, C.C. and Surman, S.B. (1998) Consideration of some implications of the resistance of biofilms to biocides. *Int. Biodet. Biodegr.* **41**, 247–259.

Nielsen, P.H., Frølund, B. and Keiding, K. (1996) Changes in the composition of extracellular polymeric substances in activated sludge during anaerobic storage. *Appl. Microbiol. Biotech.* **44**, 823–830.

Nielsen, P.H., Jahn, A. and Palmgren, R. (1997) Conceptual model for production and composition of exopolymers in biofilms. *Wat. Sci. Tech.* **36**(1), 11–19.

Nielsen, P.H. and Keiding, K. (1998) Disintegration of activated sludge flocs in presence of sulphide. *Wat. Res.* **32**, 313–320.

Nielsen, P.H. and Jahn, A. (1999) Extraction of EPS. In *Microbial Extracellular Polymeric Substances* (ed. J. Wingender, T. Neu and H.-C. Flemming), pp. 49–72. Springer, Berlin.

Nivens, D.E., Ohman, D.E., Williams, J. and Franklin, M.J. (2001) Role of alginate and its O acetylation in formation of *Pseudomonas aeruginosa* microcolonies and biofilms. *J. Bacteriol.* **183**, 1047–1057.

Obayashi, A.W. and Gaudy, A.F. (1973) Aerobic digestion of extracellular microbial polysaccharides. *J. Wat. Pollut. Control Fed.* **45**, 1584–1594.

Olofsson, A.-C., Zita and A., Hermansson, M. (1998) Floc stability and adhesion of green-fluorescent-protein-marked bacteria to flocs in activated sludge. *Microbiology* **144**, 519–528.

Ophir, T. and Gutnick, D.L. (1994) A role for exopolysaccharides in the protection of microorganisms from desiccation. *Appl. Environ. Microbiol.* **60**, 740–745.

Pavoni, J.L., Tenney, M.W. and Echelberger, W.F. (1972) Bacterial exocellular polymers and biological flocculation. *J. Wat. Pollut. Control Fed.* **44**, 414–431.

Pereira, M.O. and Vieira, M.J. (2001) Effects of the interactions between glutaraldehyde and the polymeric matrix on the efficiency of the biocide against *Pseudomonas fluorescens* biofilms. *Biofouling* **17**, 93–101.

Platt, R.M., Geesey, G.G., Davis, J.D. and White, D.C. (1985) Isolation and partial chemical analysis of firmly bound exopolysaccharide from adherent cells of a freshwater bacterium. *Can. J. Microbiol.* **31**, 657–680.

Poxon, T.L. and Darby, J.L. (1997) Extracellular polyanions in digested sludge: measurement and relationship to sludge dewaterability. *Wat. Res.* **31**, 749–758.

Ridgway, H.F. and Flemming, H.-C. (1996) Biofouling of membranes. In *Water Treatment Membrane Processes* (ed. J. Mallevialle, P.E. Odendaal and M.R. Wiesner), pp. 6.1–6.62. McGraw-Hill, New York.

Rudd, T., Sterritt, R.M. and Lester, J.N. (1983) Extraction of extracellular polymers from activated sludge. *Biotech. Lett.* **5**, 327–332.

Schmitt, J., Nivens, D., White, D.C. and Flemming, H.-C. (1995) Changes of biofilm properties in response to sorbed substances – an FTIR–ATR study. *Wat. Sci. Tech.* **32**, 149–155.

Schmitt, J. and Flemming, H.-C. (1999) Water binding in biofilms. *Wat. Sci. Tech.* **39**, 77–82.

Shin, H.-S., Kang, S.-T. and Nam, S.-Y. (2001) Effect of carbohydrate and protein in the EPS on sludge settling characteristics. *Wat. Sci. Tech.* **43**(6) 193–196.

Skjåk-Bræk, G., Zanetti, F. and Paoletti, S. (1989) Effect of acetylation on some solution and gelling properties of alginates. *Carbohydr. Res.* **185**, 131–138.
Smith, S.E. and Simpson, J.A. (1990) The contribution of *Pseudomonas aeruginosa* alginate to evasion of host defence. In *Pseudomonas Infection and Alginates. Biochemistry, Genetics and Pathology* (ed. P. Gacesa and N.J. Russell), pp. 135–159. Chapman and Hall, London.
Späth, R., Flemming, H.-C. and Wuertz, S. (1998) Sorption properties of biofilms. *Wat. Sci. Tech.* **37**(4–5) 207–210.
Spaeth, R. and Wuertz, S. (2000) Extraction and quantification of extracellular polymeric substances from wastewater. In *Biofilms. Investigative Methods and Applications* (ed. H.-C. Flemming, U. Szewzyk and T. Griebe), pp. 51–68, Technomic Publishing Co., Lancaster, Pennsylvania, USA.
Stewart, P.S., Roe, F., Rayner, J., Elkins, J.G., Lewandowski, Z., Ochsner, U.A. and Hassett, D.J. (2000) Effect of catalase on hydrogen peroxide penetration into *Pseudomonas aeruginosa* biofilms. *Appl. Environ. Microbiol.* **66**, 836–838.
Sutherland, I.W. (1996) Extracellular polysaccharides. In *Biotechnology. Volume 6: Products of Primary Metabolism* (ed. H.-J. Rehm and G. Reed), pp. 615–657. Verlag Chemie, Weinheim.
Sutherland, I.W. (1999a) Biofilm exopolysaccharides. In *Microbial Extracellular Polymeric Substances* (ed. J. Wingender, T. Neu and H.-C. Flemming) pp. 73–92. Springer, Berlin.
Sutherland, I.W. (1999b) Polysaccharases in biofilms – sources – action – consequences! In *Microbial Extracellular Polymeric Substances* (ed. J. Wingender, T. Neu and H.-C. Flemming), pp. 201–216. Springer, Berlin.
Sutherland, I.W. (2001) Exopolysaccharides in biofilms, flocs and related structures. *Wat. Sci. Tech.* **43** (6), 77–86.
Swift, S., Throup, J.P., Williams, P., Salmond, G.P.C. and Stewart, G.S.A.B. (1996) Quorum sensing: a population-density component in the determination of bacterial phenotype. *Trends Biochem. Sci.* **21**, 214–219.
Turakhia, M.H., Cooksey, K.E. and Characklis, W.G. (1983) Influence of a calcium-specific chelant on biofilm removal. *Appl. Environ. Microbiol.* **46**, 1236–1238.
Uhlinger, D.J. and White, D.C. (1983) Relationship between physiological status and formation of extracellular polysaccharide glycocalyx in *Pseudomonas atlantica*. *Appl. Environ. Microbiol.* **45**, 64–70.
Unz, R.F. and Farrah, S.R. (1976) Exopolymer production and flocculation by *Zoogloea* MP6. *Appl. Environ. Microbiol.* **31**, 623–626.
Urbain, V., Block, J.C. and Manem, J. (1993) Bioflocculation in activated sludge: an analytical approach. *Wat. Res.* **27**, 829 – 838.
Vallom, J.K. and McLoughlin, A.J. (1984) Lysis as a factor in sludge flocculation. *Wat. Res.* **18**, 1523–1528.
Vandevivere, P. and Kirchman, D.L. (1993) Attachment stimulates exopolysaccharide synthesis by a bacterium. *Appl. Environ. Microbiol.* **59**, 3280–3286.
Veiga, M.C., Jain, M.K., Wu, W.-M., Hollingsworth, R.I. and Zeikus, J.G. (1997) Composition and role of extracellular polymers in methanogenic granules. *Appl. Environ. Microbiol.* **63**, 403–407.
Vincke, E., Boon, N. and Verstraete, W. (2001) Analysis of the microbial communities on corroded concrete sewer pipes – a case study. *Appl. Microbiol. Biotech.* **57**, 776–785.

Volpe, G., Christiansen, J.A., Wescott, J., Leger, R. and Rumbaugh, E. (1998) Use of a slime producing organism to enhance biomass settleability in activated sludge and ASB systems. *Tappi Journal* **81**, 60–67.

Watanabe, M., Sasaki, K., Nakashimada, Y., Kakizono, T., Noparatnaraporn, N. and Nishio, N. (1998) Growth and flocculation of a marine photosynthetic bacterium *Rhodovulum* sp. *Appl. Microbiol. Biotech.* **50**, 682–691.

Watanabe, M., Suzuki, Y., Sasaki, K., Nakashimada, Y. and Nishio, N. (1999) Flocculating property of extracellular polymeric substance derived from a marine photosynthetic bacterium, *Rhodovulum* sp. *J. Biosci. Bioengng* **87**, 625–629.

Wimpenny, J. (2000) An overview of biofilms as functional communities. In *Community Structure and Co-operation in Biofilms* (ed. D.G. Allison, P. Gilbert, H.M. Lappin-Scott and M. Wilson, pp. 1–24. Cambridge University Press, Cambridge.

Whitfield, C. and Valvano, M.A. (1993) Biosynthesis and expression of cell-surface polysaccharides in Gram-negative bacteria. *Adv. Microb. Physiol.* **35**, 135–246.

Wingender, J. (1990) Interactions of alginate with exoenzymes. In *Pseudomonas Infection and Alginates. Biochemistry, Genetics and Pathology* (ed. P. Gacesa and N.J. Russell) pp. 160–180, Chapman and Hall, London.

Wingender, J., Neu, T. and Flemming, H.-C (1999a) What are extracellular polymeric substances? In *Microbial Extracellular Polymeric Substances* (ed. J. Wingender, T. Neu and H.-C. Flemming), pp. 1–19, Springer, Berlin.

Wingender, J., Jaeger, K.-E. and Flemming, H.-C. (1999b) Interaction between extracellular enzymes and polysaccharides. In *Microbial Extracellular Polymeric Substances* (ed. J. Wingender, T. Neu and H.-C. Flemming), pp. 231–251. Springer, Berlin.

Wingender, J., Grobe, S., Fiedler, S. and Flemming, H.-C. (1999c) The effect of extracellular polysaccharides on the resistance to *Pseudomonas aeruginosa* to chlorine and hydrogen peroxide. In *Biofilms in Aquatic Systems* (ed. C.W. Keevil, A.F. Godfree, D. Holt and C. Dow) pp. 93–100. Royal Society of Chemistry, Cambridge.

Wingender, J. and Jaeger, K.-E. (2002) Extracellular enzymes in biofilms. In *Encyclopedia of Environmental Microbiology* (ed. G. Bitton), vol. 3, pp. 1207–1223. John Wiley and Sons, New York.

Withers, H., Swift, S. and Williams, P. (2001) Quorum sensing as an integral component of gene regulatory networks in Gram-negative bacteria. *Curr. Opin. Microbiol.* **4**, 186–193.

Wolfaardt, G.M., Lawrence, J.R., Robarts, R.D and Caldwell, D.E. (1998) In situ characterization of biofilm exopolymers involved in the accumulation of chlorinated organics. *Microb. Ecol.* **35**, 213–223.

Wolfaardt, G.M., Lawrence, J.R. and Korber, D.R. (1999) Function of EPS. In *Microbial Extracellular Polymeric Substances* (ed. J. Wingender, T. Neu and H.-C. Flemming), pp. 171–200. Springer, Berlin.

Wuertz, S., Spaeth, R., Hinderberger, A., Griebe, T., Flemming, H.-C. and Wilderer, P.A. (2001) A new method for extraction of extracellular polymeric substances from biofilms and activated sludge suitable for direct quantification of sorbed metals. *Wat. Sci. Tech.* **43**(6), 25–31.

Wuertz, S. (2002) Gene exchange in biofilms. In *Encyclopedia of Environmental Microbiology* (ed. G. Bitton), vol. 3, pp. 1408–1420. John Wiley and Sons, New York.

Xun, L., Mah, R.A. and Boone, D.R. (1990) Isolation and characterization of disaggregatase from *Methanosarcina mazei* LYC. *Appl. Environ. Microbiol.* **56**, 3693–3698.

Zhang, X., Bishop, P.L. and Kupferle, M.J. (1998) Measurement of polysaccharides and proteins in biofilm extracellular polymers. *Wat. Sci. Tech.* **37**(4–5), 345–348.

Zhang, X. and Bishop, P.L. (2001) Spatial distribution of extracellular polymeric substances in biofilms. *J. Environ. Engng* **127**, 850–856.

# 9
# The importance of physicochemical properties in biofilm formation and activity

*Rosario Oliveira, Joana Azeredo and Pilar Teixeira*

## 9.1 INTRODUCTION

Microbial adhesion to solid surfaces is the *sine qua non* condition for the formation of biofilms. This process is mainly governed by the physicochemical properties of both microbial cells and solid surfaces. However, other factors can be also involved and might ultimately have a strong influence on the overall process. For instance, high levels of shear stress have been shown to reduce bacterial adhesion (Gjaltema *et al.* 1997). Thus, parameters like the porosity or surface roughness can have a determinant role in shielding the cells from the effects of shear.

© 2003 IWA Publishing. *Biofilms in wastewater treatment.* Edited by S. Wuertz, P.L. Bishop and P.A. Wilderer. ISBN: 1 84339 007 8.

The importance of the studies on initial microbial adhesion on biofilm formation has been questioned, because the number of cells in a mature biofilm after growth can be several times higher than the number involved in initial adhesion. Those studies are only justified as a contribution to improve the understanding of the adhesion phenomenon (Marshall 1985; Absolom *et al.* 1983). From another point of view, this approach overlooks the importance of the microorganisms that initially adhere as being the link between the colonised surface and the biofilm (Busscher *et al.* 1995a).

Those working with biofilm reactors are aware that the reactor performance is dependent on the stability and activity of the biofilm. This chapter reflects on the role of surface properties on biofilm formation and on the possibility of using knowledge of them to predict the stability and eventually the activity of a biofilm.

## 9.2 HOW ADHESION HAS BEEN PREDICTED

The process of bacterial colonization on surfaces immersed in aqueous media has been extensively studied because of the importance of attached bacteria in fields like medicine, biotechnology, biofouling and geochemistry. Different approaches have been used to describe and to simultaneously predict bacterial adhesion to solid surfaces. It can be argued that bacterial colonization of immersed surfaces is always to be expected, because it is considered to be ubiquitous. However, in practical terms it is important to determine the density of cell attachment and the ease of removal of the attached cells. A detailed outline of the most common methodologies used to describe bacterial adhesion is presented below.

### 9.2.1 Thermodynamic approach

The interaction between a microbial cell and a solid substratum is only possible from a thermodynamic point of view if it leads to a decrease in the surface Gibbs free energy (Absolom *et al.* 1983; Busscher *et al.* 1984). This means that adhesion is favorable if the variation of the total Gibbs energy is negative ($\Delta G < 0$).

The interfacial free energy of interaction of two surfaces 1 (microbial cell) and 2 (substratum) immersed in a medium 3 is given by the Dupré equation (van Oss 1991):

$$\Delta G_{132} = \gamma_{12} - \gamma_{13} - \gamma_{23} \tag{9.1}$$

$\gamma_{12}$ is the interfacial tension between surfaces 1 and 2 and the other parameter $\gamma$ have a similar meaning, according to the interacting surfaces indicated in subscript.

The calculation of the surface tension of a solid has been a controversial subject, mainly owing to it being considered a whole entity (Spelt and Neumann 1987) or being formed by components, and due to questions about the nature of those components (van Pelt et al. 1983; van Oss et al. 1988). Most authors now accept the approach of van Oss et al. (1988), in which the surface tension is the sum of the apolar electrodynamic Lifshitz – van der Waals (LW) interactions and the polar interactions, owing to electron-acceptor/electron-donor interactions, also designated as (Lewis) acid–base (AB). Thus for a given substance, i, the total surface tension is given by:

$$\gamma_i^{Tot} = \gamma_i^{LW} + \gamma_i^{AB} \tag{9.2}$$

The polar (AB) component comprises two non-additive constituents: the electron acceptor ($\gamma^+$) and the electron donor ($\gamma^-$) parameters.

As the surface tension components, LW, and, AB, are additive, the Dupré equation can be written in a different form to obtain the interfacial free energy of adhesion among the interacting entities ($\Delta G_{132}^{adh}$):

$$\Delta G_{132}^{adh} = \Delta G_{132}^{LW} + \Delta G_{132}^{AB} \tag{9.3}$$

The relation of both $\Delta G_{132}^{LW}$ and $\Delta G_{132}^{AB}$ with the individual surface tension components ($\gamma_i$) is fully described in the literature (van Oss 1991).

Calculation of the surface tension of a solid can be obtained via contact angle determination, employing at least one apolar and two polar liquids of well known surface tensions and solving Young's equation three times (Adamson 1982; van Oss 1991). When the solids are in granular form (e.g. sand), the contact angle can be indirectly determined by the thin layer wicking technique, using the Washburn equation (Constanzo et al. 1990; Teixeira et al. 1998).

From all the literature where the thermodynamic approach has been considered, we can conclude that this criterion cannot be used in a straightforward fashion. It happens that in many situations where $\Delta G_{132}^{adh} > 0$, adhesion of bacterial cells does occur. This can be shown by the results obtained using different approaches for the calculation of the surface free energy.

*Example 1.* Studies on the adhesion of three different strains of oral streptococci to glass, polymethylmethacrylate and polytetrafluorethylene, considering that the surface tension comprises a dispersion and a polar component, showed that adhesion occurs even when $\Delta G_{132}^{adh} > 0$ (Busscher and Weerkamp 1987). However, the evaluation of the number of adhering bacteria yielded the conclusion that, for each bacterial strain, adhesion increased with decreasing $\Delta G_{132}^{adh}$.

*Example 2*. Studies done to select the most suitable carrier for nitrification in an air-lift reactor gave similar results (Teixeira and Oliveira 1998). In this case, the surface tension calculations followed the approach of van Oss (1991), considering the LW and AB components. The carriers under evaluation were particles of sand, pumice stone, poraver (foam glass), basalt and limestone. The nitrifying consortium was composed of *Nitrosomonas* and *Nitrobacter*. All of the obtained interfacial free energies of adhesion were positive. However, the two materials giving the lower values of $\Delta G_{132}^{adh}$ (limestone and basalt) were the ones showing higher nitrifying activity. This can be attributed to a higher number of adhered bacteria, giving rise to a more efficient biofilm. It must be stressed that the evaluation of the nitrifying efficiency was followed during 20 days in an airlift reactor. The biofilm had previously formed around the particles of each type of carrier during one month in flasks that had been inoculated with the consortium and with the medium having been replenished every seven days.

Summarizing, the thermodynamic criterion cannot be generalized, but can point to a trend for adhesion of a given bacterial strain to different types of support.

## 9.2.2 DLVO theory

Since the advent of DLVO theory (named after Derjaguin, Landau, Verweu and Overbeek (Oliveira 1992)), formulated to explain the stability of lyophobic colloids 50 years ago, several attempts have been made to use this approach to describe bacterial adhesion, considering the cells as 'living colloids' (Marshall *et al.* 1971; van Loosdrecht *et al.* 1988; Azeredo *et al.* 1999). According to this theory the total energy of interaction arises from the balance between the LW forces and electrostatic forces, effective between a cell and the substratum, changing with the distance from the surface. Most of the substances, with the exception of some metallic hydroxides, display a net negative charge when immersed in a liquid medium with pH near neutrality. So, the inter-penetration of the electrical double layers of like charge of two approaching surfaces gives rise to a repulsive force (EL), whereas the LW forces are generally attractive. The total free energy of interaction is given by:

$$\Delta G^{TOT} = \Delta G^{LW} + \Delta G^{EL} \tag{9.4}$$

The energy profile of such an interaction can have two minima separated by an energy barrier (Oliveira 1992; van Loosdrecht *et al.* 1988). The minimum more distant from the surface (secondary minimum) is usually not very deep (the decrease in the total energy of interaction is small), and the cells accumulating in this minimum (stabilizing in this energy state) are only loosely associated with the surface. In this case adhesion is considered reversible; this means that the cells can be easily removed. Those cells that are able to overcome the energy barrier and progress to the primary minimum (close to the surface and theoretically of infinite depth) become more firmly attached; they attain a very low energy state and adhesion is irreversible.

It must be noted that the mentioned energy profile is strongly dependent on the ionic strength of the medium. A low ionic strength can promote a very high energy barrier, while a high ionic strength levels the profile; the secondary minimum and the energy barrier disappear and a firm adhesion is facilitated.

Both DLVO forces depend on the geometry of the interacting surfaces. The interaction between a sphere and a smooth flat surface is made solely by one point, whereas two smooth flat surfaces enter into entire contact. Apart from geometrical assumptions, others have to be made, namely for the calculation of *Hamaker* constants (Visser 1972; van Oss 1994) and for the determination of the electrical surface potential of the interacting entities (Hunter 1981). But, probably, the most difficult to establish is the electrical behavior of both surfaces while approaching. This is to say that they can keep both their surface potentials constant, the surface charge constant, or they can interact in a mixed mode with one at constant surface potential and the other at constant charge (Oliveira 1992). The last situation is considered the most probable in biological systems (Rajagopalan and Kim 1981).

So far, it has been difficult to find in the literature a situation where bacterial adhesion can be fully explained by DLVO theory, because other types of interaction, rather than DLVO forces, can play an important role in the overall process. This happens even if the different modes of double-layer interactions are considered.

*Example 3*. The experimental tests done to select a suitable carrier for *Alcaligenes denitrificans* in an inverse fluidized bed reactor studied several polymeric materials: high density polyethylene (HDPE), polypropylene (PP), polyvinyl chloride (PVC) and polymethylmethacrylate (PMMA) showed that adhesion occurs to the greatest extent to PP followed by PVC, HDPE and lastly to PMMA (Teixeira and Oliveira 2000). The corresponding DLVO energy profiles for the two extreme situations, adhesion to PP and to PMMA, are presented in Figure 9.1. As can be seen, considering both interacting surfaces either at constant charge (Figure 9.1a) or at constant potential (Figure 9.1b), the interaction is always repulsive ($\Delta G^{TOT} > 0$) and a high-energy barrier is formed at close contact, which would prevent adhesion. Furthermore the energy barrier is more pronounced for PP than for PMMA, which is not in accordance with the experimental results. If a mixed mode of interaction is assumed (Figure 9.1c), the formation of a secondary minimum is notorious. A plausible explanation relies on reversible adhesion taking place at distances of more than 1 nm. However, the energy barrier for PMMA is smaller than for PP, and the reversibility of adhesion is also questionable. All the supports were vigorously rinsed in distilled water after the adhesion experiments and before cell enumeration. Hence it can be assumed that only the cells strongly attached to the surface remained.

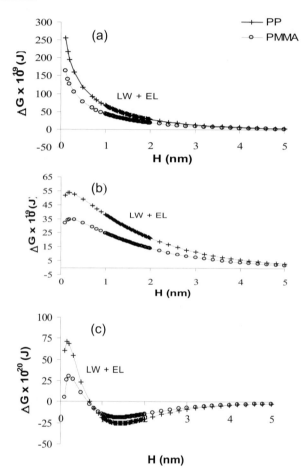

**Figure 9.1.** Variation of the DLVO total free energy of interaction between *Alcaligenes denitrificans* and the polymeric supports (PP and PMMA) as a function of the separation distance (H) under the condition of constant charge (a), constant potential (b), and mixed case (c).

## 9.2.3 XDLVO theory

Considering that in many non-metallic condensed materials, liquids and solids, polar interactions of the hydrogen bond type (AB) often occur, van Oss (1994) proposed an extension of the DLVO theory; it also accounts for the interactions due to Brownian movement forces (BR), and is generally known as the XDLVO theory. Accordingly, the total free energy of interaction is expressed by:

$$\Delta G^{TOT} = \Delta G^{LW} + \Delta G^{EL} + \Delta G^{AB} + \Delta G^{BR} \qquad (9.5)$$

When LW, EL and AB interactions are measured individually, the total free energy of interaction ($\Delta G^{TOT}$) can be obtained by summing the values of those entities plus + 1 kT for $\Delta G^{BR}$ (for systems with two degrees of freedom). If the $\Delta G^{TOT}$ is measured as a whole (by interfacial tension determination), then $\Delta G^{BR}$ is already included (van Oss 1994).

The omission of the AB forces, which are generally one or two orders of magnitude greater than the EL and LW forces, is the origin of most of the anomalies that were observed if the DLVO theory was used to interpret interfacial interactions in polar media (van Oss *et al.* 1990).

*Example 4.* Three mutants (TR, CV, and F72) of the gellan (polysaccharide) producer *Sphingomonas paucimobilis* were isolated and used in adhesion experiments. TR is the highest producer of exopolymer (EPS), followed by CV, whereas in F72 this ability is almost repressed. The adhesion assays were performed in two types of medium: in phosphate buffer saline (PBS) and in solutions of the excreted and isolated exopolymers of each mutant (Azeredo *et al.* 1999). This means that in some of the experiments the glass slides were preconditioned with the exopolymer of the mutant being tested. The number of cells per square millimeter adhering to bare glass slides (in PBS) and coated glass slides (in the EPS solutions) is shown in Table 9.1.

**Table 9.1.** Number of cells per square millimeter (± standard deviation) of each mutant of *Sphingomonas paucimobilis* adhered to glass in phosphate buffer saline (PBS) and in the corresponding solution of extracellular polymeric substances (EPS).

| Mutant | Adhesion medium | |
|---|---|---|
| | PBS | EPS |
| TR | 323±36 | 2513±215 |
| CV | 539±72 | 1508±144 |
| F72 | 646±72 | 826±72 |

Figure 9.2 presents the total free XDLVO energy of interaction as a function of the distance of separation for both types of experiment. The energy profiles for the interactions in PBS show a secondary minimum at 12–15 nm from the surface. The stabilization of the cells at this minimum is due to weak forces and an irreversible adhesion would not be expected, on account of the energy barrier at the minimum distance of separation. This may be an explanation for the small number of adhered cells (Table 9.1), which is in accordance with the depths of the secondary minima. An increase in the number of adhered cells corresponds to an increase in the secondary minimum depth.

The energy profiles obtained for adhesion in the EPS solutions clearly show a discrepancy with the practical results. This anomaly can be explained if adhesion is considered to be preferentially mediated by polymeric interactions. The polymers adsorbed to the glass surface can bind to the polymers of the EPS layer surrounding the cells. A fluorescence microscopic observation of the three mutants after binding calcofluor white and a lectin (conA) showed that mutant TR had a very thick EPS layer, which was not so large around mutant CV and was almost non-existent in mutant F72.

Although the inclusion of the AB forces is considered a drastic correction of the DLVO theory (van Oss *et al.* 1990), failures are to be expected when dealing with 'living colloids'. Apart from other phenomena, they may possess attached portions of exopolymers or surface appendages that may be able to overcome the energy barrier establishing a stable interaction with the substratum, which are difficult to account for in the calculations. This is why both DLVO and XDLVO theories are most suited to explain adhesion *a posteriori*.

## 9.3 SURFACE PROPERTIES RELEVANT FOR ADHESION

### 9.3.1 Hydrophobicity

The most well-known work on the effect of substratum wettability on bacterial adhesion is attributed to Dexter *et al.* (1975), who studied bacterial attachment in marine systems. The following studies, either in marine or medical systems, also indicate that the wettability of solid surfaces influences adhesion of bacteria, eukaryotic cells and proteins (Busscher and Weerkamp 1987; Margel *et al.* 1993; Prime and Whitesides 1993; Wiencek and Fletcher 1997; Taylor *et al.* 1997).

The wettability of a surface is now more generally referred to as hydrophobicity. In aqueous medium, adhesion is favored between hydrophobic surfaces, which can enter into closer contact by squeezing the water layer between them.

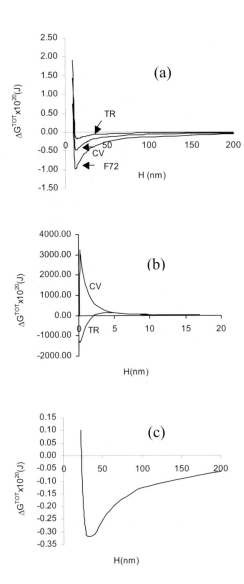

**Figure 9.2.** Variation of the XDLVO total free energy of interaction as a function of the distance of separation (H) for the interaction: (a) between the mutants TR, CV and F72 and glass in PBS medium; (b) between the mutants TR, CV and glass in EPS solutions; (c) *ibid.* for the mutant F72.

In biological systems, hydrophobic interactions are usually the strongest of all long-range non-covalent interactions, and can be defined as the attraction between apolar or slightly polar molecules, particles or cells, when immersed in water. Their sole driving force is the hydrogen bonding (AB forces) energy of cohesion between the surrounding water molecules (van Oss 1997). This means that the AB forces, if strongly asymmetrical or monopolar, are responsible for the orientation of water molecules adsorbed on the surfaces, and, as expected, water molecules oriented on the surface of one particle will repel water molecules oriented in the same manner on the surface of an adjacent particle (Parsegian et al. 1985; van Oss 1994). If the orientation of the water molecules is sufficiently strong, the two particles will not approach each other. On the other hand, if the surface is more weakly apolar, its capacity for orienting the most closely adsorbed water molecules is less pronounced and the particles will approach each other under the influence of their net LW attraction. Accordingly, 'hydrophobic' compounds or surfaces do not repel water: they attract water with rather substantial binding energies, albeit not quite as strongly as very hydrophilic ones (van Oss 1995). It should be stressed that hydrophobic attractions can prevail between one hydrophobic and one hydrophilic site immersed in water, as well as between two hydrophobic entities.

In the words of Busscher (1995), hydrophobicity is ubiquitously accepted to be a major determinant in biointerfacial reactions, but, on closer inspection, we all give different meanings to the word 'hydrophobicity' and we all use different techniques to measure 'hydrophobicity'.

Actually, several techniques have been used to determine the degree of hydrophobicity of bacterial cells or particulate materials. For materials that can be obtained in a flat plate shape, hydrophobicity has very often been expressed in terms of the contact angle formed by a sessile drop of water. In the case of bacterial cells, one of the most frequently used techniques to assess hydrophobicity is the so-called BATH (bacterial adherence to hydrocarbons) method, proposed by Rosenberg (1984), which is now more generally known as MATH (microbial adherence to hydrocarbons). In a study to characterize the hydrophobic properties of streptococcal cell surfaces (van der Mei et al. 1987), the following methods were compared: MATH, hydrophobic interaction chromatography, salting-out aggregation and contact angle measurements. Although these methods are commonly used in hydrophobicity determinations, the results obtained led the authors to the conclusion that it was not possible to define the surface 'hydrophobicity' of a bacterium, other than on a comparative level with closely related strains. Other authors (van Loosdrecht et al. 1987), studying the role of bacterial cell-wall hydrophobicity in adhesion, have also used different methods – contact angle measurements and partitioning of cells in two-phase systems (water-hexadecane and PEG-DEX) – to determine the degree of hydrophobicity of 23 bacterial strains. As a conclusion they proposed the water contact angle as the best method to quantify cell hydrophobicity, because they found some drawbacks in the utilization of two-phase systems. Later, it was observed that the zeta potentials of the hydrocarbons could be highly negative in the various solutions commonly used in MATH (Busscher et al. 1995b). So, MATH may measure a complicated interplay of long-range van der Waals and electrostatic forces and of various short-range interactions (van der Mei et al. 1995), rather than pure hydrophobic interactions.

As was already mentioned, with the techniques described above it is only possible to assess hydrophobicity in qualitative terms. However, according to van Oss (1997), it is possible to determine the absolute degree of hydrophobicity of any given substance (i) compared with water (w), which can be precisely expressed in applicable S.I. units. When the free energy of interaction, $\Delta G_{iwi}$, between two entities (i) immersed in water (w) has a positive value, i is hydrophilic, and when $\Delta G_{iwi}$ has a negative value, i is hydrophobic. More precisely (in cases of a negligible LW interaction), $\Delta G_{iwi}$ expresses the degree to which the polar attraction of entities i to water is greater (hydrophilicity) or smaller (hydrophobicity) than the polar attraction that water molecules have for each other. When the net free energy of interaction between two entities i immersed in water is sufficiently attractive (i.e., $\Delta G_{iwi}<0$), the surfaces of i are genuinely hydrophobic. The more negative $\Delta G_{iwi}$, the more hydrophobic that entity is; the more positive $\Delta G_{iwi}$, the more hydrophilic.

$\Delta G_{iwi}$ is simply related to the interfacial tension between i and water, $\gamma_{iw}$, as:

$$\Delta G_{iwi} = -2\gamma_{iw} \qquad (9.6)$$

where $\gamma_{iw}$ can be determined by contact angle measurements or thin layer wicking.

Using this last criterion, it has been easier to relate hydrophobicity with the capacity of bacteria to colonize different types of surface.

*Example 5.* Table 9.2 summarizes the results obtained in the study referred to in Example 3, which are relevant to discuss the effect of surface hydrophobicity of the polymeric materials on the attachment of *Alcaligenes denitrificans*. This bacterial strain has a $\Delta G_{iwi} = 18.2$ mJ/mm$^2$, and this means that the interaction occurred between hydrophilic bacterial cells and hydrophobic polymeric materials.

**Table 9.2.** Surface tension components ($\gamma^{LW}$ and $\gamma^{AB}$) and surface free energy of interaction between two surfaces of material i immersed in water ($\Delta G_{iwi}$), in mJ/m², and the average number of adhered bacterial cells per square millimeter (± standard deviation) (adhesion in citrate minimal medium).

| Material | $\gamma^{LW}$ (mJ/m²) | $\gamma^{AB}$ (mJ/m²) | $\Delta G_{iwi}$ (mJ/m²) | Average cell number (mm² × 10⁻³) |
|---|---|---|---|---|
| PP | 40.3 | 0 | −67.2 | 32.1±1.6 |
| HDPE | 39.5 | 3.8 | −59.2 | 20.0±1.1 |
| PVC | 37.5 | 0 | −22.0 | 13.7±1.0 |
| PMMA | 43.5 | 0 | −16.8 | 3.1±0.1 |

Table 9.2 also shows that an increase in the hydrophobicity of the polymeric supports promotes an increase in the number of adhered cells. In a closer inspection, if only the supports with $\gamma^{AB} = 0$ are considered it is possible to draw a linear correlation between the degree of hydrophobicity ($\Delta G_{iwi}$) and surface colonization. HDPE is out of this correlation because of the finite value of $\gamma^{AB}$, which is a measure of the degree of residual hydration (van Oss 1997). Thus, in spite of the intermediate hydrophobicity of HDPE, bacterial adhesion is not favored, because the bound water layer has to be removed before complete contact can occur.

*Example 6.* Four porous microcarriers, clay, foam glass, pozzolana and sepiolite, used in an anaerobic fluidized bed reactor, were compared for their ability for biomass colonization (Alves et al. 1999). The results showed that sepiolite had the greatest microbial retention capacity, followed by clay, pozzolana and finally foam glass, when expressed as mass of volatile solids per internal porous volume (gVS/$L_{internal\ porous\ volume}$). In a further development from this study, the surface tension of the carriers was determined by the thin-layer wicking technique and the $\Delta G_{iwi}$ value for each type of material was calculated. The relation between $\Delta G_{iwi}$ and the amount of attached biomass is represented in Figure 9.3. In this case all the $\Delta G_{iwi}$ values are positive, meaning that all the assayed carriers have a hydrophilic nature. However, a decrease in $\Delta G_{iwi}$ corresponds to an increase in hydrophobicity. Thus, it can be said that Figure 9.3 expresses the linear correlation between support hydrophobicity and biomass retention capacity.

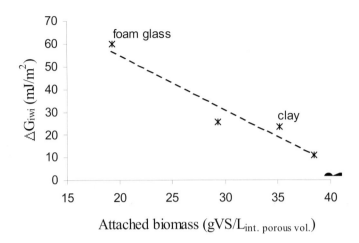

**Figure 9.3.** Relation between the attached biomass and the degree of hydrophobicity, expressed as $\Delta G_{iwi}$.

## 9.3.2 Surface charge and cation bridging

The electrical charge of the interacting surfaces is very often mentioned in adhesion studies. However, rather than surface charge, what is usually experimentally determined is the zeta potential, but both quantities can be directly related. When a particle moves through an electrolyte solution, under the influence of an electric field, part of the diffuse electrical double-layer of ions moves with the particle, while the outer region remains with the bulk phase of the environment. The interface between these two ionic regions is known as the hydrodynamic slip-plane, and the potential at this plane, with respect to an electrode at an infinite distance away in the environment, is the zeta potential ($\zeta$). The surface potential ($\psi_o$), assuming a linear Poisson–Boltzmann distribution of ions, is related to the zeta potential through

$$\psi_o = \frac{1}{4\pi\varepsilon} \zeta [1 + (d/R)] \exp(\kappa d) \qquad (9.7)$$

where $d$ is the distance between the particle surface and the slip-plane, $R$ is the particle radius and $\kappa$ is the reciprocal double-layer thickness or Debye–Hückel parameter (Oliveira 1992). The total surface charge ($Q$) can be related to the surface potential by

$$Q = 4\pi\varepsilon R (1 + \kappa R) \psi_o \qquad (9.8)$$

where $\varepsilon$ is the dielectric constant of the medium. This demonstrates that a given material has a very negative surface charge when its measured zeta potential has a high negative value.

As was briefly alluded to before, in the usual operational pH range in biofilm reactors, most of the materials (bacteria + supports) display a negative zeta potential (negative surface charge). This means that, in most the cases, bacterial adhesion is not mediated by electrostatic interactions, because these assume a repulsive effect. Nevertheless, it is possible to modify the surface of certain types of support to render them positively charged, which can be important in laboratory studies but is almost economically prohibitive for large-scale applications.

*Example 7.* The adhesion of *Pseudomonas putida* to different types of glass bead was assayed, namely: normal beads (Tamson ballotini 31/10), beads etched with 5% HF to roughen the surface, silanized beads with increased hydrophobicity, and beads positively charged by treatment with diethylaminoethyl-dextran (Gjaltema *et al.* 1997). The tests, performed under batch conditions with occasional gentle rocking, showed that the cells adhered best to the hydrophobic or positively charged beads.

The bridging effect of divalent cations, especially $Ca^{2+}$, in biological systems has been frequently referred to in the literature. These cations have a hydrophobizing effect on negatively charged particles (van Oss 1994), and according to some authors are related to increased reversible adhesion (McEldowney 1994; Takeuchi *et al.* 1997).

Referring to Example 2, the higher efficiency of limestone particles, followed by basalt, which was attributed to a higher number of adhered cells, can be a consequence of the hydrophobizing effect of $Ca^{2+}$, strongly present in limestone and in smaller quantities in the plagioclases of basalt. In Example 6, the support displaying a higher amount of adhered biomass was sepiolite, which is the material richest in surface divalent cations, in this case $Mg^{2+}$, as determined by electron dispersion spectroscopy (EDS).

Another interesting effect is the paradoxically facilitated adhesion between very negatively charged surfaces when divalent cations are present in the liquid medium.

*Example 8.* Referring once more to the study mentioned in Examples 3 and 5, the adhesion assays were done in two different liquid media: sodium phosphate buffer saline (NaPBS), containing only monovalent cations and citrate minimal medium (containing $Ca^{2+}$, $Fe^{2+}$ and $Mg^{2+}$). The ionic strength was the same in both cases. The values of the zeta potential for bacteria and the polymeric materials used as supports are presented in Table 9.3. They did not show significant deviations from one liquid medium to another.

**Table 9.3.** Zeta potential (± standard deviation) of *Alcaligenes denitrificans* and of the polymeric materials used as carriers at pH 7.3.

| Material | Zeta potential (mV) |
|---|---|
| Bacteria | −34.8±2.2 |
| Polypropylene (PP) | −45.0±1.9 |
| High-density polyethylene (HDPE) | −38.3±1.6 |
| Polyvinylchloride (PVC) | −37.7±2.1 |
| Polymethyl-methacrylate (PMMA) | −29.8±1.5 |

The average number of bacterial cells per square millimeter adhering to each type of polymeric material, in NaPBS and in citrate minimal medium, is shown in Figure 9.4.

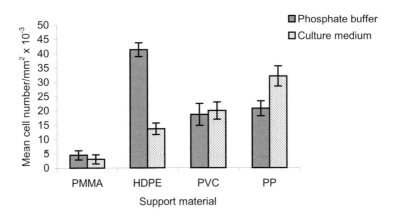

**Figure 9.4.** Average number of cells of *Alcaligenes denitrificans* per square millimeter adhering to the polymeric supports, in PBS and culture medium.

In the presence of divalent cations (citrate medium), adhesion to the more negatively charged materials is more favorable. In this situation, the number of adhered cells is linearly dependent on the increasingly negative values of zeta potential (Figure 9.5).

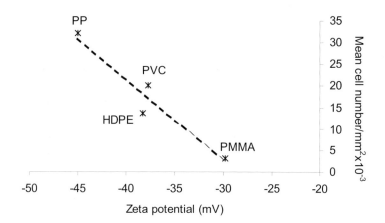

**Figure 9.5.** Relation between the number of adhering cells of *Alcaligenes denitrificans* per square millimeter and the zeta potential of the polymeric supports.

In the absence of divalent cations (NaPBS), this tendency disappears and it is only possible to directly relate bacterial adhesion with the decrease of the total XDLVO energy of interaction.

Although the properties mentioned above are determinant for adhesion, the amount of biomass present in a carrier also depends on the available surface area for microbial attachment, which is related to other physical properties.

### 9.3.3 Surface roughness and porosity

There are many reports in the literature about the advantages of using porous and rough supports for biofilm development. Apart from displaying a high surface area, a rough surface and/or internal pore space may provide a more hydrodynamically quiescent environment, thereby reducing the detachment of immobilized cells by hydraulic shearing forces (Bryers 1987; Characklis 1990; Quirynen and Bollen 1995).

The accumulation of microorganisms in porous structures is dependent upon the cell dimensions, the mode of reproduction and the pore diameter of the material (Messing and Oppermann 1979a). Accordingly, to achieve high accumulation of microbes that reproduce by fission, at least 70% of the pores of an inorganic carrier should have pore diameters in the range of one times the smallest major dimension through five times the largest major dimension of the cell. If the microbes reproduce by budding, the highest accumulation is achieved if at least 70% of the pores have diameters in the range of one times the smallest dimension of the cell to less than four times the largest cell dimension. The microorganisms that form spores and exhibit mycelial growth are considered in another report (Messing and Oppermann 1979b). For those, the highest accumulation is considered to be attained when 70% of the pores have diameters in the range of one times the smallest dimension of the fungal spore to less than about 16 times the largest dimension of that spore. Shimp and Pfaender (1982) also reported that microbe size crevices favor surface colonization. Later, Wang and Wang (1989) mathematically calculated the theoretical maximum cell retention capacities of microcarriers with different pore sizes and concluded that a mean pore diameter within a range of 2–5 times the mean cell diameter would yield the maximum immobilized cell densities. In methanogenic fluidized bed reactors, under similar start-up conditions, porous microcarriers were capable of reducing the start-up times by more than 50% compared with sand (Yee et al. 1992).

Some authors consider that surface roughness is even more important for colonization than internal surface area (Petrozzi et al 1991). Fox et al. (1990) concluded that surface roughness is critical in biofilm development during the start-up of an expanded bed reactor. The biofilm began in crevices that were protected from shear forces.

*Example 9.* The four porous microcarriers of Example 6 were characterized in terms of roughness, pore size, attached biomass and specific methanogenic activity (SMA). The latter was expressed as mass of volatile fatty acids (as COD) removed per mass of adhered volatile solids per day. The corresponding values are presented in Table 9.4. As was already mentioned, sepiolite had the highest hydrophobicity, a surface with the higher concentration of divalent cations and good cell crevice size. It is the most favorable carrier for biomass attachment and consequently for biofilm development. Despite that, it shows a paradoxical behavior, displaying the smallest methanogenic activity.

**Table 9.4.** Surface characteristics, attached biomass (g $VS/L_{internal\ porous\ volume}$) and specific methanogenic activity (SMA) of the attached biomass (g VFA-COD/g $VS_{attached}$·day) (± standard deviation).

| Material | Roughness | Pore size | Attached biomass | SMA |
|---|---|---|---|---|
| Sepiolite | ++++ | cell size crevices | 38.4±2.4 | 0.173±0.007 |
| Clay | +++ | 10–100 µm | 35.1±1.0 | 0.329±0.003 |
| Pozzolana | ++ | 10–300 µm | 29.3±1.3 | 0.340±0.038 |
| Foam glass | + | 20–1000 µm | 19.3±1.4 | 0.289±0.010 |

+, Lowest degree of roughness; ++++, highest degree of roughness.

This example suggests that a great accumulation of biomass, or the accumulation inside pores of very small size, can give rise to strong limitations in mass transfer through and from the inner biomass. This will lead to a decrease in the expected activity. It is important to be aware of this fact, especially when working with expanded bed reactors, where by an adequate manipulation of the operating conditions the unwanted excessive growth of biofilms can be controlled.

## 9.4 CONCLUDING REMARKS

In biofilm reactors, the selection of the supports for biomass immobilization is of great importance to obtain a stable biofilm leading to high overall reactor efficiency. The support must favor microbial adhesion, must be hard if subjected to high hydrodynamic shear stress, must have a low cost and must be easily available. It should also be noted that a quick, strong and uniform attachment of microorganisms to the support surface is essential to lower the start-up time of the reactor. The physicochemical properties of the support generally required to promote a stable adhesion are: a high degree of hydrophobicity; the existence of divalent cations ($Ca^{2+}$ and $Mg^{2+}$) at the surface; and a certain degree of roughness. To account for biomass detachment due to hydrodynamic shear or abrasion between carriers, roughness and porosity are of great importance. In faster-growing and higher-yielding biofilm cultures, the biofilm efficiency is less dependent on these two surface characteristics. On the contrary, slow-growing and low-yielding biofilm cultures should benefit from immobilization on rough and porous supports. However, bacteria retained in the internal surface area or inside a niche may experience diffusional resistance to the flux of substrates and products. The use of supports with large pores can overcome this problem, because in this situation the transport of metabolites can also be mediated by internal convective flow.

## 9.5 REFERENCES

Absolom, D.R., Lamberti, F.V., Policova Z., Zing W., van Oss C.J. and Neuman, A.W. (1983) Surface thermodynamics of bacterial adhesion. *Appl. Environ. Microbiol.* **46**, 90–97.

Adamson, A.W. (1982) The physical chemistry of surfaces. John Wiley, New York.

Alves, M.M., Pereira, M.A., Novais, J.M., Polanco, F.F. and Mota, M. (1999) A new device to select microcarriers for biomass immobilization: application to an anaerobic consortium. *Wat. Environ. Res.* **1**, 209–217.

Azeredo, J., Visser, J. and Oliveira, R (1999) Exopolymers in bacterial adhesion: interpretation in terms of DLVO and XDLVO theories. *Coll. Surf. B: Bioint.* **14**, 141–148.

Bryers, J.D. (1987) Biologically active surfaces: processes governing primary biofilm formation. *Biotech. Prog.* **3**, 57–68.

Busscher, H.J., Weerkamp, A.H., van der Mei, H.C., van Pelt, A.W.J., de Jong, H.P. and Arends, J. (1984) Measurement of the surface free energy of bacterial cell surfaces and its relevance for adhesion. *Appl. Environ. Microbiol.* **48**, 980–983.

Busscher, H.J. and Weerkamp, A.H. (1987) Specific and non-specific interactions in bacterial adhesion to solid substrata. *FEMS Microbiol. Rev.* **46**, 165–173.

Busscher, H.J., Bos, R. and van der Mei, H.C. (1995a) Initial microbial adhesion is determinant for the strength of biofilm adhesion. Hypothesis. *FEMS Microbiol. Lett.* **128**, 229–234.

Busscher, H.J., van de Belt-Gritter, B. and van der Mei, H.C. (1995b) Implications of microbial adhesion to hydrocarbons for evaluating cell surface hydrophobicity. 1. Zeta potentials of hydrocarbon droplets. *Coll. Surf. B: Bioint.* **5**, 111–116.

Busscher, H.J. (1995) Preface. *Coll. Surf. B: Bioint.* **5**, iii.

Characklis, W.G. (1990) Biofilm processes. In *Biofilms* (ed. W.G. Characklis and K. C. Marshall), pp 195–231. John Wiley, New York.

Constanzo, P.M., Giese, R.F. and van Oss, C.J. (1990) Determination of the acid base characteristics of clay mineral surfaces by contact angle measurements. Implications for the adsorption of organic solutes from aqueous media. *J. Adhesion Sci. Tech.* **4**, 267–275.

Dexter, S.C., Sullivan Jr, J.D., William III, J. and Watson, S.W. (1975) Influence of substrate wettability on the attachment of marine bacteria to various surfaces. *Appl. Microbiol.* **30**, 298–308.

Fox, P., Suidan, M.T. and Bandy, J.T. (1990) A comparison of media types in acetate fed expanded-bed anaerobic reactors. *Wat. Res.* **24**, 827–835.

Gjaltema, A., van der Marel, N., van Loosdrecht, M.C.M., Heijnen, J.J. (1997) Adhesion and biofilm developmenton suspended carriers in air-lift reactors: hydrodynamic conditions versus surface characteristics. *Biotech. Bioengng* **55**, 880–889.

Hunter, R.J. (1981) *Zeta potential in colloid science*. Academic Press, London and San Diego.

Margel, S., Vogel, E.A., Firment, L., Watt, T., Haynie, S. and Sogah, D.Y. (1993) Peptide, protein and cellular interactions with self-assembled monolayer model surfaces. *J. Biomed. Mater. Res.* **27**, 1463–1476.

Marshall, K.C., Stout, R. and Mitchell, R. (1971) Mechanisms of the initial events in the sorption of marine bacteria to surfaces. *J. Gen. Microbiol.* **68**, 337–348.

Marshall, K.C. (1985) Mechanisms of bacterial adhesion at solid–water interfaces. In *Bacterial Adhesion* (ed. D. C. Savage and M. Fletcher), pp 133–161. Plenum Press, New York and London.

Messing, R.A. and Oppermann, R.A. (1979a) Pore dimensions for accumulating Biomass. I. Microbes that reproduce by fissing or by budding. *Biotech. Bioengng* **21**, 49–58.

Messing, R.A. and Oppermann, R.A. (1979b) Pore dimensions for accumulating Biomass. II. Microbes that form spores and exhibit mycelial growth. *Biotech. Bioeng.* **21**, 59–67.

McEldowney, S. (1994) Effect of cadmium and Zinc on Attachment and detachment interactions of *Pseudomonas fluorescens* H2 with glass. *Appl. Environ. Microbiol.* **60**, 2759–2765.

Oliveira, R. (1992) Physico-chemical aspects of adhesion. In *Biofilms–Science and Technology* (ed L.F. Melo, T.R. Bott, M. Fletcher and B. Capdeville), pp 45–58. NATO ASI Series, Kluwer Academic Publishers, Dordrecht, Boston, and London.

Parsegian, V.A., Rand, R.P. and Rau, D.C. (1985) Hydration forces: what next? *Chemica Scripta* **25**, 28–31.

Petrozzi, S., Kut, O.M and Dunn, I.J. (1991) Protection of biofilms against toxic shocks by the adsorption and desorption capacity of carriers in anaerobic fluidized bed reactors. *Bioprocess Engng* **9**, 47–59.

Prime, K.L. and Whitesides, G.M. (1993) Adsorption of proteins onto surfaces containing end-attached oligo (ethylene oxide): a model system using self-assembled monolayers. *J. Amer. Chem. Soc.* **115**, 10714–10721.

Quirynen, M. and Bollen, C.M.L. (1995) The influence of surface roughness and surface-free energy on supra- and subgingival plaque formation in man. *J. Clin. Periodontol.* **22**, 1–14.

Rajagopalan, R. and Kim, J.S. (1981) Adsorption of Brownian particles in the presence of potential barriers. Effect of different modes of double-layer interaction. *J. Coll. Inter. Sci.* **83**, 428–448.

Rosenberg, M. (1984) Bacterial adherence to hydrocarbons: a useful technique for studying cell surface hydrophobicity. *FEMS Microbiol. Lett.* **22**, 289–295.

Shimp, R.J. and Pfaender, F.K. (1982) Effect of surface area and flow rate on marine bacterial growth in activated carbon columns. *Appl. Environ. Microbiol.* **44**, 471–477.

Spelt, J.K. and Neumann, A.W. (1987) Solid surface tension: the equation of state approach and the theory of surface tension components. Theoretical and conceptual considerations. *Langmuir* **3**, 588–591.

Takeuchi, T., Ohshima, H. and Makino, K. (1997) Effects of multivalent cations on aggregation behavior of poly(methacrylic acid)-grafting nylon microcapsules. *Coll. Surf. B: Bioint.* **9**, 225–231.

Taylor, G.T., Zheng, D., Lee, M., Troy, P.J., Gyananaht, G. and Sharma, S.K. (1997) Influence of surface properties on accumulation of conditioning films and marine bacteria on substrata exposed to oligotrophic waters. *Biofouling* **11**, 31–57.

Teixeira, P. and Oliveira, R. (1998) The importance of surface properties in the selection of supports for nitrification in air-lift bioreactors. *Biop. Engng* **19**, 143–147.

Teixeira, P. and Oliveira, R. (2000) Adhesion of *Alcaligenes denitrificans* to polymeric materials: the effect of divalent cations. *J. Adhesion* **73**, 87–97.

Teixeira, P., Azeredo, J., Oliveira, R. and Chibowski, E. (1998) Interfacial interactions between nitrifying bacteria and mineral carriers in aqueous media determined by contact angle measurements and thin layer wicking. *Coll. Surf. B: Bioint.* **12**, 69–75.

van der Mei, H.C., Weerkamp, A.H. and Busscher, H.J. (1987) A comparison of various methods to determine hydrophobic properties of streptococcal cell surfaces. *J. Microbiol. Meth.* **6**, 277–287.

van der Mei, H.C., van de Belt-Gritter, B. and Busscher, H.J. (1995) Implications of microbial adhesion to hydrocarbons for evaluating cell surface hydrophobicity. 2. Adhesion mechanisms. *Coll. Surf. B: Bioint.* **5**, 117–126.

van Loosdrecht, M.C.M., Lyklema, J., Norde, W., Schraa, G. and Zhender, A.J.B. (1987) The role of bacterial cell wall hydrophobicity in adhesion. *Appl. Environ. Microbiol.* **53**, 1893–1897.

van Loosdrecht, M.C.M., Lyklema, J., Norde, W. and Zhender, A.B. (1988) Bacterial adhesion: a physico-chemical approach. *J. Gen. Microbiol.* **68**, 337–348.

van Oss, C.J., Good, R.J. and Chaudhury, M.K. (1988) Additive and nonadditive surface tension components and the interpretation of contact angles. *Langmuir* **4**, 884–891.

van Oss, C.J., Giese, R.F. and Costanzo, P.M. (1990) DLVO and non-DLVO interactions in hectorite. *Clays Clay Minerals* **38**, 151–159.

van Oss, C.J. (1991) The forces involved in bioadhesion to flat surfaces and particles– their determination and relative roles. *Biofouling* **4**, 25–35.

van Oss, C.J. (1994) *Interfacial forces in aqueous media*. Marcel Dekker, New York.

van Oss, C.J. (1995) Hydrophobicity of biosurfaces – origin, quantitative determination and interaction energies. *Coll. Surf. B: Bioint.* **5**, 91–110.

van Oss C.J. (1997) Hydrophobicity and hydrophilicity of biosurfaces. *Curr. Opin. Coll. Int. Sci.* **2**, 503–512.

van Pelt A.W.J., de Jong, H.P., Busscher, H.J. and Arends, J. (1983) Dispersion and polar surface free energies of human enamel. *J. Biom. Mat. Res.* **17**, 637–641.

Visser, J. (1972) On Hamaker constants: a comparison between Hamaker constants and Lifshitz–van der Waals constants. *Adv. Coll. Int. Sci.* **3**, 331–363.

Wang, S. and Wang, D.I.C. (1989) Pore dimension effects on cell loading of a porous carrier. *Biotech. Bioengng* **33**, 915–917.

Wiencek, K.M. and Fletcher, M. (1997) Effects of substratum wettability and molecular topography on initial adhesion of bacteria to chemically defined substrata. *Biofouling* **11**, 293–311.

Yee, C.J., Hsu, Y. and Shieh, W.K. (1992) Effects of microcarrier pore characteristics on methanogenic fluidized bed performance. *Wat. Res.* **8**, 1119–1125.

# 10
# Influence of population structure on the performance of biofilm reactors

*Axel Wobus, Frank Kloep, Kerstin Röske and Isolde Röske*

## 10.1 INTRODUCTION

Biofilm reactors are widely used for advanced wastewater treatment, i.e. the elimination of the nutrients N and P, or the removal of xenobiotics from municipal or industrial wastewater, respectively. As a prerequisite for a successful application, microorganisms with the specific metabolic capacity have to evolve and be maintained in the bioreactor. As they are retained in the form of a biofilm, even slow-growing bacteria, i.e. nitrifiers or xenobiotic-degrading bacteria, may propagate and persist independently of the hydraulic load.

© 2003 IWA Publishing. *Biofilms in wastewater treatment*. Edited by S. Wuertz, P.L. Bishop and P.A. Wilderer. ISBN: 1 84339 007 8.

There is no doubt that the performance of biofilm reactors depends on their biotic structure, especially on the distribution and activity of the various microbial species. However, abiotic factors like transport phenomena also influence the rate at which different compounds are removed from the bulk fluid. Moreover, the biofilm development is significantly affected by reactor hydraulics and the mode of operation. The transport of dissolved substrates from bulk liquid to the cells plays a central role in the modeling of biofilm processes (Siegrist and Gujer 1985; Lewandowski *et al.* 1995; Horn and Hempel 1998). The three-dimensional, porous biofilm architecture of microbial clusters separated by interstitial voids, as visualized by confocal laser scanning microscopy (CLSM), resulted in the search for more complex models of mass transport in biofilms (de Beer and Stoodley 1995; Lewandowski *et al.* 1995).

However, the complex structure of biofilms does not only affect the transport processes but is also reflected in a heterogeneous distribution of active microorganisms in the system. To manage biofilm reactors and to develop more sophisticated models, detailed information on the different microorganisms, their distribution and activity within the biofilm is necessary. In recent years, new techniques like the fluorescent *in situ* hybridization (FISH) with rRNA-targeted oligonucleotide probes combined with CLSM or with microelectrode measurements have led to a deeper insight into the complex structure of biofilms (Amann *et al.* 1992; Poulsen *et al.* 1993; de Beer and Stoodley 1995; Manz *et al.* 1999). In some studies, information about community structure was related to the metabolic function of biofilms (Møller *et al.* 1996; Schramm *et al.* 1996). By this attempt, it seems to be possible to explain and predict the performance of biofilm reactors using the data on their biotic structure.

This chapter discusses long-term studies of two laboratory-scale membrane biofilm reactors for the treatment of wastewater containing xenobiotics and of a fluidized-bed nitrification reactor. The microbial communities on the support material were characterized by fluorescently labelled rRNA-targeted oligonucleotide probes, but also cultivation techniques were applied. Biofilm architecture was investigated by microscopic techniques, i.e. CLSM and scanning electron microscopy (SEM), respectively. As biofilms are complex biotic systems in which food-web relationships are relevant, the protozoa and metazoa in the biofilms were also examined.

## 10.2 INVESTIGATION OF THE BIOTIC STRUCTURE OF BIOFILMS: SURVEY OF METHODS

### 10.2.1 Methods for the investigation of the microbial composition of biofilm communities

The composition of microbial communities can be investigated in two different ways. It is possible to characterize microorganisms subsequent to their cultivation or to perform a qualitative analysis without cultivation by molecular techniques. In the latter case, PCR-amplified DNA-fragments are separated by gel electrophoresis (DGGE: denaturing gradient gel electrophoresis; or TGGE: temperature gradient gel electrophoresis) resulting in distinguishable fingerprints. By applying specific gene probes, special members of a microbial community or specific physiological properties can be detected. An overview of molecular methods used in microbial ecology is given in the manual of Akkermans *et al.* (1995), by Muyzer and Ramsing (1995), by Pickup and Saunders (1996), or by Loy *et al.* (2002).

To enumerate particular bacterial populations and to localize specific bacteria in biofilms, *in situ* investigations by means of immunofluorescence techniques or by hybridization with fluorescently labeled oligonucleotide probes (FISH) are used. In contrast to the classic bacteriological methods, these methods are less time consuming, and a selection of particular microbial species by the composition of the culture medium or by the cultivation conditions is avoided. Moreover, there is only a low percentage of cultivable cells in native microbial communities. Only the easily cultivable microorganisms are recovered, whereas the most prominent species cannot be found by standard cultivation procedures (Wagner *et al.* 1993; Manz *et al.* 1994). However, selective media can be used to enrich microorganisms with a specific metabolic potential and to investigate their physiological properties.

As the microorganisms in biofilms are embedded in a polymer matrix, the diffusion of antibodies into the biofilm and the contact between the antibody and the target cells may be prevented (Szwerinski *et al.* 1985). Another disadvantage of immunological methods is the high strain or species specificity of antibodies. Moreover, the cultivation of the microorganisms is a prerequisite for the preparation of a specific antiserum.

In recent years, FISH with rRNA-targeted oligonucleotide probes has become the most popular method for the investigation of the taxonomic structure of microbial communities of natural and man-made environments (for an overview, see Amann *et al.* (1995)). This is due to the general acceptance of the 16S and 23S rRNA molecules as phylogenetic markers (Olsen *et al.* 1986; Woese 1987; Head *et al.* 1998). Therefore, it is possible to identify microorganisms at different taxonomic levels, i.e. one particular strain in a complex community can be detected, as well as the proportions of the principal groups of Bacteria can be examined. Moreover, fluorescently labeled probes are important tools to localize specific microorganisms in complex environments like biofilms. A further advantage of this technique is the opportunity to use newly isolated and sequenced rRNAs from the environment of interest for probe design. The potential of the rRNA approach for the phylogenetic identification and *in situ* detection of microorganisms results from the combination of sequencing (the estimation of the genetic diversity of an environmental sample) and probing (to identify and enumerate whole fixed cells in the sample) (Amann *et al.* 1995).

To get first information about the microbial composition of biofilm communities, it is useful to apply probes specific to the α-, β- and γ subclass of Proteobacteria, to the Gram-positive bacteria with a high GC (guanine and cytosine) content and to the *Cytophaga–Flavobacterium* group. These groups were shown by different investigators (Wagner *et al.* 1993; Manz *et al.* 1994) to be the dominant bacterial groups in wastewater communities. However, the main groups of bacteria comprise a very broad range of physiologically quite different taxa. Thus changes in community structure relevant to the metabolic potential of the biofilms could not be detected. The examination of biofilm samples with genus or species-specific probes will also allow more detailed information on the composition of microbial communities concerning functional aspects. For example, different probes for the detection of ammonia- and nitrite-oxidizing bacteria were developed to characterize nitrifying microbial communities in wastewater treatment plants (Wagner *et al.* 1995, 1996; Mobarry *et al.* 1996; Schramm *et al.* 1996).

Notwithstanding the high potential of FISH to examine the microbial diversity in different environments, some limitations still exist in probing highly complex samples (Amann *et al.* 1995). Frequently encountered problems are background fluorescence due to autofluorescing particles, i.e. particles of algae or humic substances, and low signal intensity after *in situ* hybridization. The latter can be caused by too few copies or insufficient accessibility of the target molecules, i.e. the rRNA molecules. Therefore, slowly growing or inactive bacteria will not be detected. Thick cell walls (of Gram-positive bacteria) or higher-order structures in the ribosomes may also prevent probe hybridization.

For the investigations presented in this chapter, the biofilms were removed from the carrier material by sonication to homogenize the samples and to diminish the number of aggregates that can hamper the quantification of cells by epifluorescence microscopy. By sampling at intervals and in different reactor depths, it becomes possible to detect temporal and spatial (longitudinal) gradients in taxonomic composition of the microbial communities of different biofilm reactors. The oligonucleotide probes (MWG Biotech, Ebersberg, Germany) used for examination of biofilm samples in this study are listed in Table 10.1. The hybridization procedure is as described by Manz *et al.* (1992).

**Table 10.1.** Oligonucleotide probe data.

| Specificity | Probe | Sequence (5'–3') | rRNA target | Reference |
|---|---|---|---|---|
| Most Bacteria | EUB338 | 5'-GCTGCCTCCCGTAGGAGT-3' | 16S | Amann *et al.* 1990 |
| α-subclass of Proteobacteria | ALF1b | 5'-CGTTCGYTCTGAGCCAG-3' | 16S | Manz *et al.* 1992 |
| β-subclass of Proteobacteria | BET42a | 5'-GCCTTCCCACTTCGTTT-3' | 23S | Manz *et al.* 1992 |
| γ-subclass of Proteobacteria | GAM42a | 5'-GCCTTCCCACATCGTTT-3' | 23S | Manz *et al.* 1992 |
| Gram-positive bacteria with high G+C content of DNA | HGC69a | 5'-TATAGTTACCACCGCCGT-3' | 23S | Roller *et al.* 1994 |
| Cytophaga–Flavobacteria group | CF319a | 5'-TGGTCCGTGTCTVAGTAC-3' | 16S | Wagner *et al.* 1994 |
| Fluorescent pseudomonads | Ps56aX | 5'-GCTGGCCTAGCCTTC-3' | 23S | Schleifer *et al.* 1992 |
| Halophile and halotolerant members of the genus *Nitrosomonas* | Neu23a | 5'-CCCCTCTGCTGCACTCTA-3' | 16S | Wagner *et al.* 1995 |
| *Nitrosomonas europaea*, *N. eutropha*, *N.* sp. C56, *Nitrosococcus mobilis* | Nsm156 | 5'-TATTAGCACATCTTTCGAT-3' | 16S | Mobarry *et al.* 1996 |
| *Nitrosolobus multiformis*, *Nitrosospira briensis*, *Nitrosovibrio tenuis* | Nsv443 | 5'-CCGTGACCGTTTCGTTCCG-3' | 16S | Mobarry *et al.* 1996 |
| Ammonia-oxidizing β-Proteobacteria | Nso190 | 5'-CGATCCCTGCTTTTCTCC-3' | 16S | Mobarry *et al.* 1996 |

**Table 10.1.** *continued*

| Specificity | Probe | Sequence (5′–3′) | rRNA target | Reference |
|---|---|---|---|---|
| Ammonia-oxidizing β-Proteobacteria | Nso1225 | 5′-CGCGATTGTATTACGTGTGA-3′ | 16S | Mobarry *et al.* 1996 |
| *Nitrobacter* sp. | NIT1 | 5′-CACCTCTCCCGAACTCAA-3′ | 16S | Wagner *et al.* 1996 |
| *Nitrobacter* sp. | NIT2 | 5′-CGGGTTAGCGCACCGCCT-3′ | 16S | Wagner *et al.* 1996 |
| *Nitrobacter* sp. | NIT3 | 5′-CCTGTGCTCCATGCTCCG-3′ | 16S | Wagner *et al.* 1996 |
| *Paracoccus* sp. | PAR651 | 5′-ACCTCTCTCGAACTCCAG-3′ | 16S | Neef *et al.* 1996 |
| Competitor probes | | | | |
| Beta | Cgam | 5′-GCCTTCCCACATCGTTT-3′ | 23S | Manz *et al.* 1992 |
| Gamma | Cbeta | 5′-GCCTTCCCACTTCGTTT-3′ | 23S | Manz *et al.* 1992 |
| Neu23a | CTE | 5′-TTCCATCCCCCTCTGCCG-3′ | 16S | Wagner *et al.* 1995 |
| NIT3 | CNIT3 | 5′-CCTGTGCTCCAGGCTCCG-3′ | 16S | Wagner *et al.* 1996 |

## 10.2.2 Utilization of protozoa and metazoa as indicators of the ecological conditions in and around biofilms

Biofilms are complex biotic communities consisting of protozoa and metazoa in addition to bacteria. As the interactions between protozoa, metazoa and bacteria are complex, both positive and negative effects on the biofilm architecture as well as biotic structure and activity as a whole are possible. It is important to know which types of protozoa and metazoa are dominant, as their life strategies may strongly differ and consequently influence the behavior of the biofilm in different ways.

Thickness and composition of biofilms are not only controlled by substrate concentration and shear stress, but also by predator–prey interactions between eukaryotes and bacteria. Protozoa and metazoa in biofilm reactors exert a direct influence on the composition and performance of the microbial community. Protozoa feeding on biofilm may decrease the microbial biomass and possibly eliminate slow-growing bacteria, but filter-feeding species also improve the quality of the bulk fluid (the effluent quality) by the removal of suspended bacteria. Therefore, the interactions between bacteria and eukaryotic organisms are of particular significance to control the performance of a biofilm reactor. Consequently, the major goal of this contribution was to identify the protozoa and metazoa, and their spatial and temporal distribution.

As a prerequisite for quantification, the biomass has to be removed from the carrier material (in membrane biofilm reactors, from the silicone tubings, pieces with a length of 20–50 mm) and suspended in a small amount of mineral water. This suspension can be concentrated to a volume of 1–3 ml by centrifugation.

For counting larger organisms, 10 µl of this suspension were examined by using a Thoma chamber. A Kolkwitz chamber (volume 1 ml) may be used to examine large metazoa. By homogenization of the suspension with an Ultra-Turrax at a very low speed, the enumeration of small protozoa in a Fuchs–Rosenthal counting chamber became possible.

Further calculations are based upon the total number of individuals from all of the subsamples. Thus, no arithmetic means are presented. The numbers of individuals may be related to the corresponding area of the carrier material and may be multiplied by the volumes of the different species (according to Foissner (1995)), or to average values obtained from microscopic measurement) to calculate the biomass (fresh mass).

In terms of feeding behavior, the protozoa and metazoa may be classified in a simplified manner as follows:
- carnivorous and cannibalistic (organisms that feed on other protozoa and metazoa);
- burrowers (organisms feeding in tunnels within a biofilm or a floc);
- browsers (grazing on the surface of the biofilm);
- filter feeders (organisms that feed on suspended particles, mainly bacteria, by ciliary activity).

They are either mobile at or near the surface of the biofilm or are attached.

## 10.2.3 Microscopic techniques for the examination of biofilm architecture

Microbial biofilms associated with surfaces form a heterogeneous architecture with many microniches, which vary continually in time, depending on physicochemical and biological conditions of the environment. Whereas the traditional light microscope techniques are meaningful tools to identify protozoa and metazoa (refer to section 10.2.2), and bacterial cells are routinely enumerated using fluorescence microscopy, only a few methods are suitable to study a biofilm as a three-dimensional specimen.

SEM provides one well-developed method to get special qualitative information about morphological structures (e.g. the development of biofilms, the occurrence of extracellular polymers, or the presence of protozoans suggesting predator–prey interactions). A major disadvantage of SEM is the required preparation of the samples. The usual drying procedures based on extensive dehydration lead to artifacts of the well-hydrated natural biofilm (Richards and Turner 1984; Stewart *et al.* 1995). Additionally, SEM provides only an image of the surface of the biofilm without revealing information about the inner architecture.

To examine the fully hydrated biofilm *in vivo*, CLSM allows the study of native biofilms and the optical penetration under the biofilm surface without drying procedures (Lawrence *et al.* 1998). The principal advantage of CLSM is the possibility to scan different optical sections through the whole thickness of a stained or hybridized biofilm followed by digital interpretation with image analysis software. Therefore, many possibilities of representation (e.g. 3-D pictures) provide a powerful tool to investigate hitherto unknown structures of living biofilms. Recent investigations exploiting this technique have shown a heterogeneous structure with water channels in its open architecture (Lawrence *et al.* 1991; Wolfaardt *et al.* 1994; de Beer and Stoodley 1995; Massol-Deya *et al.* 1995; Neu and Lawrence 1997). By applying CLSM to the investigation of the nitrifying biofilms described in section 10.3.2, the multiform morphologies of bacteria resulting from different influencing factors became visible (Kloep *et al.* 2000).

Furthermore, the combination of CLSM and FISH with rRNA-targeted oligonucleotide probes provides another powerful tool to show microenvironments within the biofilm. It is possible to study microbial interactions, e.g. between ammonia oxidizers and the nitrite oxidizers, whose growth occurs in aggregates very close to each other (Schramm *et al.* 1996, 1998). The application of CLSM is limited because of photobleaching of fluorescent probes, which allows only relatively short studying times of the samples. Nevertheless, the CLSM could be applied for the systematic study of very different biofilms.

## 10.3 CASE STUDIES

### 10.3.1 Membrane-grown biofilms for the treatment of wastewater containing xenobiotics

In this study, membrane biofilm reactors (MBR) were used for the removal of chlorophenols, which serve as a model for poorly degradable xenobiotics from wastewater. Gas permeable silicone tubing was laid out in a helical fixed bed; it served both for oxygen supply and as support material for the biofilm (as proposed by Kniebusch *et al.* (1990)). The membrane-grown 'inverted' biofilm is characterized by opposite vertical gradients of oxygen and substrate concentration (Figures 10.1 and 10.2).

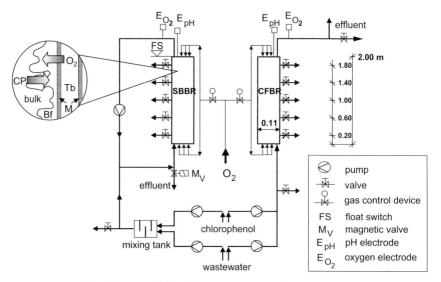

**Figure 10.1.** Flow diagram of the laboratory plant with membrane biofilm reactors for the treatment of wastewater containing xenobiotics and schematic representation of a biofilm grown on gas permeable silicone tubings; the arrows in this scheme indicate the opposite diffusive transport of oxygen and chlorophenols resulting in opposite concentration gradients in the biofilm. Tb, silicone tubing; M, semipermeable membrane; Bf, biofilm; CP, chlorophenols; the proportions in these diagrams are not in scale.

In comparing two identical MBRs, one operated discontinuously as a Sequencing Batch Biofilm Reactor (SBBR) (as introduced by Wilderer (1992), the other with continuous flow (Continuous Flow Biofilm Reactor, CFBR), the influence of different operation strategies on the biotic structure and reactor performance was investigated. A schematic representation of the laboratory plant (liquid volume about 10 l), which has been previously described (Wobus *et al.* 1995), is given in Figure 10.1. The wastewater passed through the CFBR in the upflow direction. To have a standard for comparison, the 6 hour residence time of the water in the CFBR corresponded to the duration of one cycle of the SBBR, which included a fill, a reaction and a draw period. The SBBR was filled from the bottom to the top within half an hour. After circulating the water during the reaction period of 5 h, the reactor was drained against the direction of filling for half an hour. To characterize the biotic structure of biofilms, the longitudinal distribution of biomass and microbial activity, biofilm samples were taken from both reactors and from different sections of these reactors.

Both reactors were operated with an addition of 4-monochlorophenol (MCP) over a period of about 10 months using constant and fluctuating influent conditions. After an adaptation period of approximately 10 days, more than 95% of the added MCP was eliminated in both reactors, most of the time (Figure 10.3). However, at influent concentrations in excess of 50 mg/l MCP, higher effluent concentrations (mean: 3.45 mg/l, between the 15th and 180th day of operation) were measured more frequently in the CFBR than in the SBBR (mean: 0.88 mg/l ). If the reactors were subjected to periodically changing influent concentrations or to shock loads, the effluent quality deteriorated, especially in the CFBR (Wobus and Röske 2000).

**Figure 10.2.** Three coils of one of the membrane biofilm reactors with tubings laid out in the helical fixed bed (after about 6 weeks of operation).

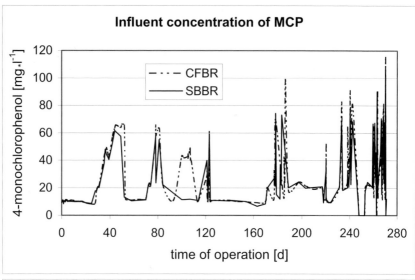

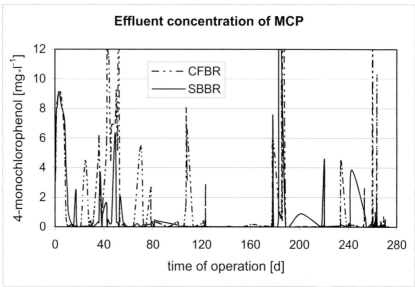

**Figure 10.3.** Influent and effluent concentrations of 4-monochlorophenol (MCP) in two differently operated membrane biofilm reactors over a period of about 10 months.

Whereas the degradation of chlorophenols (and other wastewater compounds) proceeds along the reactor depth of the CFBR, the elimination process in the SBBR takes place on a temporal scale. Therefore, concentration profiles along the vertical axis in the CFBR and at corresponding intervals on the timescale of the SBBR were compared (Figure 10.4). Continuous flow conditions resulted in longitudinal gradients of chlorophenol and a decrease of removal rates per area with increasing reactor depth. According to the short filling period, the distribution of chlorophenols and, therefore, of the biomass across the SBBR was more homogeneous. In the first two hours of the reaction time, about 95% of the MCP load was eliminated.

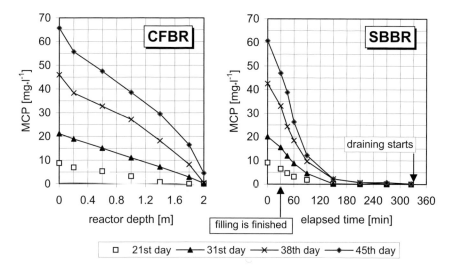

**Figure 10.4.** Characteristic profiles of the 4-monochlorophenol (MCP) concentration along the reactor depth in the CFBR, and on temporal scale in the SBBR (between the 21st day and the 45th day of operation, the influent concentration of MCP was stepwise increased).

## *10.3.1.1 Results of microbiological examination*

By the application of fluorescently labelled, rRNA-targeted oligonucleotide probes, generally more than 60% of the total cell count was detected, and the identification at the level of principal phylogenetic groups of Bacteria became possible. Because a distinct amount of ribosomes per cell is necessary to obtain a fluorescence signal (Poulsen *et al.* 1993), the high percentage of hybridized cells indicates that most bacteria in the biofilms were active (also in the thick biofilms from the inflow section of the CFBR). An example of the typical longitudinal biomass distribution in both reactors is given in Figure 10.5. The distribution of fresh biomass of procaryotes and eukaryotes, as calculated from the number and volume of bacterial and eukaryote cells, was in good agreement with the longitudinal gradient of the DNA content of biofilm extracts.

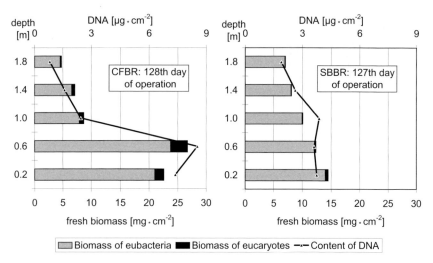

**Figure 10.5.** Comparison of the spatial (longitudinal) distribution of biomass of bacteria and eukaryotes and of DNA content across the two different operated biofilm reactors (after 127/128 days of operation with addition of 4-monochlorophenol).

Neither principal shifts in the proportions of principal groups of Bacteria nor conspicuous differences between the CFBR and the SBBR were observed over an experimental period of 9 months (Figure 10.6). The composition of the microbial communities did not significantly differ from the population structure found in other wastewater treatment systems (Wagner *et al.* 1993; Manz *et al.* 1994). The representatives of the β subclass of Proteobacteria (BET42a positive cells) amounted to 10–50% of the total number of bacterial cells and were dominant over the total length of the experimental period.

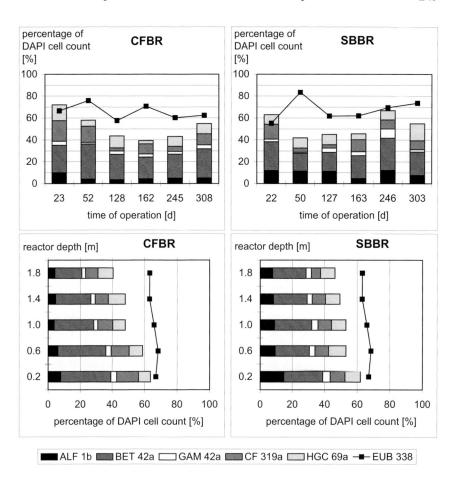

**Figure 10.6.** Above: abundances of probe-positive cells in relation to the total cell number (by 4′,6-diamidino-2-phenylindole (DAPI) staining) in two differently operated membrane biofilm reactors for the elimination of xenobiotics on a temporal scale (days of operation with addition of 4-monochlorophenol); the bars represent averages of samples from five segments. Below: distribution of probe-positive cells (related to the DAPI cell count) across the longitudinal axis in both reactors (averages from seven sampling dates). The names of group-specific probes are given in Table 10.1.

According to their proportion of hybridizable cells, the *Cytophaga–Flavobacteria* cluster (CF319a) and the group of Gram-positive bacteria with a high G+C content (HGC69a) were the next most important groups. Members of both groups, like *Arthrobacter*, *Rhodococcus* or *Flavobacterium* are known for their capacity to decompose chlorinated phenols (Häggblom 1990; Commandeur and Parsons 1994). The population dynamics showed strong fluctuations of the proportions of both groups temporally (Figure 10.6). The percentage of cells hybridizing with the HGC69a probe decreased between days 22 and 50 of MCP addition in the SBBR according to the stepwise increase of the inflow concentration of 4-monochlorophenol (MCP) from 10 mg/l (until day 28) up to 65 mg/l (from day 42 to 52). A similar population shift was observed in case of the change of a regime of continuous feeding to shock loading experiments with up to 60 mg/l MCP (between day 80 and 123) in the CFBR. Thus, no conclusions can be drawn from these investigations with regard to the principal role of members of one of these groups in the MCP removal. Possibly, the members of the HGC group were more sensitive or less competitive at high MCP concentrations than other groups of bacteria in the biofilm reactors.

The capacity to degrade chloroaromatics is widespread among the different groups of Bacteria. For example, various strains of pseudomonads (belonging to the β- or γ-subclass of Proteobacteria) are capable of biodegrading chlorophenols (Knackmuss and Hellwig 1978; Häggblom 1990; Commandeur and Parsons 1994; Schlömann 1994). Thus, changes in community structure relevant to the metabolic potential of the biofilms could not be detected by taxonomic probes specific to the principal groups of Bacteria, which comprise a very broad range of physiologically quite different taxa.

In a further experiment, a mineral medium with addition of MCP was used to select bacteria capable of using MCP as the sole carbon source. The recovery from biofilm samples from both reactors was 4% at the highest total cell count. By probing the colonies growing on this medium with group-specific oligonucleotides, it was found that less than 20% of the cells were affiliated with the β-subclass and more than 80% with the γ-subclass of Proteobacteria. These isolates could be identified as members of the fluorescent pseudomonads by means of the specific oligonucleotide probe Ps56aX (Schleifer et al. 1992). Thus, the question arises whether the MCP degradation in the reactors was performed by a low number of very active bacteria belonging to the genus *Pseudomonas*, or whether these bacteria obtained a selective advantage by the composition of the culture medium.

Another process performed in the membrane biofilm reactors was nitrification. As the wastewater was characterized by a relatively low organic load (besides the MCP) and by an ammonia concentration between 15 and 30 mg N $l^{-1}$, and an efficient oxygen supply to the biofilm was allowed by the semipermeable membranes, the conditions in which nitrifiers can grow with less competition from heterotrophic, oxygen-consuming bacteria were realized. Contrary to the capacity to degrade chloroaromatics, only a distinct group of closely related Proteobacteria is distinguished by their capability to oxidize ammonia to nitrite and nitrite to nitrate. Therefore, more detailed information on the nitrifying capacity of the biofilms can be obtained by the use of genus-specific probes, i.e. a probe specific to *Nitrosomonas* (Neu23a) (Wagner et al. 1995). In the first months of reactor operation, less than 4% of the total cell count hybridized with the Neu23a probe, but in the following period it increased to 6–14% (corresponding to 10–20% of EUB338). Accordingly, a substantial part of the ammonia was oxidized (Figure 10.7). A high proportion of *Nitrosomonas* in the upper segment of the CFBR, detected with the specific oligonucleotide probe, was coincident with a rapid decrease of the $NH_4$ concentration (i.e. a high nitrification rate) in this part of the reactor. Thus, the performance of the biofilm reactors can obviously be related to data on the composition of the microbial communities with regard to nitrification.

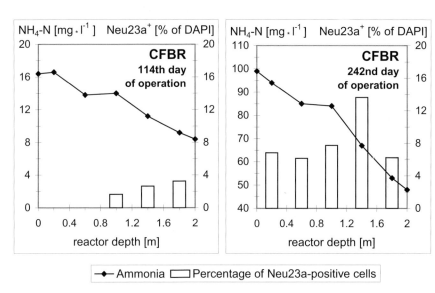

**Figure 10.7.** Comparison between the vertical profile of ammonia concentration and the longitudinal distribution of ammonia-oxidizing bacteria (as detected by the NEU23a oligonucleotide probe) in relation to the DAPI cell count in the CFBR in an earlier (left) and a later (right) operational period (4-monochlorophenol was added during the whole period).

## 10.3.1.2 Protozoan and metazoan communities of the biofilms

Biofilms are complex biotic systems in which food-web relationships are relevant. Though the biomass of the protozoa and metazoa is usually low, compared with the mass of bacteria (as Figure 10.5 also shows), they may significantly affect the stability of the ecosystem 'biofilm reactor' (Röske *et al.* 2001). The specific goals of studying the protozoan and metazoan communities were: (i) to relate species composition and spatial distribution to the reactor performance; and (ii) to obtain information about the microenvironments in the biofilms from the occurrence of indicator organisms.

In our experiments with addition of MCP to both reactors, a total of 51 protozoan species was found (Table 10.2). The SBBR had a lower diversity. Nearly 50%, 24 of 41 ciliate species, were present in both reactors. Four species were found in the SBBR, which were absent in the CFBR.

**Table 10.2.** Total number of species of the meiofauna in the membrane biofilm reactors.

| Group | CFBR | SBBR |
|---|---|---|
| Ciliates | 37 | 29 |
| Flagellates | >9 | >8 |
| Rotifers | 4 | 4 |
| Other invertebrates | >4 | >4 |

Comparing the composition of the protozoan and metazoan communities and their distribution along the reactor depth, significant effects of the different modes of operation on the biotic structure became evident (Figures 10.8 and 10.9). Operating a biofilm reactor with continuous flow (CFBR), i.e. steady-state conditions, resulted in a spatial succession towards a more complex community with several trophic levels, according to the longitudinal gradients of wastewater compounds. For example, the proportion of metazoa, especially rotifers, increased with the distance from the inlet, which coincided with an increase in dissolved oxygen and a decrease in the MCP concentration. Among the protozoa, the ciliates normally were dominant in the CFBR.

Owing to the discontinuous mode of operation, temporal shifts in the dominance of eukaryotes in the SBBR occurred, with alternating maxima of different groups of protozoa (Figure 10.8). For example, there were mass growths of amoebae followed by a maximum of the colorless flagellate *Peranema*. The high abundance of heterotrophic flagellates and of amoebae has to be considered as a phenomenon characteristic of the starting phase in the operation of wastewater treatment plants or of overloading (Curds 1982). In the case of the SBBR, the more unstable biotic structure may be attributed both to the discontinuous feeding pattern and to the hydraulic stress during the filling and draining period.

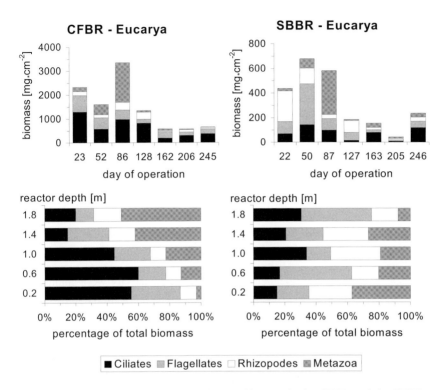

**Figure 10.8.** Above: composition of eukaryote biomass in the CFBR and the SBBR at different sampling dates during the operation with addition of 4-monochlorophenol (MCP) (averages of samples from five segments). Below: changes in percentage composition of eukaryote biomass across the longitudinal axis in both reactors (averages from seven sampling dates).

Though the protozoan and metazoan communities in the SBBR seemed to be more susceptible to disturbances, a high stability of effluent quality (also in case of shock loads) was attributed to this mode of operation (Wobus and Röske 2000). Protozoa and metazoa may exert influence on the performance of biofilm reactors in different ways. Considering the relevance of a predator–prey relationship, a vigorous grazing of these organisms may decrease the number of bacteria crucial to the transformation of wastewater compounds (as described for nitrifying biofilms by Lee and Welander (1994)). On the other hand, the quality of the effluent will be improved by the removal of suspended bacteria by the filtering activity, especially of peritrich ciliates (Curds 1982; Pauli *et al.* 2001).

Most of the invertebrates found in the reactors are burrowers that force their body through the interstices of the biofilm. In this study, a mass growth of oligochaetes (genus *Nais*) resulted in a critical reduction of active bacterial biomass in both reactors.

Among the eukaryotes, the ciliates were most interesting because of their high proportion. Furthermore, different species normally differ from one another in their environmental requirements and, in particular, type of feeding and food, to such an extent that they represent good bio-indicators (Foissner and Berger 1996). In the CFBR, a spatial (longitudinal) succession of ciliates has been observed, from the dominance of bacteria-feeders towards a more complex community with omnivorous species and predators (of protozoa, mainly other ciliates) like *Podophrya fixa* and *Prodiscophrya collini* in the uppermost segments (Figure 10.9). With the exception of the carnivorous *Acineria*, this group was much less represented in the SBBR. In this reactor, species that exclusively or preferably feed upon bacteria dominated the ciliate community, indicating a treatment system in the starting phase of operation or exposed to disturbances.

Species that feed on suspended bacteria by ciliary activity were abundant in both reactors, but the number of free-swimming species was much higher in the CFBR. However, stalked (i.e. sessile) members of this ecological group (*Carchesium*, *Vorticella*, *Epistylis*, *Opercularia*) were found with eight species and high numbers in the SBBR, two more than in the CFBR. Browsers of the order Hypotrichida mainly feed by grazing sessile bacteria on or in biofilms. Five species were abundant in the CFBR and only three in the SBBR, but two of these achieved very high numbers.

From the abundance of ciliates, which have high demands on the oxygen concentration, conclusions can be drawn about the oxygen supply in the biofilms. For example, the indicator organism *Aspidisca cicada* increased with the distance from the inlet of the CFBR. On the other hand, ciliates (like *Metopus*, *Trimyema*, and *Dexiotricha*), which tolerate strictly anaerobic conditions, were only present in the CFBR, especially in the inflow section. These species, which are typical inhabitants of anaerobic mud and overloaded and/or oxygen-deficient activated sludge (Foissner and Berger 1996), indicate an intermittent oxygen demand. In the SBBR, the presence of a sufficient oxygen supply in the whole reactor was supported by the dominance of the α-meso-saprobic *Vorticella convallaria*.

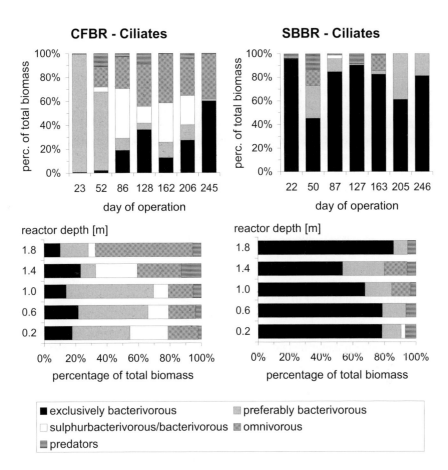

**Figure 10.9.** Above: temporal changes in percentage composition of ciliate biomass (by the type of feeding) in both reactors during operation with MCP addition (averages of samples from five segments). Below: percentage composition of the ciliate communities in different segments of the CFBR and the SBBR (averages from seven sampling dates).

## 10.3.2 Nitrifying biofilms in a fluidized bed reactor

### *10.3.2.1 Microbial composition of the biofilms*

Knowledge about microbial structure of nitrifying communities in wastewater biofilms is still limited. Microbial nitrification, a two-step process of the oxidation of ammonium ($NH_4^+$) to nitrate ($NO_3^-$) via nitrite ($NO_2^-$), is done by the chemolithoautotrophic ammonia-oxidizers and nitrite-oxidizers. By the utilization of a fluidized bed biofilm reactor, which provides a qualified operation for the treatment of nitrogen-containing wastewaters (Röske *et al.* 1995), a high concentration of nitrifiers may be achieved irrespective of their low growth rate and independent of the hydraulic loading. For an accurate recording of the process performance in wastewater treatment biofilms, it is necessary to analyze the microbial biofilm composition. Thus, the employment of FISH with 16S rRNA-targeted oligonucleotide probes seems to be promising. Previously developed probes for ammonia- and nitrite-oxidizing bacteria were used to describe nitrifying populations of activated sludge (Wagner *et al.* 1995; 1996) and biofilms (Mobarry *et al.* 1996; Schramm *et al.* 1996; Okabe *et al.* 1999), and provide thereby the microbial *in situ* characterization of our nitrifying biofilm reactor in space and time.

In this we measured the development of a microbial biofilm on small polypropylene tubules (5 mm length) as carrier material in a fluidized bed reactor during an experimental time of 10 months. The nitrification reactor was arranged as the final treatment step subsequent to the tanks for the elimination of organic carbon and phosphorus (Kloep *et al.* 2000). An $NH_4^+$-N elimination of more than 95% at influent concentrations of up to 80 mg $NH_4^+$-N/l was observed during a period with 75% carrier material (v/v), with an average surface loading of 0.632 g $NH_4^+$–N/$m_F^2$d ($m_F$: effective surface area of carrier) and a hydraulic retention time of 1.8 h.

Figure 10.10 shows the abundances of probe-positive cells relative to those visualized by DAPI staining versus time after hybridization of biofilm samples with oligonucleotide probes for Bacteria and group-specific probes to identify α-, β-, and γ-Proteobacteria, the *Cytophaga–Flavobacterium* group, and Gram-positive bacteria with a high G+C content of the DNA. As several bacteria showed growth in dense cell clusters, these samples could only be examined after some modification. The clusters could not be efficiently disintegrated by ultrasonication. Thus, the detection of bacteria grown in clusters was first done with exclusion of the clusters (Figure 10.10). In addition, the clusters were enumerated on the basis of their areal coverage per microscopic field of vision. This was only used as a relative method, with results shown in Figure 10.11.

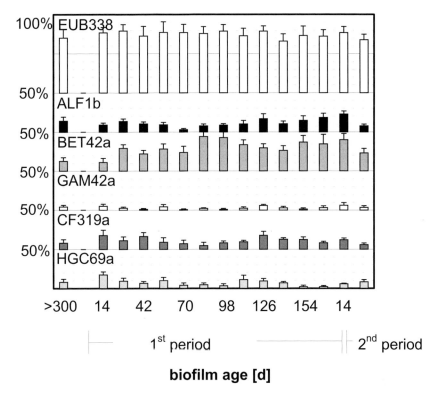

**Figure 10.10.** *In situ* probing of the nitrifying biofilm in two operational periods. Abundances of probe-positive cells are relative to those visualized by DAPI staining (error bars indicate standard deviations of 10 counts, respectively). The names of the group-specific probes may be found in Table 10.1. Biofilm age >300 belongs to a prior treatment period.

During the whole study, 66–79% of the total number of cells could be hybridized with the Bacteria-specific probe EUB338. The proportion of α-Proteobacteria ranged from 3% to 23% of the DAPI cell number, with an average of 10%. The proportion of cells affiliated to the *Cytophaga–Flavobacterium* group was about 10% of the DAPI number. The amount of Gram-positive bacteria with a high G+C content of the DNA detected with probe HGC69a was lower (<10%). With regard to the low amount below 5%, the γ-Proteobacteria were of small importance to the nitrifying biofilm community. This observation is consistent with previous investigations of wastewater communities (Wagner *et al.* 1993; Manz *et al.* 1994), which indicated the low proportion of the γ-Proteobacteria as well. The β-Proteobacteria achieved the highest proportion (11–44%) of the DAPI cell number. This amount, however, does not include the cells in the clusters detected with the probe for β-Proteobacteria. Thus, the real number, as the sum of the β-Proteobacteria shown in Figure 10.10 and the clusters hybridized with BET42a (Figure 10.11), was still higher (15–55%). During the study beginning in winter, only the proportion of the β-Proteobacteria demonstrated a slight increase towards the summer months (Figure 10.11).

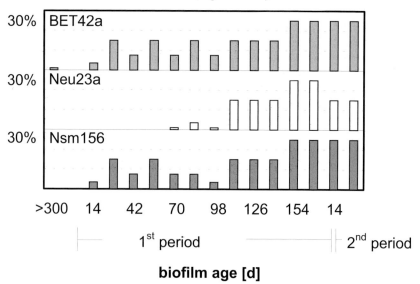

**Figure 10.11.** Percentage of nitrifiers in cell clusters (enumerated on the basis of their area coverage per microscopic field of vision) in relation to the total cell number after hybridization with two specific probes. The gene probe BET42a was included as a control surmising all cells hybridizing to the two probes specific for ammonia-oxidizing bacteria. BET42a also binds to non-nitrifying bacteria. Biofilm age >300 belongs to a prior treatment period.

The formation of clusters is a characteristic growth feature of ammonia-oxidizing bacteria, in particular, *Nitrosomonas* sp. (Wagner *et al.* 1995; Mobarry *et al.* 1996), and could also be confirmed by our observations. Most of the clusters were smaller than 10 μm; only a few clusters were larger than 15 μm in diameter. With the exception of two marine species of the genus *Nitrosococcus*, all ammonia-oxidizers belong to the β-Proteobacteria. In the present study, probe Nsm156 hybridized with the clusters over the whole experiment with a similar distribution as probe BET42a (Figure 10.11), suggesting that all bacteria in clusters are ammonia-oxidizing bacteria. Bacteria affiliated to probe Neu23a had been first detected at day 70 of the 1st operational period. After the first appearance, the amount of the clusters hybridized with Neu23a remained similar to that of the clusters hybridized with probe Nsm156. This points to a shift in the population structure of the ammonia-oxidizing bacteria. It is evident from Table 10.1 that both probes cover ammonia-oxidizing β-Proteobacteria, but with target sites in different regions of the 16S rRNA (Mobarry *et al.* 1996). Obviously, at around day 70 of the 1st experimental period, the ammonia-oxidizing community changed from a Nsm156$^+$/Neu23a$^-$ genotype to a Nsm156$^+$/Neu23a$^+$ genotype. Interestingly, this time was preceded by a sharp temperature increase, from about 11 °C to 15 °C, within one week.

Members of the bacteria affiliated to probe Nsv443 could not be detected. This minor importance of the other known ammonia-oxidizing genera (*Nitrosospira*, *Nitrosovibrio*, and *Nitrosolobus*) supports the suggestion that the ammonia-oxidizing community in the present biofilm was only dominated by different types of the genus *Nitrosomonas*. In contrast, the failure of hybridization signals of probes Nso190 and Nso1225 for *Nitrosomonas*. (Mobarry *et al.* 1996) in our study was unexpected, as all target organisms of both probes Neu23a and Nsm156, which hybridized cell clusters, are covered by these probes for ammonia-oxidizing β-Proteobacteria (see also Table 10.1). Obviously, the ammonia oxidizers in the investigated biofilm reactor were Nso190$^-$- and Nso1225$^-$-genotypes, but were probe Neu23a- and Nsm156-positive cells. Therefore, this biofilm consisted of hitherto not yet described bacteria of the genus *Nitrosomonas*, which could extend our knowledge about the phylogenetic relationship of ammonia-oxidizing microorganisms. Furthermore, the detection of clusters, firstly with bacteria of the same morphology and secondly exclusively hybridized as Neu23a$^+$- and Nsm156$^+$-cells, points to a strongly related ammonia-oxidizing community. For practical applications this suggests that the population was very sensitive to external factors.

Using probes NIT1, NIT2, and NIT3 for *Nitrobacter* sp. as nitrite-oxidizing bacteria, no hybridization signal could be observed in any of the samples. This is in accordance with several recent studies (Hovanec *et al.* 1998; Juretschko *et al.* 1998), which indicated a more general importance of *Nitrospira* spp. in nitrifying systems.

## 10.3.2.2 Properties of the carrying material and the biofilm

The polypropylene tubules offered a large specific surface per volume unit (1.41 $m^2/l$) because of their high coefficient of roughness (Figure 10.12 above). Thus, enough area should be available to the bacteria to grow to an optimum biomass concentration. Tubules consisted of two different habitats: the outer side with direct contact to the surrounding environment and the inner side with steep concentration gradients of dissolved oxygen and ammonia from the openings towards the central part of the lumen. Because of these contrasts, we assume different successions of the development of the biofilm structure. Additionally, we increased the concentration of the carrier material in the biofilm reactor, providing an increasing elimination of ammonia with higher microbial biomass (Kloep *et al.* 2000). In the experimental period with the highest concentration of the carrier material (75%), the balance between biofilm loss and growth processes on the outside of the carrier material was dominated by shear forces. This resulted in the formation of a thin and compact biofilm. In a period with a low amount of carrier material (10%) in the biofilm reactor, the collisions between the tubules obviously were not significant. Under these conditions a smooth and thick biofilm could be observed by means of a light microscope.

Unlike the outer surface of the tubules, the lumen was protected from shear forces. The microscopical examination revealed dense growth of stalked peritrich ciliates in the lumen (Figure 10.12 below). This indicates that the molecular diffusion of oxygen and ammonia to the microbial films in the lumen may have been forced through the 'bioturbation', i.e. the microcurrents generated by the peristome currents of the peritrichs. Regarding the size of the microbial cells in the biofilm and the thickness of the Prandtl boundary layer, these microcurrents may be relevant to the oxygen and ammonia flux and thus to the performance of the reactor. Owing to the very low intensity of shear forces, after several weeks the lumen was completely filled with a complex biofilm consisting of sessile and free-living ciliates, as well as some species of rotifers. Nematodes could only be observed during a period of high $NH_4^+$-loading and corresponding oxygen deficiency. Furthermore, filamentous bacteria dominated the architecture of the biofilm in the lumen. Unlike the outside of the tubules, the biofilm losses in the lumina of the carrier material may be attributed to grazing activities of protozoa and metazoa. The outer surface of the carrier material was free of these organisms. The different kinds of biofilm losses indicate that these processes are very relevant to the net growth of the biofilms.

Population structure and biofilm reactor performance 257

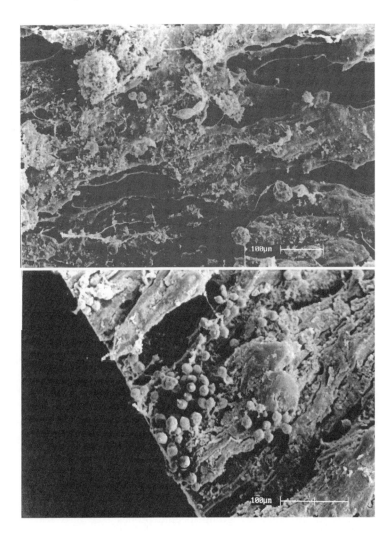

**Figure 10.12.** Scanning electron micrographs of polypropylene tubules used as carrier material in a fluidized bed biofilm reactor. Above: interior surface of a developing young biofilm; below: interior surface with protozoans.

## 10.4 CONCLUDING REMARKS

In the past decade, several new tools have become available for the examination of biofilm structure, i.e., the three-dimensional architecture of biofilms as well as their microbial composition. This leads to new ideas on biofilm structure and development. Regarding the performance of biofilm reactors in wastewater treatment, the main question is how data on biofilm structure can be related to the metabolic function of the biofilm.

In this contribution, the compositions of microbial communities of two laboratory-scale membrane biofilm reactors for the treatment of wastewater containing xenobiotics and of a fluidized-bed nitrification reactor were investigated. No significant differences between the two biofilm systems were observed by means of fluorescently labeled 16S rRNA-targeted oligonucleotide probes specific to the principal divisions of Bacteria. Although a high proportion of the microbial population could be related to these phylogenetic groups, no conclusions can be drawn as to specific activities of the biofilms.

If the desired metabolic capability is associated with only one distinct taxonomic group of bacteria, *in situ* probing at a lower taxonomic level can reveal more information. For example, ammonia oxidizers form such a closely related group of β-Proteobacteria for which a range of probes specific to different members of this group is available (Wagner *et al.* 1995; Mobarry *et al.* 1996). Consequently, a correlation between the proportion of detected cells and the nitrification of the membrane biofilm reactors could be observed by means of a oligonucleotide probe specific to members of the genus *Nitrosomonas* (Neu23a). Moreover, a seasonal shift in the population structure of the nitrifying biofilms in the fluidized bed reactor was detected by using two different probes covering different proportions of ammonia oxidizing bacteria. Therefore, it is possible to explain differences in the nitrification efficiency (owing to variations of one or more operating parameters) by such a succession of the predominant members of the nitrifying bacteria. It becomes possible to detect the potential for nitrification in a reactor subjected to low or fluctuating influent concentrations of ammonia by group-specific probes (see also Kloep *et al.* 2000).

On the elimination of xenobiotics, there have been only limited opportunities for FISH to identify the members of the microbial community responsible for the biodegradation. For example, the capacity to degrade chlorophenols is widespread among different groups of bacteria. Moreover, for many taxa (especially for those that are difficult to cultivate or are non-cultivable) the metabolic potential is unknown. Special degradative capabilities are also often encoded by broad-host-range plasmids and may only be found in some strains of a given taxon. Therefore, probes directed towards catabolic genes may help to overcome the limitations of the rRNA-directed, taxonomic probes (Ka et al. 1994). However, the convergent evolution of catabolic pathways, like the chlorocatechol pathway for the degradation of chloroaromatics, has to be taken into account if the degradative potential of microbial communities is examined (Schlömann 1994; Schlömann and Eulberg 1998).

Biochemical assays for the determination of the activity of enzymes involved in a specific metabolic process can be done as another way for the examination of a specific metabolic potential of a biofilm. As described elsewhere (Wobus et al. 2000), different proportions of specialized microorganisms at different reactor depths became evident by means of a catechol-1,2-dioxygenase assay.

The investigation of the protozoan and metazoan communities may be useful for an efficient control of biofilm processes for two reasons: (i) microbial activities that are crucial to the reactor performance may be negatively affected by the grazing activity of eukaryotes (Lee and Welander 1994); and (ii) information about the micro-environment in the biofilms may be gathered from the occurrence of indicator organisms (Pauli et al. 2001). In this study it was shown that different operating conditions (like continuous/discontinuous loading or shear stress) are reflected by the composition of the eukaryotic communities of the biofilms. However, no strong effects of the different colonizations on the reactor performance were detected. Nevertheless, the possible role of predators has to be considered if conversions performed by slow-growing bacteria are limited, though abiotic factors are optimal. The interactions between bacteria and eukaryotes may be of interest to optimize and to control the (active) biomass, i.e. the biofilm thickness, in biofilm reactors.

## 10.5 REFERENCES

Akkermans, A.D.L., VanElsas, J.D. and deBrujn, F.J. (1995) *Molecular Microbial Ecology Manual.* Kluwer Academic, Dordrecht, Boston, and London.

Amann, R.I., Krumholz, L. and Stahl, D.A. (1990) Fluorescent-oligonucleotide probing of whole cells for determinative, phylogenetic, and environmental studies in microbiology. *J. Bacteriol.* **172**, 762–770.

Amann, R.I., Stromley, J., Devereux, R., Key, R. and Stahl, D.A. (1992) Molecular and microscopic identification of sulfate-reducing bacteria in multispecies biofilms. *Appl. Environ. Microbiol.* **58**, 614–623.

Amann, R.I., Ludwig, W. and Schleifer, K.-H. (1995) Phylogenetic identification and *in situ* detection of individual microbial cells without cultivation. *Microbiol. Rev.* **59**, 143–169.

Commandeur, L.C.M. and Parsons, J.R. (1994) Biodegradation of halogenated aromatic compounds. In *Biochemistry of Microbial Degradation* (ed. C. Ratledge), pp. 423–458, Kluwer Academic, Dordrecht.

Curds, C.R. (1982) The ecology and role of protozoa in aerobic sewage treatment processes. *Annu. Rev. Microbiol.* **36**, 27–46.

de Beer, D. and Stoodley, P. (1995) Relation between the structure of an aerobic biofilm and transport phenomena. *Wat. Sci. Tech.* **32**(8), 11–18.

Foissner, W., Berger, H., Blatterer, H. and Kohmann, F. (1995) *Taxonomische und ökologische Revision der Ciliaten des Saprobiensystems*, Band IV: *Gymnostomatea, Loxodes, Suctoria, Inform.* Ber. Bayer. Landesamt Wasserwirtschaft, München.

Foissner, W., and Berger, H. (1996) A user-friendly guide to the ciliates (Protozoa, Ciliophora) commonly used by hydrobiologists as bioindicators in rivers, lakes, and waste waters, with notes on their ecology. *Freshwat. Biol.* **35**, 375–482.

Häggblom, M.M. (1990) Mechanisms of bacterial degradation and transformation of chlorinated monoaromatic compunds. *J. Basic Microbiol.* **30**, 115–141.

Head, I.M., Saunders, J.R. and Pickup, R.W. (1998) Microbial evolution, diversity, and ecology: A decade of ribosomal RNA analysis of uncultivated microorganisms. *Microb. Ecol.* **35**, 1–21.

Horn, H. and Hempel, D.C. (1998) Modeling mass transfer and substrate utilization in the boundary layer of biofilm systems. *Wat. Sci. Tech.* **37**(4–5), 139–147.

Hovanec, T.A., L.T. Taylor, L.T., Blakis, A. and Delong, E.F. (1998) *Nitrospira*-like bacteria associated with nitrite oxidation in freshwater aquaria. *Appl. Environ. Microbiol.* **64**, 258–264.

Juretschko, S., Timmermann, G., Schmid, M., Schleifer, K.H., Pommerening-Roser, A., Koops, H.P. and Wagner, M. (1998) Combined molecular and conventional analyses of nitrifying bacterium diversity in activated sludge: *Nitrosococcus mobilis* and *Nitrospira*- like bacteria as dominant populations. *Appl. Environ. Microbiol.* **64** (8), 3042–3051.

Ka, J.O., Holben, W.E. and Tiedje, J.M. (1994) Use of gene probes to aid in recovery and identification of functionally dominant 2,4-dichlorophenoxyacetic acid-degrading populations in soil. *Appl. Environ. Microbiol.* **60**, 1116–1120.

Kloep, F., Röske, I. and Neu, T.R. (2000) Performance and microbial structure of a nitrifying fluidized-bed reactor. *Wat. Res.* **34**(1), 311–319.

Knackmuss, H.-J. and Hellwig, M. (1978) Utilization and cooxidation of chlorinated phenols by *Pseudomonas* sp. B13. *Arch. Microbiol.* **117**, 1–7.

Kniebusch, M.M., Wilderer, P.A. and Behling, R.D. (1990) Immobilization of cells at gas permeable membranes. In *Physiology of Immobilized Cells* (ed. J.A.M. De Bont, J. Visser, B. Mattiasson and J. Tramper), pp. 149–160. Elsevier Science, Amsterdam.

Lawrence, J.R., Korber, D.R., Hoyle, B.D., Costerton, J.W. and Caldwell, D.E. (1991) Optical sectioning of microbial biofilms. *J. Bacteriol.* **173**, 6558–6567.

Lawrence, J.R., Wolfaardt, G.M. and Neu, T.R. (1998) The study of biofilms using confocal laser scanning microscopy. In *Digital Image Analysis of Microbes: Imaging, Morphometry, Fluorometry, and Motility Techniques and Applications* (ed. M.H.F. Wilkinson and F. Schut), pp. 431–465, John Wiley, Chichester and Weinheim.

Lee, N.M. and Welander, T. (1994) Influence of predators on nitrification in aerobic biofilm processes. *Wat. Sci. Tech.* **29**(7), 355–363.

Lewandowski, Z., Stoodley, P. and Altobelli, S. (1995) Experimental and conceptual studies on mass transport in biofilms. *Wat. Sci. Tech.* **31**(1), 153–162.

Manz, W., Amann, R.I., Ludwig, W., Wagner, M. and Schleifer, K.-H. (1992) Phylogenetic oligodeoxynucleotide probes for the major subclasses of proteobacteria: Problems and solutions. *System. Appl. Microbiol.* **15**, 593–600.

Manz, W., Wagner, M., Amann, R.I. and Schleifer, K.-H. (1994) *In situ* characterization of the microbial consortia active in two wastewater treatment plants. *Wat. Res.* **28**, 1715–1723.

Manz, W., Wendt-Potthoff, K., Neu, T.R., Szewzyk, U. and Lawrence, J.R. (1999) Phylogenetic composition, spatial structure, and dynamics of lotic bacterial biofilms investigated by fluorescent *in situ* hybridization and confocal laser scanning microscopy. *Microb. Ecol.* **37**, 225–237.

Massol-Deya, A.A., Whallon, J., Hickey, R.F. and Tiedje, J.M. (1995), Channel structure in aerobic biofilms of fixed-film reactors treating contaminated groundwater. *Appl. Environ. Microbiol.* **61**, 769–777.

Mobarry, B.K., Wagner, M., Urbain, V., Rittmann, B.E. and Stahl, D.A. (1996) Phylogenetic probes for analyzing abundance and spatial organization of nitrifying bacteria. *Appl. Environ. Microbiol.* **62**, 2156–2162.

Møller, S., Pedersen, A.R., Poulsen,L.K., Arvin, E., and Molin,S. (1996) Activity and three-dimensional distribution of toluene-degrading *Pseudomonas putida* in a multispecies biofilm assessed by quantitative *in situ* hybridization and scanning confocal laser microscopy. *Appl. Environ. Microbiol.* **62**, 4632–4640.

Muyzer, G. and Ramsing, N.B. (1995) Molecular methods to study the organization of microbial communities. *Wat. Sci. Tech.* **32**(8), 1–9.

Neef, A., Zaglauer, A., Meier, H., Amann, R.I., Lemmer, H. and Schleifer, K.-H. (1996) Population analysis in a denitrifying sand filter: Conventional and *in situ* identification of *Paracoccus* spp. in methanol-fed biofilms. *Appl. Environ. Microbiol.* **62**, 4329–4339.

Neu, T.R., and Lawrence, J.R. (1997) Development and structure of microbial biofilms in river water studied by confocal laser scanning microscopy. *FEMS Microbiol. Ecol.* **24**, 11–25.

Okabe, S., Satoh, H. and Watanabe, Y. (1999) *In situ* analysis of nitrifying biofilms as determined by *in situ* hybridization and the use of microelectrodes. *Appl. Environ. Microbiol.* **65**, 3182–3191.

Olsen, G.J., Lane, D.J., Giovannoni, S.J., Pace, N.R. and Stahl, D.A. (1986) Microbial ecology and evolution: a ribosomal RNA approach. *Ann. Rev. Microbiol.* **40**, 337–365.

Pauli, W., Jax, K. and Berger, S. (2001) Protozoa in wastewater treatment: function and importance. In *The Handbook of Environmental Chemistry* (ed. B. Beek), vol. 2, part K: *Biodegradation and Persistance*, pp. 203–252, Springer-Verlag, Berlin and Heidelberg.

Pickup, R.W. and Saunders, J.R. (1996) *Molecular Approaches to Environmental Microbiology*. Ellis Horwood, London.

Poulsen, L.K., Ballard, G. and Stahl, D.A. (1993) Use of fluorescence *in situ* hybridization for measuring the activity of single cells in young and established biofilms. *Appl. Environ. Microbiol.* **59**, 1354–1360.

Richards, S.R. and Turner, R.J. (1984) A comparative study of techniques for the examination of biofilms by scanning electron microscopy. *Wat. Res.* **18**, 767–773.

Roller, C.M., Wagner, M., Amann, R.I., Ludwig, W. and Schleifer, K.-H. (1994) *In situ* probing of gram-positive bacteria with high DNA G+C content by using 23S rRNA-targeted oligonucleotides. *Microbiology* **140**, 2849–2858.

Röske, I., Uhlmann, D. and Kermer, K. (1995) Demands on the performance of domestic wastewater treatment plants in view of the quality of river waters. (In German.) *Korrespond. Abwass.* **42**, 356–364.

Röske, I., Wobus, A. and Röske, K. (2001) Biotic diversity and ecological stability of biofilms in reactors for advanced waste water treatment. *Verh. Internat. Verein. Limnol.* **27**, 3248–3253.

Schleifer, K.-H., Amann, R.I., Ludwig, W., Rothemund, C., Springer, N. and Dorn, S. (1992) Nucleic acid probes for the identification and *in situ* detection of *Pseudomonas*. In *Pseudomonas: Molecular Biology and Biotechnology*, (ed. E. Galli, S. Silver and B. Withold), pp. 127–134. American Society for Microbiology, Washington, D.C.

Schlömann, M. (1994) Evolution of chlorocatechol catabolic pathways. Conclusions to be drawn from comparisons of lactone hydrolases. *Biodegradation* **5**, 301–321.

Schlömann, M. and Eulberg, D. (1998) Convergent evolution of chlorocatechol catabolism - a problem for the design of functional gene probes. In *Microbiology of Polluted Aquatic Ecosystems* (ed. P.M. Becker), UFZ Report 10/98, pp. 109–117. UFZ Centre for Environmental Research, Leipzig-Halle.

Schramm, A., Larsen, L.H., Revsbech, N.P., Ramsing, N.B., Amann, R.I. and Schleifer, K.-H. (1996) Structure and function of a nitrifying biofilm as determined by *in situ* hybridization and the use of microelectrodes. *Appl. Environ. Microbiol.* **62**, 4641–4647.

Schramm, A., de Beer, D., Wagner, M. and Amann, R.I. (1998) Identification and activity *in situ* of *Nitrosospira* and *Nitrospira* spp. as dominant populations in a nitrifying fluidized bed reactor. *Appl. Env. Microbiol.* **64**, 3480–3485.

Siegrist, H. and Gujer, W. (1985) Mass transfer mechanisms in a heterotrophic biofilm. *Wat. Res.* **19**, 1369–1378.

Stewart, P.S., Murga, R., Srinivasan, R. and de Beer, D. (1995) Biofilm structural heterogeneity visualized by three microscopic methods. *Wat. Res.* **29**, 2006–2009.

Szwerinski, H., Gaiser, S. and Bardtke, D. (1985) Immunofluorescence for the quantitative determination of nitrifying bacteria: interference of the test in biofilm reactors. *Appl. Microbiol. Biotechnol.* **21**, 125–128.

Wagner, M., Amann, R.I., Lemmer, H. and Schleifer, K.-H. (1993) Probing activated sludge with oligonucleotides specific for proteobacteria: inadequacy of culture-dependent methods for describing microbial community structure. *Appl. Environ. Microbiol.* **59**, 1520–1525.

Wagner, M., Erhart, R., Manz, W., Amann, R.I., Lemmer, H., Wedi, D. and Schleifer, K.-H. (1994) Development of an rRNA-targeted oligonucleoide probe specific for the genus *Acinetobacter* and its application for *in situ* monitoring in activated sludge. *Appl. Environ. Microbiol.* **60**, 792–800.

Wagner, M., Rath, G., Amann, R.I., Koops, H.-P. and Schleifer, K.-H. (1995) *In situ* identification of ammonia-oxidizing bacteria. *System. Appl. Microbiol.* **18**, 251–264.

Wagner, M., Rath, G., Koops, H.-P., Flood, J. and Amann, R.I. (1996) *In situ* analysis of nitrifying bacteria in sewage treatment plants. *Wat. Sci. Tech.* **34**(1–2), 237–244.

Wilderer, P.A. (1992) Sequencing batch biofilm reactor technology. In *Harnessing Biotechnology for the 21st Century* (ed. M.R. Ladisch and A. Bose), pp. 475–479, American Chemical Society, Washington, D.C..

Wobus, A. and Röske, I. (2000) Reactors with membrane-grown biofilms: their capacity to cope with fluctuating inflow conditions and with shock loads of xenobiotics. *Wat. Res.* **34**, 279–287.
Wobus, A., Ulrich, S. and Röske, I. (1995) Degradation of chlorophenols by biofilms on semi-permeable membranes in two types of fixed bed reactors. *Wat. Sci. Tech.* **32**(8), 205–212.
Wobus, A., Röske, K. and Röske, I. (2000) Investigation of spatial and temporal gradients in fixed-bed biofilm reactors for wastewater treatment. In *Biofilms: Investigative Methods and Applications* (ed. H.-C. Flemming, U. Szewzyk, T. Griebe), pp. 165–194. Technomic Publishing Co., Lancaster, PA.
Woese, C.R. (1987) Bacterial evolution. *Microbiol. Rev.* **51**, 221–271.
Wolfaardt, G.M., Lawrence, J.R., Robarts, R.D., Caldwell, S.J. and Caldwell, D.E. (1994) Multicellular organization in a degradative biofilm community. *Appl. Environ. Microbiol.* **60**, 443–446.

# 11

# Detachment: an often-overlooked phenomenon in biofilm research and modeling

*Eberhard Morgenroth*

## 11.1 INTRODUCTION

Detachment is the removal of small or larger parts of biomass from a biofilm. In the operation of biofilm reactors, detachment is required to prevent clogging of the reactor due to excessive biomass growth. However, the significance and the mechanisms of detachment vary for different reactor configurations. Trickling filters have a large overall porosity and are less sensitive to biomass accumulation. But new, fine-grained material in submerged biofilters offers a large surface area with a small porosity, and for those systems efficient removal of excess biomass through backwashing has become an important operational

© 2003 IWA Publishing. *Biofilms in wastewater treatment*. Edited by S. Wuertz, P.L. Bishop and P.A. Wilderer. ISBN: 1 84339 007 8.

parameter (Lazarova and Manem 2000). In water distribution systems, biofilm growth should be minimized and efficient detachment is required during cleaning. However, during operation the contamination of water through the detachment of large biofilm particles should be avoided. Thus, detachment rates and detachment dynamics are important parameters for the operation of water and wastewater systems.

The purpose of this chapter is to look beyond the influence of detachment on overall biofilm accumulation and clogging of reactors. A theoretical background and two case studies will be presented on how detachment has a significant influence on the microbial ecology of biofilms and on the overall performance of biofilm reactors. Section 11.2 describes detachment mechanisms and how they are implemented in current mathematical models. Section 11.3 then lists two examples of the influence of detachment on the competition between heterotrophic and autotrophic bacteria in a biofilm with dynamic detachment (section 11.3.1) and the influence of dynamic detachment on enhanced biological phosphorus removal (EBPR) in biofilm reactors (section 11.3.2).

## 11.2 DETACHMENT MECHANISMS

### 11.2.1 Categories

Under steady-state conditions, biofilm growth is balanced by detachment from the biofilm. Different processes are responsible for detachment of biomass from the biofilm, and four categories of detachment processes can be distinguished (Bryers 1988): (1) abrasion; (2) erosion; (3) sloughing; (4) predator grazing. Abrasion and erosion refer to the removal of small groups of cells from the surface of the biofilm. Those detachment processes are differentiated by their mechanism. Erosion is caused by forces resulting from the moving fluid in contact with the biofilm surface, while abrasion is caused by the collision of particles, e.g., during backwashing. Sloughing, in contrast, refers to the detachment of relatively large particles whose characteristic size is comparable to or greater than the thickness of the biofilm itself. In Figure 11.1, a simplified schematic of different detachment processes is shown. During sloughing, a fraction of the biofilm is removed down to the substratum, but detachment is often not effective over the entire surface of the biofilm. Erosion and abrasion are assumed to be effective over the entire surface of the biofilm. However, bacteria are not removed from the base of the biofilm through erosion or abrasion.

The detachment of biofilm parts occurs when external forces exceed the internal strength within the biofilm. For example, during the backwashing of submerged biofilm reactors, external forces are intentionally increased by pumping air and water into the reactor at high flow rates to fluidize the filter bed. During fluidization, particles collide resulting in a scouring of the filter media, detachment of biofilms through abrasion and a washout of detached biofilm. On the other hand, detachment can result from a decreasing internal strength of the biofilm. For example, in biofilm reactors operated for denitrification, $N_2$ gas bubbles may form within the biofilm, resulting in sloughing events (Harremoës et al. 1980). With an increased biofilm thickness in trickling filters, the biofilm structure becomes unstable, resulting in the sloughing of parts of the biofilm. However, even though mechanistic explanations can be provided for some of the observed phenomena of detachment, others cannot be explained well. For example, no sufficient explanation can be given for major sloughing events in trickling filters that occur seasonally. Besides mechanistic explanations based on external forces and internal processes influencing biofilm strength, other researchers have suggested that cell-to-cell communication could be used to explain certain selective detachment patterns (Davies et al. 1998). Selective detachment could be an explanation for the maintenance of an internal porosity within a biofilm that increases the overall mass transfer into the biofilm by allowing advective transport. Biofilm detachment is still an emerging research field and actual mechanisms controlling detachment are far from being well understood (Rittmann and Laspidou 2002).

The overall process of detachment will be a combination of physical, biological, and chemical processes (Rittmann 1989; Bryers 2000). However, depending on the type of operation, one detachment mechanism may be dominant. In submerged biofilters, backwashing is regularly applied to remove excess biomass to prevent clogging. Abrasion is likely to be the dominant detachment process in those types of biofilm reactor. On the other hand, sloughing (and maybe erosion) will be the dominant detachment process(es) in trickling filters or rotating biological contactors. Thus, when applying mathematical models or when doing laboratory experiments, care should be taken to implement the appropriate mode of detachment of the biofilm system of interest.

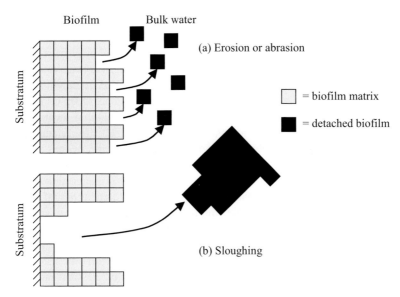

**Figure 11.1.** (a) Erosion and abrasion are assumed to be continuous processes that remove small biofilm parts evenly from the surface of the biofilm. (b) Sloughing is a local event that removes large pieces of biofilm and biofilm is often removed all the way to the substratum.

## 11.2.2 Experimental determination

Experiments to quantify detachment are usually based on suspended solids concentrations measured in the effluent of a biofilm reactor. In most cases, these data only allow the determination of average detachment rates without spatial or temporal resolution (Trulear and Characklis 1982; Chang and Rittmann 1988; Peyton and Characklis 1993; Peyton *et al.* 1995). For example, Peyton *et al.* (1995) measured the mass flux of detached biofilm in the effluent of a reactor with controlled shear stress (using a rotating annular reactor). For the calculation of the average detachment rate, it was assumed that the area where detachment occurred was equal to the surface area of the substratum. Then the rate of biomass detachment ($r_{d,M}$ with units of M L$^{-2}$ T$^{-1}$) can be calculated as:

$$r_{d,M} = \frac{\text{mass of biofilm detached per unit time}}{\text{surface area of substratum}} \qquad (11.1)$$

However, even though the overall rate of detachment, $r_{d,M}$, may be constant, local detachment rates may very well fluctuate with time. For example, van Benthum *et al.* (1995) suggested that biofilm formation in an airlift reactor is determined by a repeated formation of valleys in the biofilm architecture caused by local detachment. They stated that biofilms are dynamic structures, in which formation of cracks and fissures is continuously taking place. Open crevices are rapidly filled with fresh biomass and new crevices form at another location.

In section 11.3, it will be shown that those local fluctuations can have a significant influence on population dynamics and on overall process performance. Although the local distribution of detachment is generally assumed to be a random process, Stewart *et al.* (1997) suggested different detachment rates for different bacterial cells. Thus, different types of bacteria may produce an increased adhesive strength resulting in a selective advantage compared to other bacteria that are easier to remove by detachment.

For the calculation of average detachment rates (Equation 11.1), the assumption of using the area of the substratum as an approximation for the area where detachment is taking place must be critically evaluated. Currently, there is only very limited information available on how detachment occurs in heterogeneous biofilm architectures. The actual surface area of the heterogeneous biofilm contour can be significantly larger, compared with the area of the substratum, and therefore average detachment rates could overestimate local detachment rates. On the other hand, in these heterogeneous architectures, local detachment may not be randomly distributed over the entire biofilm contour, but detachment may be dominant, for example, at the peaks within the biofilm.

In conclusion, it can be stated that detachment mechanisms in biofilms have to be further investigated. There are some quantitative investigations of the overall detachment rate. However, very little information is available on local detachment rates and dynamics, even though it will be shown below that dynamics of local detachment are an important factor influencing microbial ecology in biofilms.

## 11.2.3 Mathematical description

When developing mathematical biofilm models, one is faced with the dilemma of having to include a quantitative description of detachment regardless of the currently rather limited understanding of the process. As will be shown below, the rate and the category of detachment can have a significant influence on the performance of biofilm reactors. When developing mathematical biofilm models, a range of different approaches to include detachment have been taken, which can be summarized considering the work of Wanner and his co-workers

(Wanner and Gujer 1984, 1986; Fruhen *et al.* 1991; Wanner *et al.* 1994; Wanner and Reichert 1996).

Wanner and Gujer (1984) and Wanner and Reichert (1996) modeled carbon oxidation and nitrification. They assumed a constant biofilm thickness, and thus the rate of detachment was equal to the rate of growth. Wanner and Gujer (1986), however, assumed a detachment rate that was a function of biofilm thickness and the total biomass within the biofilm. Wanner *et al.* (1994) modeled a biofilm system with a variable detachment rate. The value of the detachment rate was adjusted to match simulation results with experimental data. Fruhen *et al.* (1991) did not include detachment at all in their mathematical model for organic carbon oxidation and nitrification. In the time frame of the simulations, detachment was assumed to be negligible. The mathematical model was evaluated only for a period of 22 days, where net biofilm growth occurred before steady state.

The inherent problems of detachment modeling, lack of understanding of the fundamental mechanisms of detachment and inability to predict the location of detachment were clearly stated by Kissel *et al.* (1984). Their mathematical model did not include detachment, although these processes (loss of mass by periodic sloughing or continuous shearing) were considered to be very significant under some conditions. However, detachment processes were still poorly understood. Some information on detached masses of solids was available, but the location within a film at which detachment would occur could not be predicted. The location of detachment is important in the case of sloughing because the composition of the film remaining after sloughing may be very different from that originally present.

Summarizing different approaches to include detachment in a mathematical model, it can be seen that numerous expressions have been proposed for specific experimental applications (Table 11.1). Reported detachment rate equations can be categorized according to the major factor controlling detachment: biofilm thickness, shear, growth rate or substrate utilization rate, backwashing. Many models do not specify a mechanism of detachment and assume no detachment or a constant biofilm thickness. Although each has given reasonable results for the specific experimental condition, none are general enough to extrapolate to other environmental conditions. The abundance of detachment rate equations partly reflects the failure of any one expression to model the detachment rate over a wide range of conditions (Peyton and Characklis 1993). Most detachment rates reported in Table 11.1 apply to a continuous operation of a biofilm system, and reported detachment processes are also considered to be continuous processes. For biofilm reactors with distinctly different operating conditions over time, such as most biofilters that use backwashing as the main mechanism for biofilm removal, continuous models for detachment cannot be used. Picioreanu et al. (2001) developed a two-dimensional mathematical model where detachment occurred if local stress exceeded the internal biofilm strength. To describe discrete detachment events, Morgenroth and Wilderer (1999) and Rittmann et al. (2002) used a backwashing mechanism where all biomass above a defined base thickness of the biofilm was removed.

**Table 11.1.** Detachment rate expressions (modified from Peyton and Characklis (1993) and Tijhuis et al. (1995))[1]

| Mechanism of detachment related to … | Reported detachment rate expression, $r_{d,M}$ [M L$^{-2}$ T$^{-1}$] | Reference |
|---|---|---|
| None specified | 0 | Kissel et al. 1984; Fruhen et al. 1991 |
|  | Constant biofilm thickness | Wanner and Gujer 1984 |
| Biofilm thickness | $k_d (\rho_F L_F)^2$ | Trulear and Characklis 1982; Bryers 1984 |
|  | $k_d \rho_F L_F^2$ | Wanner and Gujer 1986 |
|  | $k_d \rho_F L_F$ | Kreikenbohm and Stephan 1985; Chang and Rittmann 1987; Rittmann 1989 |
| Shear | $k_d \rho_F \tau$ | Bakke et al. 1984 |
|  | $k_d \rho_F L_F \tau^{0.58}$ | Rittmann 1982 |
| Growth rate or substrate utilization rate | $L_F (k_d' + k_d'' \mu)$ | Speitel and DiGiano 1987 |
|  | $k_d \cdot r_S \cdot L_F$ | Robinson et al. 1984; Peyton and Characklis 1993; Tijhuis et al. 1995 |
| Backwashing down to a predefined base thickness | $k'_d L_F$ normal operation | Morgenroth and Wilderer (1999); Rittmann et al. 2002 |
|  | $k''_d (L_F - L_{\text{base thickness}})$ backwashing |  |

[1] Explanation of symbols: $k_d$, $k_d'$, $k_d''$ = detachment rate coefficients, $\rho_F$ = biofilm volumetric mass density [M L$^{-3}$], $L_F$ = biofilm thickness [L], $L_{\text{base thickness}}$ = predefined biofilm thickness after backwashing [L], $\mu$ = specific growth rate [T$^{-1}$], $r_S$ = substrate utilization rate [M L$^{-2}$ T$^{-1}$], $\tau$ = shear stress [M L$^{-1}$ T$^{-2}$].

## 11.2.4 Location of detachment

The detachment expressions in the previous section provide rate information representing averaged values over the surface area of the biofilm and over time. When applying these detachment rate expressions in a mathematical model, the location and dynamics of detachment need to be specified. Basically, there are three different approaches to include detachment in mathematical models that differ with respect to location of detachment and detachment dynamics (Figure 11.2). Similarities and differences between these three approaches are discussed.

**Surface detachment ($r_{d,S}$)** assumes that detachment is active only at the outer surface of the biofilm (Figure 11.2a). Bacterial growth within the biofilm results in a volume expansion. The change of the overall biofilm thickness ($L_F$) results from balancing volume expansion with surface detachment

$$\frac{dL_F}{dT} = u_F(z = L_F) - r_{d,S} \qquad (11.2)$$

where $u_F(z = L_F)$ is the advective velocity of the biofilm matrix caused by growth and volume expansion at the surface of the biofilm (L T$^{-1}$) and $r_{d,S}$ is the rate of surface detachment (L T$^{-1}$). Surface detachment was used in the studies of Wanner and co-workers that were described in the previous section, and surface detachment is implemented in the computer program AQUASIM (Wanner and Reichert 1996).

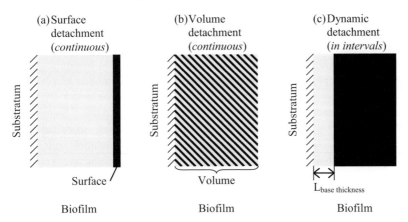

**Figure 11.2.** Different modes of detachment where the location of detachment is shown shaded. (a) Surface detachment is active only at the surface of the biofilm; (b) volume detachment is modeled as a continuous decay process that is active over the entire thickness of the biofilm; (c) dynamic detachment is modeled as a dynamic process where all the biomass above a defined base thickness ($L_{base}$ thickness) is removed in intervals.

**Volume detachment** ($r_{d,V}$) assumes that the process of detachment is distributed over the entire thickness of the biofilm (Figure 11.2b). Essentially, detachment is modeled by introducing an additional decay process into the rate expressions. The change of the overall biofilm thickness can be calculated from the volume detachment rate as

$$\frac{dL_F}{dT} = u_F(z = L_F) - r_{d,V} \cdot L_F \tag{11.3}$$

where $u_F(z = L_F)$ is again the advective velocity of the biofilm matrix caused by growth (but *not* taking into account detachment) and $r_{d,V}$ is the rate of volumetric detachment ($T^{-1}$). Volume detachment has been used in some early biofilm models (Saez and Rittmann 1992) and is also the basis for formulating the pseudo-analytical solution of a biofilm model (Rittmann and McCarty 2001).

**Dynamic detachment** differs from both surface and volume detachment in that detachment is modeled not as a continuous process but as discrete events occurring in certain intervals (Figure 11.2c). The resulting change of overall biofilm thickness can then be calculated as

$$\frac{dL_F}{dT} = u_F(z = L_F) - \begin{cases} 0 & \text{no detachment} \\ r_{d,S} & \text{during sloughing event} \end{cases} \tag{11.4}$$

where $r_{d,S}$ can be defined so that all biofilm above a pre-defined base thickness is removed (see Table 11.1 and Morgenroth and Wilderer (1999) for details).

This dynamic detachment can be used to model backwashing events where a significant amount of biofilm is detached during backwashing, while detachment is considered to be negligible during normal operation. This approach of dynamic detachment has been used by Morgenroth and Wilderer (1999) and Morgenroth and Wilderer (2000) to evaluate competition in biofilms. A similar approach to model dynamic detachment was used by Rittmann *et al.* (2002).

Whether or not these different approaches to model detachment affect model predictions depends on the application. If the model assumes a homogeneous distribution of bacteria over the thickness of the biofilm, then surface and volume detachment will yield identical results and detachment rates can be related to one other and to the experimentally determined detachment rate ($r_{d,M}$).

$$r_{d,S} = r_{d,V} \cdot L_F \tag{11.5}$$

$$r_{d,S} = \frac{r_{d,M}}{\rho_F} = \frac{\text{mass of biofilm detached per unit area per unit time}}{\text{density of biofilm}} \tag{11.6}$$

$$r_{d,V} = \frac{r_{d,M}}{\rho_F \cdot L_F} = \frac{\text{mass of biofilm detached per unit area per unit time}}{\text{mass of biofilm per unit area}} \tag{11.7}$$

However, for simulations describing competition between different groups of bacteria in a heterogeneous mixed culture biofilm, the type of detachment has a significant influence on model predictions (Morgenroth and Wilderer 2000). Assuming surface detachment, slower growing bacteria preferentially grow towards the base of the biofilm, because they are protected from detachment.

In Figure 11.3, advective velocities over the thickness of a biofilm are shown for two cases assuming surface or volume detachment, respectively. Velocity profiles shown were calculated using AQUASIM (Wanner and Reichert 1996). For the case of surface detachment, volume expansion is the result of bacterial growth, whereas decay of bacteria result in volume reduction. From the velocity profiles it can be seen that at every location, growth exceeds decay resulting in advective velocities away from the substratum. For example, the volume expansion of bacteria in layers 1–3 causes the bacteria in layer 4 to be pushed away from the substratum. For the case of volume detachment, volume reduction is also the result of volume-averaged detachment. It can be seen that for the case of volume detachment, the advective velocity is directed towards the substratum. Towards the base of the biofilm, decay and volume detachment dominate over growth resulting in an overall volume reduction. However, towards the surface of the biofilm, growth exceeds decay and detachment, and the overall thickness of the biofilm remains constant (i.e., the velocity at the surface of the biofilm is zero). In the given example, in layers 1–3 volume detachment exceeds growth resulting in a volume reduction and an advective velocity towards the substratum. It needs to be further evaluated how these different approaches of surface and volume detachment influence the predictions of mathematical biofilm models.

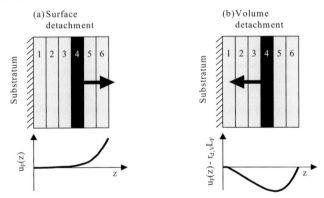

**Figure 11.3.** (a) Surface and (b) volume detachment result in different predictions of the advective velocity within the biofilm ($u_F(z)$).

## 11.2.5 Residence time distribution

Competition of bacteria within a biofilm will depend on the competition for both substrate and space. As a result, different types of organism will find their ecological niche at different locations within the biofilm. Substrate availability within the biofilm is determined by the balance between mass transport into the biofilm and substrate utilization rates. In many cases, limited mass transport will result in substrate limitations towards the base of the biofilm. Competition for space within the biofilm is governed by detachment processes that are balancing growth within the biofilm. Organisms growing at the surface of the biofilm are more likely to be removed by detachment and slower growing bacteria often find their ecological niche at the base of the biofilm where they are better protected from detachment. As a tool to discuss competition for space within a biofilm, a methodology to calculate residence time distributions over the thickness of biofilm is developed for the cases of surface and dynamic detachment.

Residence times within the biofilm will depend on surface detachment rates and on advective transport of bacteria within the biofilm. If detachment takes place only at the surface of the biofilm, it cannot be concluded that bacteria or particles at the base of the biofilm are never removed. Growth of bacteria at the base of the biofilm will lead to an increase in volume and a subsequent displacement of the bacteria above. Thus, depending on the net growth rate, bacteria are consecutively transported from the base of the biofilm towards the surface where abrasion or erosion can occur (Figure 11.3a). Detachment of a bacterial cell and of particulate matter from a biofilm depends on its specific location within the biofilm. As a result, the cell retention time is dependent on the location within the biofilm and will differ for cells at the base of the biofilm that are protected from detachment compared to cells at the surface of the biofilm. A local cell retention time ($\theta_{X,local}(z)$) is a function of the distance from the substratum and can be defined as follows (Morgenroth and Wilderer 2000):

**Definition:** *If the movement of a particle (e.g., a bacterial cell) within the biofilm is traced, then the* local cell retention time ($\theta_{X,local}(z)$) *is defined as the mean time that the traced particle will remain in the biofilm before detachment.*

Detachment of a particle from a biofilm is a two-step process. First, the particle must be transported by advection past the base thickness ($L_{base\ thickness}$) where $L_{base\ thickness}$ is defined as the thickness of the biofilm after backwashing. Advective transport within the biofilm results from organism growth, subsequent volume expansion, and displacement of the biofilm matrix above.

The local cell retention time can be calculated as the sum of the average time to reach the base thickness plus the average time for subsequent detachment:

$$\theta_{X,\text{local}}(z) = [\text{average time to reach } L_{\text{base thickness}}] \\ + [\text{average time until backwashing}] \quad (11.8)$$

For biofilm systems without backwashing only the first term in Equation (11.8) is relevant and $L_{\text{base thickness}}$ is equal to the total biofilm thickness. The average time required for a traced particle to reach $L_{\text{base thickness}}$ can be evaluated from advective velocities of the rigid biofilm matrix. Let $z^*(t)$ be the solution to the initial value problem

$$\frac{dz}{dt} = v(z,t) \quad \text{and} \quad z(t=t_0) = z_0 \quad (11.9)$$

where:  $z$ = distance from the substratum;
$v(z,t)$ = advective velocity of the biofilm matrix as determined in the mathematical model;
$t_0, z_0$ = initial time and location within the biofilm.

Then the time required to reach the base thickness ($T_{\text{advection}}(z_0,t_0)$) can be calculated for any initial $t_0$ and $z_0$ from

$$z^*(t_0 + T_{\text{advection}}) = L_{\text{base thickness}}. \quad (11.10)$$

The average time for detachment after reaching the base thickness is $0.5 \cdot T_{\text{backwashing interval}}$. For a location above the base thickness ($z_0 > L_{\text{base thickness}}$), the mean time of detachment depends on the probability of the particle to be located at that position at $t = t_0$. The following expression can be used to evaluate $\theta_{X,\text{local}}(z)$ from modeling results:

$$\theta_{X,\text{local}}(z) = \begin{cases} \overline{T}_{\text{advection}}(z) + \frac{1}{2} T_{\text{backwashing interval}} & \text{for } z < z_{\text{base thickness}} \\ \frac{1}{2} T_{\text{backwashing interval}}(1 - F(z)) & \text{for } z > z_{\text{base thickness}} \end{cases} \quad (11.11)$$

where:  $\overline{T}_{\text{advection}}(z)$ = time average of $T_{\text{advection}}$ (Equation 11.10) for a location $z$ within the biofilm;
$T_{\text{backwashing interval}}$ = backwashing interval;
$F(z)$ = cumulative probability distribution for the biofilm thickness.

The definition of the local residence time distribution can be used to evaluate local growth conditions within a biofilm based on results from a numerical simulation of the biofilm. By using local residence time distributions, knowledge and experiences from suspended cultures can be translated to biofilm systems. As in suspended cultures, retention times for bacteria (calculated as sludge age or dilution rate) have been identified as one of the key parameters determining microbial competition.

In addition to calculating the local solids residence time, an overall solids residence time can be calculated relating the detached biomass to the amount of biofilm in the system

$$\theta_{X,\,overall} = \frac{\text{average mass of biofilm}}{\text{average mass of detached biofilm}} \quad (11.12)$$

For volume detachment, $\theta_{X,overall}$ is equal to the inverse of the volume detachment rate, $r_{d,V}$ (Rittmann and McCarty 2001). Using a mathematical model, Morgenroth and Wilderer (2000) showed that competition within a biofilm is related to $\theta_{X,local}$ rather than $\theta_{X,overall}$. However, the experimental determination and the significance of $\theta_{X,local}$ and $\theta_{X,overall}$ is ongoing.

### 11.2.6 Summary

Different modes of detachment can be identified that can result in local or overall dynamic variations of the biofilm thickness with time, even though average overall mass detached is constant. Detachment occurs when external forces exceed internal strengths within the biofilm. For example, in submerged biofilm reactors, detachment occurs mainly during backwashing, where the filter bed is fluidized and external forces are increased by increased flow velocities and particle collisions. However, very limited quantitative experimental data on detachment processes are available. Experimental data for detachment are usually measured under steady-state conditions and often in a laminar flow field. Those data from a very protective growth environment in a laboratory system may not be applicable to systems with different modes of detachment. The lack of knowledge is reflected in current mathematical models, where detachment is usually assumed as a constant rate or detachment is completely neglected.

## 11.3 INFLUENCE OF DETACHMENT ON COMPETITION IN BIOFILMS AND ON OVERALL PROCESS PERFORMANCE

In this section, results from two case studies on the effect of detachment dynamics on competition in biofilms and biofilm reactor performance are discussed. The first example evaluates the influence of detachment dynamics on competition between heterotrophic and autotrophic bacteria (Morgenroth and Wilderer 2000). In the second example, the influence of detachment dynamics in systems with enhanced biological phosphorus removal is discussed (Morgenroth and Wilderer 1999).

## 11.3.1 Competition between fast- and slow-growing bacteria

Wanner and Gujer (1984) showed for a constant thickness mixed culture biofilm that the slower-growing autotrophic bacteria are located further towards the base of the biofilm, whereas the faster-growing heterotrophic bacteria dominate at the surface of the biofilm. Meanwhile, such a stratification of bacterial species over the thickness of a biofilm has been shown experimentally (Rittmann and Manem 1992; Okabe *et al.* 1996). The question remains: how the model predictions of Wanner and Gujer are influenced by the assumption of a constant thickness. Using mathematical modeling, the system of Wanner and Gujer (1984) was evaluated with three different types of detachment (Table 11.2). Details of these simulations are provided in Morgenroth and Wilderer (2000).

**Table 11.2.** Cases evaluated using surface or dynamic detachment.

| Detachment process | Time interval between backwashing events ($T_{backwashing\ interval}$) | Biofilm thickness, μm | |
|---|---|---|---|
| | | Mean | Range |
| Constant biofilm thickness | 0 d | 982 | 982 |
| Backwashing | 1 d | 982 | 782–1187 |
| Sloughing | 7 d | 982 | 10–2047 |

### 11.3.1.1 Influence on overall mass bacterial species

In Figure 11.4a,b the overall masses of heterotrophic and autotrophic bacteria are compared for different modes of detachment. For sloughing, it can be seen that the heterotrophic biomass increased by a factor of almost three compared to the constant biofilm thickness simulation. The reason for the increase in the heterotrophic biomass at longer detachment intervals is that, during unrestricted growth between backwashing events, bacteria at the surface grow with high growth rates, whereas decay and inactivation at the base of the biofilm remain slow. The overall mass of autotrophic bacteria is affected differently compared with heterotrophic biomass accumulation because, in the case of sloughing, the growth of autotrophs is transport limited for the thick biofilms. For backwashing, the mass of autotrophic biomass increased slightly over the mass calculated in the constant thickness simulation. However, for sloughing, the autotrophic biomass decreased by up to an order of magnitude. As the maximum density of autotrophs is located at the base of the biofilm, the thick layer of heterotrophic biomass during unrestricted growth between sloughing events caused large mass transfer limitations, which significantly reduced autotrophic growth rates.

### 11.3.1.2 Influence on system performance

The resulting overall performance of the system is summarized in Figure 11.4c,d, where the flux of COD and ammonia into the biofilm is compared for different modes of detachment. For the degradation of COD, it can be seen that the increased mass of heterotrophic bacteria did not result in a corresponding increase of the COD flux. Thus, for the given conditions, COD oxidation was limited by mass transfer into the biofilm rather than by the mass of heterotrophic bacteria present. The flux of ammonia decreased slightly in the case of backwashing compared with the constant thickness simulation, even though the mass of autotrophic bacteria was larger. However, for the case of sloughing, ammonia flux into the biofilm was an order of magnitude lower compared with the flux in the constant thickness simulation. This decrease in performance is even larger than the corresponding decrease in the mass of autotrophic bacteria (Figure 11.4). It can be concluded that biofilm thickness dynamics has a significant influence on the competition between heterotrophic and autotrophic bacteria. High variations of the biofilm thickness dynamics caused by long backwashing intervals further increase the dominance of the faster growing bacteria.

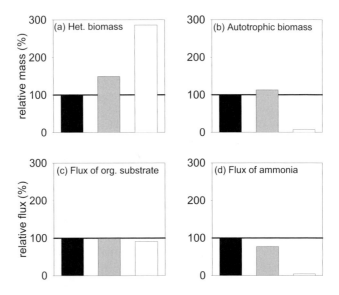

**Figure 11.4.** Mass of bacteria and substrate flux for different detachment intervals (Filled bars, constant biofilm thickness; gray bars, backwashing ($T_{backwashing\ interval}$ = 1 d); open bars, sloughing ($T_{backwashing\ interval}$ = 7 d). Relative values are given, where 100% equals the results for the constant biofilm simulation. (From Morgenroth and Wilderer 2000.)

The influence of detachment dynamics can be observed in the dynamic evaluation of organism distributions between backwashing events (Figure 11.5). The faster-growing heterotrophs can grow with little influence of the autotrophic biomass, and the maximum density of the heterotrophic biomass is always located at the surface of the growing biofilm. Below the biofilm surface, the heterotrophic mass fraction decreases owing to inactivation and respiration, as growth rates decrease as a result of mass transfer limitations at those locations.

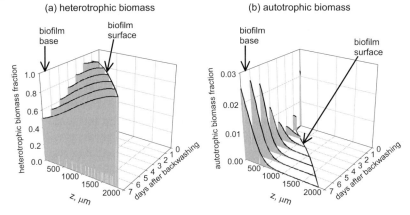

**Figure 11.5.** Evolution of biomass fractions of (a) heterotrophic and (b) autotrophic biomass in the biofilm shown for 7 days following a sloughing event. (From Morgenroth and Wilderer 2000.)

### 11.3.1.3 Local cell residence time distribution

Organism distributions for different types of detachment are shown in Figure 11.6 with local cell residence times and cumulative probability distributions of the biofilm thickness. The base thickness after backwashing and the detachment intervals were chosen to obtain identical average biofilm thickness for all three scenarios. The distribution of heterotrophic and autotrophic bacteria within the biofilm is determined by their individual strategies to compete for their common substrate (oxygen) and space within the biofilm. Comparing biomass distributions in Figure 11.6g,h,i, it can be seen that the maximum density of the faster growing heterotrophic biomass is always located at the surface of the biofilm, while the slower growing autotrophs are predominantly located below the heterotrophic layer, where they are better protected from detachment. However, the distribution of the autotrophic biomass differs significantly for simulations with different detachment mechanisms. For simulations with constant thickness and backwashing, the maximum densities of autotrophic bacteria are located 686 μm and 608 μm away from the substratum, respectively. For sloughing, the maximum density of the autotrophic biomass is located directly at the substratum.

Local cell residence time distributions ($\theta_{X,local}(z)$) are shown in Figure 11.6d,e,f. Although the overall residence times ($\theta_{X,overall}$) range from 1.9 to 3.4 days for all simulated cases, distributions of $\theta_{X,local}(z)$ vary pronouncedly from the constant thickness and backwashing simulations on the one hand, to sloughing on the other. For the constant thickness simulation, the base of the biofilm is substrate limited and towards the base the biofilm is almost entirely (>99%) composed of inert material. As a result, volume expansion at the base of the biofilm approaches zero and, for example, below $z = 518$ μm, the calculated $\theta_{X,local}(z)$ increases above 100 days. For backwashing, the overall shape of $\theta_{X,local}(z)$ is similar to the constant thickness simulation, but $\theta_{X,local}(z)$ reaches 100 days further down towards the substratum at $z = 314$ μm. For sloughing, the local residence time distribution of the entire biofilm above $z = 10$ μm was smaller than 3.5 days. Those low $\theta_{X,local}(z)$ values for sloughing resulted in low concentrations of autotrophic bacteria and lead to the significant decrease in ammonia flux into the biofilm.

Within a biofilm, different bacterial populations have to find their optimal niche. Faster-growing heterotrophs are dominant at the surface with corresponding low $\theta_{X,local}(z)$, where they can take full advantage of high substrate concentrations. Autotrophic bacteria, with a slower growth rate relative to their competing heterotrophic bacteria, find a niche within the biofilm where they can balance limitations caused by their removal from the system by advective transport and subsequent detachment on one hand, and substrate limitations on the other hand. For sloughing, the optimal location for the autotrophs was pushed all the way down to the substratum. The general consequence for the competition between heterotrophic and autotrophic growth in a biofilm with dynamic detachment is that, with increasing amplitude of biofilm thickness variations, slower-growing bacteria find their niche closer to the substratum.

### 11.3.1.4 Summary

It can be concluded that the simplifying assumptions of a constant biofilm thickness and continuous detachment in most mathematical biofilm models need to be re-evaluated. It appears that overall reactor performance and biofilm architecture are significantly affected by detachment patterns. The need to take into account the mode of detachment applies both to mathematical models and to laboratory experiments. Results from experiments using a laminar flow channel (where detachment may be dominated by continuous erosion) may not

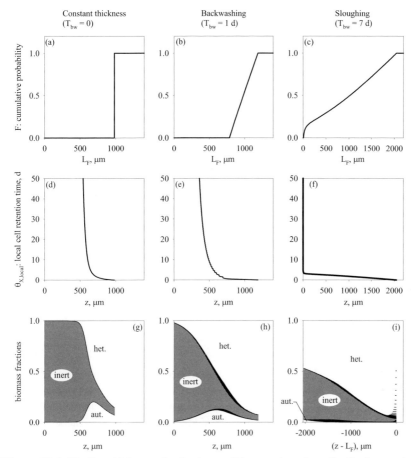

**Figure 11.6.** Biofilm thickness distribution, solids retention time distribution, and the organism distribution over the thickness of the biofilm for different mechanisms of detachment. (a)–(c): Cumulative probabilities of having a biofilm thickness smaller than LF. Solids retention time and bacterial distributions over the thickness of the biofilm for different mechanisms of detachment. (d)–(f): Solids retention time distribution as defined in Equation (11.11). (g)–(i): Relative distribution of bacteria and inert biological material. The heterotrophic fraction plotted above the inert fraction (shaded area). The autotrophic biomass is plotted below the inert fraction. For backwashing and sloughing, multiple curves for different times in the simulation were superimposed in each diagram. (Reprinted from Morgenroth and Wilderer 2000.)

be appropriate to explain the performance of submerged biofilm reactors (where detachment may be dominated by backwashing) or trickling filters (where detachment may be dominated by sloughing). The local cell retention time ($\theta_{X,local}(z)$) varies over the thickness of the biofilm. At the surface of the biofilm, the local retention time approaches zero, while the local retention time can increase infinitely at the base of a biofilm if the growth rate of organisms approaches zero at that particular location. An overall retention time ($\theta_{X,overall}$), calculated from the total mass of the biofilm divided by the mass of the detached biofilm, is not meaningful. Considering heterotrophic and autotrophic competition for substrate and space, it was shown that the local cell retention time ($\theta_{X,local}(z)$) varies significantly between constant thickness simulations or daily backwashing on the one hand and weekly sloughing on the other. With increasing detachment intervals, the mass of heterotrophic biomass increased by almost a factor of three whereas the flux of COD into the biofilm remained constant. Both the mass of autotrophs and the flux of ammonia into the biofilm were an order of magnitude lower in the case of sloughing compared with constant thickness simulations.

## 11.3.2 Detachment as a relevant part of the process

The key to enhanced biological phosphorus removal (EBPR) is the activity of certain microorganisms that take up phosphate under aerobic or anoxic conditions, accumulating it as polyphosphate. Under anaerobic conditions the stored polyphosphate is a source of energy for the microorganisms, where phosphate is released and organic substrate is accumulated. Phosphorus removal is achieved when bacteria with a high content of stored phosphorus are removed from the system. EBPR is a common technology that is being implemented in a growing number of full-scale activated sludge systems around the world. The potential of EBPR in biofilm reactors has so far mainly been demonstrated in laboratory and pilot-scale systems (Gonzalez-Martinez and Wilderer 1991; Goncalves and Rogalla 1992; Morgenroth and Wilderer 1999; Arnz et al. 2001). These biofilm systems can be advantageous when the footprint of the plant or the settleability of sludge are of concern.

One of the major differences between phosphorus removal in activated sludge and biofilm systems is that, in biofilms, efficient detachment of bacteria can limit the overall performance. In Figure 11.7, two outcomes for the fate of polyphosphate accumulated in bacteria are shown. A net phosphorus removal is achieved only if a cell is detached from the biofilm. Thus, the role of detachment is significantly different for biofilm reactors operated for biological phosphorus removal, on the one hand, and COD or nitrogen removal on the other. In COD or nitrogen removal, bacteria act only as a catalyst in transforming a chemical component into a gaseous product ($CO_2$ or $N_2$) that is subsequently released to the atmosphere.

# Biofilm detachment

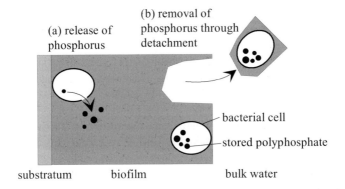

**Figure 11.7.** Biological phosphorus removal in biofilms. (a) If bacteria are not removed from the biofilm, then phosphorus will be released again due to endogenous respiration or cell lysis. (b) Phosphorus is removed only through detachment of bacteria with a high polyphosphate content. (Reprinted from Morgenroth and Wilderer 1999.)

## *11.3.2.1 Short local residence times are beneficial*

EBPR is linked to the detachment of biomass, and as a result, bacteria that grow in locations that are exposed to detachment (i.e., at the surface of the biofilm) have a larger contribution to the overall phosphorus removal (Figure 11.7). If a fraction of the biomass was never removed from the system, it may still release and accumulate phosphorus during anaerobic and aerobic phases, respectively. For this accumulation and release of phosphorus, the bacteria require organic substrate—but still no net phosphorus removal is achieved. In the long run, accumulated phosphorus will slowly be released as a result of lysis or endogenous respiration (Figure 11.7a). Thus, any substrate used by that biomass fraction that is never removed from the system is wasted with respect to EBPR. As the availability of organic substrate usually limits EBPR performance, any COD wasted will reduce the overall process efficiency.

Morgenroth and Wilderer (1999) have calculated local residence time distributions for a biofilm reactor operated for biological phosphorus removal by using a mathematical biofilm model and the approach described in section 11.2.5. Advective transport within the biofilm results from organism growth and subsequent volume expansion resulting in a displacement of the biofilm matrix above. It was found that the average advective velocity distribution within the biofilm could adequately be described with the following linear velocity distribution:

$$v = 0.054 \, \text{d}^{-1} \cdot z \tag{11.13}$$

where $v$ = growth velocity, $z$ = distance from the substratum. By integration of Equation (11.3) the time ($T_{advection}$) for a particle to move from a location $z_0$ to the base thickness of the biofilm ($L_{base\ thickness}$) can be calculated as

$$T_{advection} = 18 \, \text{d} \cdot \ln(L_{base\ thickness} / z_0) \tag{11.14}$$

The mean residence time of a particle in the biofilm is the sum of the time required to reach the base thickness and the average time required for subsequent detachment:

$$\theta_X(z) = T_{advection} + 0.5 \cdot T_{backwashing\ interval} \tag{11.15}$$

where $T_{backwashing\ interval}$ is the time between two backwashing events. The resulting mean residence time distribution is shown in Figure 11.8. Residence times in the upper third of the biofilm were below 10 days. These residence times are comparable to solids retention times used in activated sludge systems operated for biological phosphorus removal (Henze et al. 2002). In the lower part of the biofilm local residence times are above 80 days. As a result, the lower part of the biofilm has only a limited contribution towards the overall phosphorus removal. Thus, it can be concluded that frequent backwashing resulting in thin and active biofilms is advantageous to achieve efficient biological phosphorus removal.

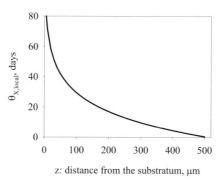

**Figure 11.8.** Theoretical mean particle cell residence time distributions (from Equation 11.5) with $T_{backwashing\ interval} = 1$ d. (Reprinted from Morgenroth and Wilderer 1999.)

## 11.3.2.2 Loss of active biomass

Backwashing with long intervals between detachment events causes significant fluctuations of the amount of phosphorus-removing bacteria. In Figure 11.9a,c the effects of different backwashing intervals on the effluent phosphorus concentration are shown. Even intensive backwashing had no effect on the effluent phosphorus concentration in the experimental investigation, if applied in short intervals (once or twice a week) (Figure 11.9a). However, with an increased backwashing interval of 15 days, a significant increase of the effluent phosphorus concentration after backwashing was observed. The reason for this increase was that, during backwashing, too much active biomass was removed from the system. After this sudden increase of effluent phosphorus, effluent concentrations decreased again within a week as a result of biofilm growth. A similar qualitative conclusion can be drawn from the mathematical simulation, where only a minor increase in the effluent phosphorus concentration was observed with a backwashing interval of 1 day (Figure 11.9b), but the effluent concentration increases to three times the influent phosphorus concentration after backwashing with a 15-day backwashing interval (Figure 11.9d).

The peculiar effect of an increase of the effluent phosphorus above influent concentrations (e.g., Figure 11.9d) can be explained by an imbalance between the organic load in the influent and the active biomass. After backwashing, a significant amount of the active biomass was lost. As phosphorus release is a fast process while uptake of phosphorus is generally slow, the remaining bacteria released a large amount of phosphorus but could not accumulate released phosphorus fast enough within the aerobic phase (Filipe et al. 2001). The increase of the effluent phosphorus concentrations above the influent concentrations was due to the release of phosphorus stored in previous cycles.

## 11.3.2.3 Summary

Detachment dynamics influence EBPR in two different ways. (1) Location and dynamics of detachment influence the local residence time distribution of bacteria within the biofilm. To achieve efficient EBPR, bacteria have to be removed from the system once they have accumulated high amounts of polyphosphate. As a result, low local residence times towards the surface of the biofilm are beneficial for EBPR. Alternatively, very long local residence times towards the base of the biofilm can be detrimental for overall phosphorus removal efficiency due to release of previously stored polyphosphate through endogenous respiration or cell lysis. (2) The frequency of backwashing and resulting detachment dynamics can result in a significant variability of active biomass over time. In EBPR, these variations of active biomass can cause an imbalance between phosphorus release and uptake kinetics that can result in increased effluent phosphorus concentrations after backwashing.

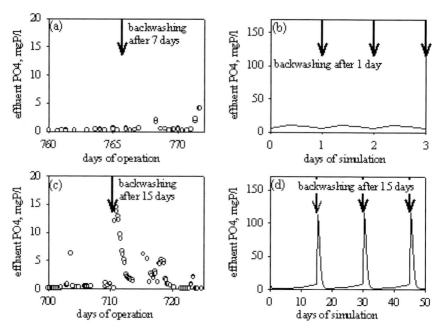

**Figure 11.9.** Influence of backwashing (arrow) on the effluent PO$_4$ concentration: (a) with normal backwashing intervals (weekly), no effect on the effluent phosphorus concentration was observed; (c) when the backwashing intervals were increased, backwashing resulted in a temporary deterioration of the phosphorus removal; (b, d) In the mathematical simulation, similar effects of the backwashing interval are observed. (Reprinted from Morgenroth and Wilderer 1999.)

## 11.4 CONCLUDING REMARKS

Microbial ecology is influenced by exposing bacteria to necessary substrates for growth and by giving enough time for the bacteria to grow and to proliferate. Changes in substrate availability or retention time of bacteria within the system will result in changes of competitive advantages of different groups of bacteria and a shift in the population dynamics. Only recently, new experimental methods allow us to study local availability of substrate (see Chapter 7) and local distributions of bacterial populations (see Chapter 10) in great detail. In this chapter on detachment, it was shown that retention of bacteria within a biofilm system can have a significant influence on microbial ecology. However, mechanisms of detachment and detachment dynamics leading to the described local retention time distributions within the biofilm have so far not been well understood. It can be concluded, however, that focusing only on studying population dynamics and substrate availability within a biofilm without taking into account detachment is like taking a snap-shot of a situation, while neglecting major factors responsible for the development of the current situation (Figure 11.10).

Current understanding on detachment indicates that both microbial ecology and overall process performance can be significantly influenced by location and dynamics of detachment. However, understanding of mechanisms and rates of detachment is still very limited. More experimental data are required to quantify detachment at the micro-scale (e.g., in a flow chamber under a confocal laser scanning microscope) and in full-scale biofilm reactor applications. A better quantitative understanding will help to improve mathematical models of biofilms (see Chapters 1–5). Finally, an improved understanding of detachment mechanisms can become the basis for developing new processes where detachment is not only included to prevent the system clogging, but where detachment processes are deliberately used to manipulate the microbial ecology.

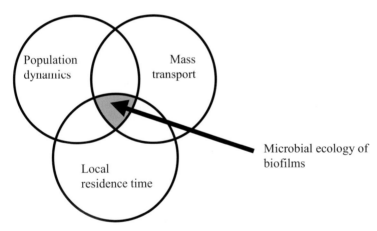

**Figure 11.10.** Information required to understand microbial ecology in biofilms.

## 11.5 REFERENCES

Arnz, P., Arnold, E. and Wilderer, P.A. (2001) Enhanced biological phosphorus removal in a semi full-scale SBBR. *Wat. Sci. Tech.* **43**(3), 167–174.

Bakke, R., Trulear, M.G., Robinson, J.A. and Characklis, W.G. (1984) Activity of *Pseudomonas aeruginosa* in biofilms: steady state. *Biotech. .Bioengng* **26**, 1418–1424.

Bryers, J.D. (1984) Biofilm formation and chemostat dynamics: Pure and mixed culture considerations. *Biotech. Bioengng* **26**, 948–958.

Bryers, J.D. (1988) Modeling biofilm accumulation. In *Physiological Models in Microbiology*, (ed. M. Bazin and J.J. Prosser), pp. 109–144. CRC Press, Boca Raton, Florida..

Bryers, J.D. (2000) Biofilm formation and persistence. In: *Biofilms II: Process Analysis and Application* (ed. J.D. Bryers), pp. 45–88, John Wiley, New York.

Chang, H.T. and Rittmann, B.E. (1987) Mathematical modeling of biofilm on activated carbon. *Environ. Sci. Tech.* **21**, 273–280.

Chang, H.T. and Rittmann, B.E. (1988) Comparative study of biofilm shear loss on different adsorptive media. *J. Wat. Pollut. Control Fed.* **60**, 362–368.

Davies, D.G., Parsek, M.R., Pearson, J.P., Iglewski, B.H., Costerton, J.W. and Greenberg, E.P. (1998) The involvement of cell-to-cell signals in the development of a bacterial biofilm. *Science* **280**, 295–297.

Filipe, C.D.M., Meinhold, J., Jorgensen, S.B., Daigger, G.T. and Grady, C.P.L. (2001) Evaluation of the potential effects of equalization on the performance of biological phosphorus removal systems. *Wat. Environ. Res.* **73**, 276–285.

Fruhen, M., Christan, E., Gujer, W. and Wanner, O. (1991) Significance of spatial distribution of microbial species in mixed culture biofilms. *Wat. Sci. Tech.* **23**(7/9), 1365–1374.

Goncalves, R.F. and Rogalla, F. (1992) Biological phosphorus removal in fixed films reactors. *Wat. Sci. Tech.* **25**(12), 165–174.

Gonzalez-Martinez, S. and Wilderer, P.A. (1991) Phosphate removal in a biofilm reactor. *Wat. Sci. Tech.* **23**(7–9), 1405–1415.

Harremoës, P., Jansen, J.L.C. and Kristensen, G.H. (1980) Practical problems related to nitrogen bubble formation in fixed film reactors. *Prog. Wat. Tech.* **12**, 253–269.

Henze,M., Harremoës, P., Jansen, J.L.C. and Arvin, E. (2002): *Wastewater Treatment: Biological and Chemical Processes,* 3rd edition. Springer, Berlin.

Kissel, J.C., McCarty, P.L. and Street, R.L. (1984) Numerical simulation of mixed-culture biofilm. *ASCE J. Environ. Engng* **110**, 393–411.

Kreikenbohm, R. and Stephan, W. (1985) Application of a two-compartment model to the wall growth of *Pelobacter acidigallici* under continuous culture conditions. *Biotech. Bioengng* **27**, 296–301.

Lazarova, V. and Manem, J. (2000) Innovative biofilm treatment technologies for water and wastewater treatment. In *Biofilms II: Process Analysis and Application*, (ed. J.D. Bryers), pp. 159–206. John Wiley, New York.

Morgenroth, E. and Wilderer, P.A. (1999) Controlled biomass removal – the key parameter to achieve enhanced biological phosphorus removal in biofilm systems. *Wat. Sci. Tech.* **39**(7), 33–40.

Morgenroth, E. and Wilderer, P.A. (2000) Influence of detachment mechanisms on competition in biofilms. *Wat. Res.* **34**, 417–426.

Okabe, S., Hiratia, K., Ozawa, Y. and Watanabe, Y. (1996) Spatial microbial distributions of nitrifiers and heterotrophs in mixed-population biofilms. *Biotech. Bioengng* **50**, 24–35.

Peyton, B.M. and Characklis, W.G. (1993) A statistical-analysis of the effect of substrate utilization and shear-stress on the kinetics of biofilm detachment. *Biotech. Bioengng* **41**, 728–735.

Peyton, B.M., Skeen, R.S., Hooker, B.S., Lundman, R.W. and Cunningham, A.B. (1995) Evaluation of bacterial detachment rates in porous-media. *Appl. Biochem. Biotech.* **51–52**, (SPR), 785–797.

Picioreanu, C., van Loosdrecht, M.C.M. and Heijnen, J.J. (2001) Two-dimensional model of biofilm detachment caused by internal stress from liquid flow. *Biotech. Bioengng* **72**, 205–218.

Rittmann, B.E. (1982) The effect of shear stress on biofilm loss rate. *Biotech. Bioengng* **24**, 501–506.

Rittmann, B.E. (1989) Detachment from biofilms. In *Structure and function of biofilms* (ed. W.G. Characklis. and P.A. Wilderer), pp. 49–58. John Wiley, New York.

Rittmann, B.E. and Laspidou, C.S. (2002) Biofilm detachment. In *Encyclopedia of Environmental Microbiology*, (ed. G. Bitton), pp. 544–550. John Wiley, New York.

Rittmann, B.E. and Manem, J.A. (1992): Development and experimental evaluation of a steady-state, multispecies biofilm model. *Biotech. Bioengng* **39**, 914–922.

Rittmann, B.E. and McCarty, P.L. (2001) *Environmental Biotechnology: Principles and Applications,* McGraw-Hill, New York.

Rittmann, B.E., Stilwell, D. and Ohashi, A. (2002) The transient-state, multiple-species biofilm model for biofiltration processes. *Wat. Res.* **36**, 2342–2356.

Robinson, J.A., Trulear, M.G. and Characklis, W.G. (1984) Cellular reproduction and extracellular polymer formation by *Pseudomonas aeruginosa* in continuous culture. *Biotech. Bioengng* **26**, 1409–1417.

Saez, P.B. and Rittmann, B.E. (1992) Accurate pseudoanalytical solution for steady-state biofilms. *Biotech. Bioengng* **39**, 790–793.

Speitel, G.E. and DiGiano, F.A. (1987) Biofilm shearing under dynamic conditions. *ASCE J. Environ. Engng* **113**, 464–475.

Stewart, P.S., Camper, A.K., Handran, S.D., Huang, C.T. and Warnecke, M. (1997) Spatial distribution and coexistence of *Klebsiella pneumoniae* and *Pseudomonas aeruginosa* in biofilms. *Microb. Ecol.* **33**, 2–10.

Tijhuis, L., van Loosdrecht, M.C.M. and Heijnen, J.J. (1995) Dynamics of biofilm detachment in biofilm airlift suspention reactors. *Biotech. Bioengng* **45**, 481–487.

Trulear, M.G. and Characklis, W.G. (1982) Dynamics of biofilm processes. *J. Wat. Pollut. Control Fed.* **54**, 1288–1301.

van Benthum, W.A.J., van Loosdrecht, M.C.M., Tijhuis, L. and Heijnen, J.J. (1995) Solids retention time in heterotrophic and nitrifying biofilms in a biofilm airlift suspension reactor. *Wat. Sci. Tech.* **32**(8), 53–60.

Wanner, O., Debus, O. and Reichert, P. (1994) Modeling the spatial distribution and dynamics of a xylene-degrading microbial population in a membrane-bound biofilm. *Wat. Sci. Tech.* **29**(10/11), 243–251.

Wanner, O. and Gujer, W. (1984) Competition in biofilms. *Wat. Sci. Tech.* **17**(2/3), 27–44.

Wanner, O. and Gujer, W. (1986) A multispecies biofilm model. *Biotech. Bioengng* **28**, 314–328.

Wanner, O. and Reichert, P. (1996) Mathematical modeling of mixed-culture biofilms. *Biotech. Bioengng* **49**, 172–184.

## Acknowledgements

Partial financial support was provided from an NSF CAREER award (BES-0134104). Discussions with Peter Wilderer, Edward Schroeder, Bruce Rittmann, and Mark van Loosdrecht are gratefully acknowledged. Rachel Michaud and Young Chul Choi also provided help in the preparation of this manuscript.

# Architecture, population structure and function: Conclusions

*Stefan Wuertz*

Predicting biofilm system performance is an art based on solid scientific and engineering principles. It takes a good understanding of biological and reactor kinetic principles to design a system that will withstand serious operational and environmental challenges. In pondering the question posed at the beginning of this section: *'Can information obtained by physicochemical analytical techniques and microbial population studies be used to explain and predict the performance of biofilm systems?'*, the answer after reading the preceding chapters can only be a resounding, 'Yes, it can!'

One of the most surprising revelations that has emerged from detailed biofilm studies is the dominant role played by extracellular polymeric substances (EPS). Measuring their abundance and composition has led to insights into sludge settleability and dewaterability as well as into the role of EPS in maintaining extracellular enzyme activity for the hydrolysis of macromolecules before cellular uptake and metabolism. There are now several published EPS extraction protocols gentle enough to minimize cell lysis, which have led to new insights regarding the functions of EPS including their role as sorbent for hydrophobic substances. What has become apparent is that sorption properties of biofilms are strongly influenced by the prevalence and distribution of hydrophilic and hydrophobic regions in EPS molecules. The effect of different compositions of EPS (mainly polysaccharides versus proteins) has not been unambiguously determined yet but, clearly, physicochemical measurements will enable us to make predictions about the ability of a particular biofilm system to remove hydrophobic contaminants based on its sorption ability. EPS must also be incorporated in multidimensional models, although their precise function – rigid hindrance, porous microstructure or viscoelastic structure – has not yet been resolved.

© 2003 IWA Publishing. *Biofilms in wastewater treatment*. Edited by S. Wuertz, P.L. Bishop and P.A. Wilderer. ISBN: 1 84339 007 8.

In terms of mass transport the availability of microsensors and confocal laser scanning microscopy (CLSM) was key to discovering biofilm heterogeneity, and its quantification has been achieved by using advanced image analysis tools. It is not yet clear to what extent this heterogeneity affects mass transport phenomena on a reactor scale and whether it will result in improved mathematical description of processes in biofilm reactors (see also part 1 for modeling aspects).

In this section we have learned that the physicochemical characteristics of the biofilm support material influence both microbial adhesion and detachment. Hence surface properties may be used to make predictions about stability and activity of biofilms. Although such knowledge certainly has its merits when choosing suitable support media in biofilm operations and reducing start-up periods, it also leads to another question: 'Can the mechanical characteristics of biofilms be measured and used to predict the performance of biofilm systems?' Systematic rheological studies have not been done on wastewater systems but could yield useful information about structural inhomogeneities.

A great deal of information is available regarding the diversity of microbial populations in a biofilm reactor. This includes the trophic levels of prokaryotes, protozoa and metazoa, with nearly all of the available molecular information from members of the Bacteria and Archaea. Information about the abundance of protozoa and metazoa should be included in reactor operation as discussed in this section. Only some of the molecular biology techniques are quantitative, such as fluorescent *in situ* hybridization (FISH). Other molecular techniques like denaturing gradient gel electrophoresis (DGGE) are more like fingerprinting methods that indicate the diversity of a population and changes therein. They are excellent reactor monitoring tools and can be expanded as needed to establish gene libraries encompassing rRNA sequences of essentially all the microbial players within the experimental limitations of the method. The information gained can be used to develop and apply specific FISH probes vital in the detection of reactor disturbances and to help identify the affected bacterial genera. A case in point is the recent discovery that many nitrifying wastewater treatment plants do not contain significant numbers of nitrite-oxidizing *Nitrobacter* bacteria but rather members of the – so far unculturable from wastewater – *Nitrospira*. These microorganisms have very different physiological requirements from *Nitrobacter*. Obviously, when adding activated sludge or biofilm from another plant, one should ensure that the desired bacterial genera are present. Biofilm systems also depend to a great extent on efficient removal of excess biomass, frequently achieved by backwashing. This operational feature has a great effect on the microbial ecology of the system by affecting the local retention times of individual populations and hence their activities.

Overall, there is mounting evidence that microbial population analysis and physicochemical characterization can assist in the successful operation of biofilm reactors. Part 3 of this book addresses the question of whether the flow of information from fundamental biofilm research has been directed towards feasible results on a practical scale.

# PART THREE

## FROM FUNDAMENTALS TO PRACTICAL APPLICATIONS

# From fundamentals to practical applications: Introduction

## Peter A. Wilderer

The first reactor ever built to achieve biological wastewater treatment was a biofilm reactor. Biological wastewater treatment technology started in the late 19th century when engineers realized that the slime forming at the surface of gravel exposed to sewage helped clean the water. At the beginning, rock media were simply submerged in the wastewater to be treated, but soon the engineers found that trickling the water over a packing of rocks resulted in far better treatment. It became clear that the slime was formed by microorganisms, and it was realized that these organisms were capable of removing organic pollutants and ammonia when supported with oxygen. The trickling filter was born.

For many decades, trickling filters were successfully designed, built and operated by engineers. Because of the lack of detailed knowledge about the very nature of the slime – the biofilm, as we call it today – design procedures and operation guidelines were mostly based on experience gained by trial and error. The treatment results were reasonable for both protection of the receiving waters and costs.

In recent years, the advances made in microbiology, molecular biology, analytical chemistry, sensor design, and numerical computation allow completely new insights into the microbial community forming a biofilm, and in the system of reactions taking place in a biofilm reactor. Mathematical models are available to describe the performance of biofilms and biofilm reactor systems. Methods of molecular biology combined with confocal laser scanning microscopy allow in-depth description of spatial and temporal microbial distribution and of the 3-dimensional structure of biofilms. By using novel physical and chemical tools such as photoacoustic spectrometry the time-dependent development of the 3-dimensional structure of biofilms can be

© 2003 IWA Publishing. *Biofilms in wastewater treatment.* Edited by S. Wuertz, P.L. Bishop and P.A. Wilderer. ISBN: 1 84339 007 8.

observed *in situ* and non-destructively. The information gained can be fed into computer programs designed to visualize the various components, transport processes and metabolic reactions. One can virtually 'fly' through biofilms.

In summary, a door has been opened into the world of biofilm systems. Engineers and practitioners are overwhelmed with detailed information and colorful images of high resolution. Will all this lead to novel technological developments, to an improved performance of biofilm reactors, to enhancement of reactor efficiency, and to a reduction of costs? Did scientists ask the right questions, and were the factors affecting the performance of biofilm reactors properly addressed? After critically reviewing the current situation, it appears that the bridge between biofilm science and biofilm reactor engineering has not been properly built yet. In the following chapters an attempt is made to lay the foundation of that urgently needed bridge.

# 12
# Deduction and induction in design and operation of biofilm reactors

*Poul Harremoës*

## 12.1 INTRODUCTION

Within the past two to three decades there has been a significant development of technology, away from the trickling filter towards biofilm reactors with a different approach to the configuration of the carrier material and the operation of the reactor with respect to biomass control.

In the same period, the engineering approach to design of wastewater treatment plants has developed from shear pragmatism to a science-based concept. The pragmatic approach is based on the experience gained from trial and error. The scientific approach is based on description of cause–effect relationships, mostly in mathematical form. Its purpose is that the relationships should be of a more universal nature than just experience, which is limited to the data set from which the specific experience has been gained. It is a general

© 2003 IWA Publishing. *Biofilms in wastewater treatment.* Edited by S. Wuertz, P.L. Bishop and P.A. Wilderer. ISBN: 1 84339 007 8.

engineering experience that a more generalized scientific description contains a better approximation to reality and may be better extrapolated to circumstances outside the narrow range of experience.

Current research outlined in this book will contribute to the scientific understanding of the phenomena involved in biofilm reactors. It remains to be seen how these new approaches will aid in the engineering design and operation of biofilm reactors.

## 12.2 TECHNOLOGICAL DEVELOPMENT

### 12.2.1 Basic principles

The basic principle of biofilm reactors is that the biological processes take place inside an attached biomass. The kinetic characteristic is that the substrates have to diffuse through the biomass to reach the bacteria in the biofilm and the products have to diffuse out (Harremoës 1978). To make a reactor work, there are two basic principles that have to perform: (1) the substrates have to be transported to the biofilm and the products have to be transported away from the biofilm; and (2) the biomass has to be controlled avoid clogging.

There are seven basic principles for the transport of substrate, consisting of electron donor and acceptor plus nutrients (Harremoës and Wilderer 1993):

1. A **three-phase** system: carrier material with biofilm, bulk water and air. The carrier material and the biofilm are *fixed*, while the water trickles past the biofilm surface and air moves upward or downward as a third phase.
2. A **three-phase** system: carrier material with biofilm, bulk water and air. The carrier material and the biofilm are *fixed*, while the water flows through the reactor together with gas bubbles.
3. A **three-phase** system: carrier material with biofilm, bulk water and air. The carrier material and the biofilm *move* around in the reactor, while the water flows through the reactor together with gas bubbles.
4. A **two-phase** system: carrier material with biofilm, and bulk water. The carrier material and the biofilm are *fixed*, while the water flows through the reactor with both electron donor and acceptor.
5. A **two-phase** system: carrier material with biofilm and bulk water. The carrier material and the biofilm *move* around in the reactor, while the water flows through the reactor with both electron donor and acceptor.
6. A **three-phase** <u>membrane</u> system: a diffusive membrane with the biofilm and water on one side and the gas on the other side.
7. A **two-phase** <u>membrane</u> system: a diffusive membrane with the biofilm and water on both sides, but electron donor on one side and electron acceptor on the other side.

The traditional approach has been system 1: the trickling filter. An equally long history is associated with one type of reactor in system 3: the rotating biological contactor (RBC). Both trickling filter and RBC suffer from severe problems with respect to control of biomass. During the past 15 years, developments have been associated with systems 2 and 4: the biological filter, where the biomass is controlled by backwashing. During the past 10 years developments have been associated with another type of reactor in systems 3 and 5: the moving bed reactor, either as fluidized bed or as a bed in which elements of carrier with biofilm are suspended in the reactor. Currently, the membrane reactors are under rapid development, because the price of membranes has decreased by an order of magnitude in recent years.

## 12.2.2 Demands on performance

The practice of design of wastewater treatment was based on recommended loading figures and simple rules of thumb no more than a generation ago. This will never be obsolete, because the simplicity will always appeal to the practitioner, but today the demands on performance of wastewater treatment plants are so great that more scientific understanding is becoming a prerequisite for engineering success.

The following list illustrates the historical development of demands on the performance of wastewater treatment plants:
- BOD (below 30 mg/L before 1970)
- Source control of inhibitors for BOD-degradation
- Source control of detergents (linear alkyl sulfonates (LAS) substituted with alkyl benzene sulfonates (ABS), 1960s)
- SS (below 30 mg/L before 1970)
- Ammonia (below 1 mg/L starting in the 1970s)
- Source control of inhibitors for nitrification (starting in the 1970s)
- Source control of metals owing to sludge contamination (1980s)
- Odor control (starting seriously in 1980s)
- Tot. P (down to 0.3–1 mg/L, starting 1970–80)
- Tot. N (down to 8–10 mg/L, starting 1990)
- Source control of organic chemicals due to sludge contamination (1980s)
- New concerns: endocrine disrupters, antibiotics, resistant bacteria (1990s)

The pressure on the technology to perform according to an ever-broader spectrum of demands will increase the need to understand the processes involved and to use that knowledge in improving engineering design and operation.

## 12.3 PRAGMATISM VERSUS THEORY-BASED MODELS

### 12.3.1 Engineering craftsmanship

Pragmatism has been the successful basis for engineering development for centuries, even millennia. The craftsmanship of structures and machinery has been a sound basis for development, but the fact is that this craftsmanship in combination with science has improved practice and has expanded the perspectives in all disciplines of engineering.

This point is best illustrated by comparing the technological development in China and Europe. Technological development in China was ahead of that in Europe until the Renaissance. In China there were cast iron, pumps, suspension bridges, porcelain, and so forth centuries before they became available in Europe (Temple 1998). But the development of science and experimentation in Europe, starting with Galilei, Kepler, and Newton, resulted in a new understanding of cause–effect relationships based on laws of logic, laws of nature, and experimentation. These advances provided the basis for the quantum leap in the following centuries of technological development in Europe, as opposed to a continued development based on correlative experience and pragmatism in China (Harremoës 2000). The development of science-based technology is still in full swing, now also in China. It will have its impacts on engineering, also on wastewater treatment design and operation; but nobody knows precisely how.

The fact is that the practice of wastewater treatment has been late in the transformation from sheer pragmatism to science-based approaches. This is most likely due to the complexity of the biological systems used in biological wastewater treatment. However, there are exceptions: The theory of aeration was established a hundred years ago. That simple theory of proportionality has become an indispensable part of pragmatism since the first development of the activated sludge treatment plant, to the extent that practitioners hardly consider it a theory, but an integral part of practice. 'There is nothing more practical than a good theory.'

**Example 12.1**
Design of rotating biological contactors (RBCs) has always been very pragmatic. This is illustrated in Figure 12.1, which shows the effluent concentration in mgBOD$_5$/L versus the loading of the filter expressed as gBOD$_5$m$^{-2}$s$^{-1}$, where m$^2$ is the surface area of the disks (Pöpel 1964). Note that the scales are logarithmic. The results come from 16 investigations, where the loading with domestic wastewater has been recorded and the effluent measured on full-scale plants varying from 30 to 44,000 PE (person equivalents). The plants consist of one to four tanks in series with total disk area varying from 142 to 54,600 m$^2$.

The suggested use of this information is presented as follows. To achieve 90% probability of successful design aiming at an effluent concentration of <25 mgBOD$_5$/L, the design load must be less than 3.5 gBOD$_5$m$^{-2}$s$^{-1}$. The assumption is that performance of RBCs can be interpreted as a random sample of design among a statistical sample of plants. Despite much research this figure still represents a valuable pragmatic tool for design.

Figure 12.2 shows an alternative interpretation of the same data (Harremoës 1977). The removal rate is shown as a function of the effluent concentration. Because RBCs tend to be totally mixed systems, the effluent concentration can be assumed to be equal to the concentration in the bulk liquid in the reactor. That is what the biofilm is exposed to. The theoretical prediction is that the reaction should be ½-order due to diffusion limitation into a partly penetrated biofilm. It is seen from Figure 12.2, that this is the case. Compared with Figure 12.1 the scatter has decreased and thus Figure 12.2 is a better description. The result is in a sense surprising: (1) some of the RBCs have more than one compartment, and then the basic assumption about totally mixed reactors breaks down; (2) the theory is derived with the assumption of dissolved substrate. That is in contradiction to the fact that most of the organic matter in wastewater is in a particulate form, to which the classical theory of substrate diffusion does not apply (Morgenroth *et al.* 2002).

The two figures illustrate the difference between two completely different concepts: (1) the **loading approach**, where the reactor is a black box and the performance can be described as the result of the load on the reactor; (2) the **conceptual approach**, where the reactions in the reactor are described by a combination of laws of nature, in this case: water and mass balance, Monod kinetics, and diffusion in the biofilm.

It is a frequent misperception that the latter approach always leads to complicated formulations, alien to the practitioner. The example above illustrates that the conceptual approach may be as simple as the loading approach.

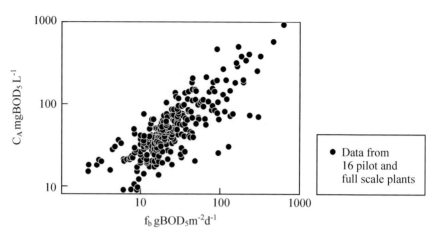

**Figure 12.1.** Empirical data from extensive measurements of effluent concentration of $BOD_5$, $C_A$, as a function of bioreactor surface loading of RBCs with domestic wastewater, $f_b$. Notice the logarithmic scales and the variability. Design criteria were suggested on a probabilistic basis (Pöpel 1964). The figure illustrates the thinking behind the loading design concept.

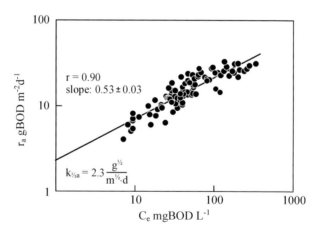

**Figure 12.2.** The same data as shown in Figure 12.1 were later re-interpreted (Harremoës 1977). The removal rate in the reactor is shown as a function of the effluent concentration. This corresponds to a conceptual interpretation, where the bulk concentration determines the removal rate in the biofilm. In its simplest form, this assumption leads to ½-order kinetics, corresponding to a slope of 0.5 in Figure 12.2. It can be seen that the scatter decreased significantly by plotting surface removal rate as a function of effluent concentration. The figure illustrates the simplest version of a conceptual design approach, based on the very same data as in Figure 12.1.

In the loading approach to design, no attempt is made to understand the phenomena that determine the effluent as a cause–effect relationship between driving forces (among which loading is only one), properties (reactor characteristics) and resulting effluent. The approach presented in Figure 12.1 is an empirical approach, supplemented with a statistical interpretation. However, the expectation would be that more knowledge about wastewater characteristics, reactor characteristics, and biofilm properties would improve the relationship and reduce the uncertainty of design. When this is discussed in the profession, it is argued that loading is the practical tool for design! One would think that a reduced uncertainty in design would be a more valid argument for a pragmatic design approach.

## 12.3.2 Science-based determinism

The complexity of wastewater treatment plants by today's standards is hard to incorporate in simple design, like rules of thumb. The combination of several different processes into one plant (organic carbon removal, nitrification, denitrification, bio-P-removal) calls for a systematization and structuring of the experience. The mathematical formulation of a model constitutes such a systematization and structure. It creates a framework of understanding and a compounding of experience.

It has to be realized that the mathematical description of transport and processes is an idealized world based on the formalism of logic, which can be brought to fit the real world only by experience. In philosophical terminology (called epistemology) a mathematical description is based on *deduction* by the logic rules of mathematics on the basis of certain assumptions, but the applicability to engineering practice can be achieved only by *induction,* i.e. by experience.

In its idealistic form this approach is called *determinism*. It is assumed that if the assumptions were met and the parameters of the model were true, then a known input of driving forces on the system will create only one solution. This basic assumption has to be kept in mind under all circumstances because there are many deviations from this idealization. The virtue of the approach is that the deterministic structure is attempted to be universal in nature and to the extent that this is the case, it is possible to expand the application into areas without experience. The practice of using models is to apply this feature and at the same time be aware of the idealizations that may in fact not fit as well to reality as the modeler tends to believe. The use of models creates a whole new set of approaches to the practice of engineering. We are still in the transition from one form of practice to another.

## 12.4 MODELS OF BIOFILM REACTORS

### 12.4.1 Model structure, variables, parameters and forcing input

See Chapters 1–5 for an extensive discussion of biofilm models and simulation techniques. The formulation of a model is based on the following components.

#### *12.4.1.1 Model structure*

The structure of a deterministic model is the mathematical formulation of all the laws of nature that are assumed to be relevant to the simulation of the performance of a treatment plant. The detailed deterministic model is based on a *reductionistic* approach, by which all the phenomena of relevance are described in detail and built into the complex model on the basis of fundamental laws of integration, like the mass balance equations for water and for each compound of the system.

#### *12.4.1.2 Models state variables*

Any model contains state variables, which are the descriptors of the model and supposedly also of the real system, like the concentration of a compound or biomass.

#### *12.4.1.3 Model parameters*

The model structure and the laws associated with the structure contain parameters, which together with the formulation of the law constitute the simulation of the performance of the system. It is attempted to choose laws of such universality that the parameters are invariant, i.e. they are constants, independent of environmental circumstances or history. However, the reality is that the structure does not always feature such universality that the parameters are universally constant. In fact, some parameters are not constant at all, because they depend on circumstances not accounted for in the laws incorporated in the formulation of the model structure. It is important to have these parameters well identified and the choice related to the pertinent circumstances.

#### *12.4.1.4 Forcing input*

The model is capable of simulation of the cause–effect relationship between an external load on the system, which is described by the model. The application of the model consists of analyzing the effects of chosen loads on the system. These external loads have to be determined or assumed. In the deterministic formulation the model calculates the output associated uniquely with the chosen input.

The practice of model application is to choose a model structure, such that it represents the key features of relevance to the engineering application in question. This can be based on *a priori* knowledge and a dedicated investigation of the local circumstances at reasonable cost. The aim is to create results, which are of adequate reliability for the application.

## 12.4.2 Model applications

Models can be applied for the following purposes.

### *12.4.2.1 Planning tool*

Treatment plants are just one component in a much larger system of water management. Water management is delineated by the hydrological catchment, in which system the treatment plant functions as the engineering facility that converts polluted water to purified water before discharge from an urban catchment to the receiving waters. This system is very complex and difficult to manage. In recent years the analysis of managerial options has been investigated with models of the whole hydrological catchment or of the local urban catchment. That calls for models of treatment plants that are integrated with other models of the urban sewer system and of the receiving waters. Only the most relevant of features can find room in models of such scale and complexity (Ljiklema *et al* 1993; Harremoës *et al*. 1994; Matos 2002).

### *12.4.2.2 Analysis of existing plants*

The reason for investigating existing plants may be either because they do not operate as intended or because they have to be upgraded to better performance, e.g., extended to include nutrient removal. The virtue of the analysis is that there is a lot of experience to deal with. A model can be adjusted to fit the data and then used to analyze options for improvement. However, that applies only to the extent that experience contains relevant data sets suitable for model fitting (e.g. Dupont and Sinkjær 1994; Sinkjær *et al*. 1996a,b; Harremoës *et al*. 1998).

### *12.4.2.3 Design of new plants*

New plants suffer from lack of information on the circumstances of operation in the future. Forecasting is inherently uncertain and the design has to account for that uncertainty. Therefore, it is frequent practice to use the traditional design rules approach for design. This can be justified because of the uncertainty of forecasting. However, in important cases it is relevant to test the performance of complex plants against the assumed forcing inputs and to test the sensitivity of the design to variability of assumed input.

### *12.4.2.4 Real-time control of plant*

It is the tradition of the profession to operate wastewater treatment plants based on fixed settings and manual control. Development has demonstrated the virtues of dynamic control of the plants. Several advantages can be achieved: better effluent and savings on energy and chemicals. Real-time control of wastewater treatment plants is still in its infancy and there are many approaches to real-time control (Lijklema *et al.* 1993; Lynggaard-Jensen and Harremoës1996; Harremoës and Rauch 1999; Spanjers *et al.* 1998).

The idea behind fixed rule-based control is that operation is controlled automatically on the basis of fixed rules on an 'if-then-else' basis. That is in fact how treatment plants have been run for years. The difference by today's standard is that information from sensors can be used as the information base for fixed rules. The virtue of the simulation model is the fact that such rules can be tested with respect to performance by running the model of the plant on the basis of information on the existing plant in an effort to find good rules.

The optimized control requires an on-line model for the optimization. The optimization is based on an objective function. Such a function consists of the relevant performance variables and the merit associated with that performance. It may be least cost or best effluent, or a combination of the two. The model of the plant can predict the performance on the basis of forecasted input. The operational options are analyzed to choose the best value of the objective function (a minimum or maximum). The model can be adjusted to account for the most recent experience with the plant operation, periodically or on-line. In this case, the model is a prerequisite for the improved performance. This system is quite complex. To make the totality of the system less overwhelming there is a need to simplify the models to suit the purpose of real-time control.

### *12.4.2.5 Models as research tools*

There are still many features of the operation of treatment plants that are not well understood and not well described. Models may serve as tools with which to analyze information gained from dedicated investigations. This calls for models chosen to describe specifically the processes under investigation to identify which interpretation fits the best, or even better: show the greatest universality.

## 12.4.3 Level of aggregation

Level aggregation means the level of detail or resolution in the description of the outcome. The level of aggregation is determined by the engineering application. In the extreme, the planning tools need a low level of aggregation whereas the research tools necessarily need a high level of aggregation on the processes to be investigated. In an integrated model of the river Rhine little detail and low resolution in time and space is required, and uncertainty is evened out in the integration of the system. On the other hand, a specific investigation of the performance of an existing treatment plant during rain may call for a high resolution in time to account for the processes during the fast changes associated with transient loading during rain, e.g. during hydraulic overload of the final clarifier.

The level of desired aggregation has implications on the level of complexity of the model to be chosen. Any model structure is a choice between simplicity and complexity. The key is to fit the choice to the application in question.

## 12.5 MODEL CALIBRATION AND PARAMETER ESTIMATION

### 12.5.1 Model structure

A model consists of a model structure, variables and parameters. The ideal of a model is achieved as an attempt to create a model with as much universality as possible. Looking at the past 500 years of scientific development it becomes clear that this choice is no simple thing. It is based on the accumulated know-how over an equal span of time. The ideal model is the model which is simple and which has been shown by induction to apply universally to the scope of engineering application in question.

Nobody with a scientific background will question the universality of the concept of mass balance as a valid tool in dealing with models. It is another question whether we can formulate the components in the equation with equal certainty.

> **Example 12.2**
> The equations for bacterial growth are well established. Though these equations are recognized as the basic formulation of growth and yield, it is well known that these equations are mere simplifications of a much more complex reality, governed by enzymatic activity in a complicated interplay between the organisms and the environment. It is implicitly assumed that the equations apply and that the three parameters (maximum growth rate, yield constant, and half saturation constant) are constants. From a scientific point of view, we know that this is a gross simplification; but from a pragmatic point of view the issue is not whether it is true or not. What counts is whether it is a good assumption, and that depends entirely on the engineering problem in question. Choice of simplicity that suits the issue in question is not science, but an engineering art.

The art of model application is to know when chosen assumptions constitute a sufficiently good approximation to the reality of the problems in question. This can be achieved by induction only, i.e. by experience. The advantage of such formulations is that the model constitutes a simplified condensation of a much more complicated experience, which is complicated because of the lack of structure otherwise associated with experience. The formulation can claim universality due to induction within the range of experience only.

The issue of universality is crucial. It is not possible to make a scientific investigation into the applicability of all the components in a model for each case of application. The claim of universality rests on *a priori* knowledge of the laws of nature and their applicability.

It is very difficult to test the applicability of the model structure, other than by trial and error and by comparison with the sum of *a priori* knowledge. However, it is important always to be on the alert for signs in each particular case with respect to applicability. A few examples will illustrate this.

Nitrifiers are sensitive to inhibitors in domestic wastewater. This knowledge may change the formulation of the growth equation and/or change the value of the constants. There are several new formulations of the growth equation available and new parameters are introduced (Sinkjær *et al.* 1996b).

The yield constant is not a constant, because the bacteria can store substrate and utilize it later. Description of these phenomena requires a new formulation of the growth equation.

Population changes in the treatment plant alter the properties of the bacteria present, and the growth equations and/or the parameters may change. There is no simple set of options for alternative formulations of equations in this case.

### 12.5.2 Parameter calibration, verification, and estimation

For any application of a model it is essential to determine the parameters that fit the problems in question. Some parameters are reasonably well known based on *a priori* knowledge. That applies to the acceleration of gravity, the viscosity of water, some stoichiometric constants, and so forth. Nobody dealing with modeling of wastewater treatments would dream of finding such parameters on the basis of a dedicated local investigation. In other cases the parameters are known to depend on local conditions and have to be determined on the basis of local data. There are several procedures for this very important process.

#### *12.5.2.1 Calibration*

The most frequently used procedure with which to make models fit to a particular situation is called *calibration*. This is based on a time series of loading of the treatment plant and of the concentrations of the effluent from the plant. This combination of the input and output data express the transformation performed by the plant, which is precisely what the model is supposed to simulate. The procedure is now to change the parameters of the model such that the best fit to the effluent data is achieved. This is frequently done empirically on a trial and error basis. The best fit can be defined as the set of parameters that gives a minimum of the standard variation of the difference between the data and the output from the model calculations.

## 12.5.2.2 Verification

In an attempt to evaluate the result of the model performance with these parameters, another time series of loading of the treatment plant and of the concentrations of the effluent from the plant is used to compare the fit without adjusting the parameters. This procedure is frequently considered to be a test of the quality of the model application. The approach is called *verification*. In practice the procedure is to use one half of a time series of data for calibration and the rest of the available data series for verification.

There is reason for concern in the interpretation of the result (Harremoës and Madsen 1999) because there is no guarantee that another parameter set could not have achieved a similar fit. In other words, the parameter set is not uniquely defined by the procedure. In fact, there may be little difficulty in getting a good fit if there are a sufficient number of parameters to manipulate. The data set may not contain information of such kind or extent as to allow the parameters to be determined on that basis. Some parameters may not be *identifiable* on the basis of the information contained in the data series (see below).

Suppose that the two data series have identical statistical properties (belong to the same statistical population), then there is no reason to expect a different fit for the verification, except for variations within the standard deviation of the fit. Then, why not use the whole series as a means for determination of the standard deviation of the fit and use that as a measure of the fit.

Suppose that the two series do not have the same statistical properties, then the whole approach is false. However, the reason may be a poor set of parameters. It might be wiser to use the whole series for calibration because the wider range of information will create a better calibration. The procedure of calibration/verification has one significant virtue:

The procedure can identify *over-parameterization*, i.e., when there are too many parameters to be determined compared with the information available in the time series. In such a case, a good fit can be achieved by calibration, but it will fail in verification. The fact is that most models are *empirically underdetermined* by information from data series because the series from practice do not contain sufficient information.

Each increase of complexity of a model increases the number of parameters to be calibrated. There are examples to show that increase in model complexity and detail can in fact decrease the fitness and the predictive quality of the model due to *over-parameterization*.

Depending on the quality of the data series, the tendency is that quite a small number of parameters and only certain parameters can be calibrated by this approach. The solution is to choose which parameters are to be calibrated and consider all other parameters to be uniquely determined by *a priori* knowledge. The art is to choose the right parameters for calibration, because the requirement is that they are identifiable on the basis of the data sets available, and that there is sufficient *a priori* knowledge by which to fix the rest of the parameters to a predetermined value.

### 12.5.3 Parameter estimation

There exist statistical procedures for parameter estimation. The virtue of these procedures is that they can be used to determine the best parameter values that satisfy chosen statistical criteria. The result includes the uncertainty with which these parameters have been determined and thus whether the parameter is identifiable on the basis of the data available. Included is also the standard deviation of the fit and information on the fitness of the model structure. These are the procedures to be recommended. The problem is that these procedures are quite complex and usage quite specialized (Harremoës and Madsen 1999).

### 12.5.4 Experimental design

The experience is that model application in practice is based on inadequate data. Without realizing it, the modeler fits his or her model to inadequate data and assumes to have calibrated the model for practical application. The consequence may be use for extrapolations far outside the range of information covered by the data and outside a reasonable expectation of model performance.

The solution is to apply experimental design dedicated to the problem in question (Harremoës 1997). The fundamental rule is that the data provided from experimentation shall contain data with such information that the parameter in mind can be identified. A simple example can illustrate the point. If the intention is to determine the $K_s$-value in the growth equation, it is quite obvious that the experiments have to include data on growth rates at concentrations in the low range of performance, say below $2*K_s$. If not, there is no way of determining $K_s$. However, in a usual calibration procedure there is no warning not to fix the $K_s$-value, unless the trial and error approach reveals that changing the $K_s$-value does not influence the fit. In a more sophisticated fashion a sensitivity analysis can be done on $K_s$. In a situation with no data in the proper range, there is no alternative to consider $K_s$ an *a priori* known value. The real alternative in case $K_s$ needs to be determined is to design an experiment such that it does contain the information required to determine $K_s$. The approach is to make $K_s$ identifiable by design. This reasoning applies to all determinations of parameter, e.g., determination of $\mu_{max}$, in relation to $K_s$ and $Y_{max}$. In fact, most experiments and monitoring make it possible to determine the ratio $\mu_{max}/Y_{max}$, not the individual value of each. The standard procedure is to measure $Y_{max}$ separately or consider $Y_{max}$ *a priori* known (Carstensen *et al.* 1995).

The point is that data series should contain *excitations* created by transient loading to cover a large range of situations. This has interesting implications with respect to operation: To get adequate information, the treatment plant should be run with as much variation as possible without disturbing the effluent compliance. This is interesting because the standard practice of operating treatment plants is to run them as invariantly as possible. Ironically, this is the way by which the least information about the kinetic properties of the treatment plant is obtained.

In conclusion, no model should be applied without proper attention to the requirements associated with the application and without deciding on a proper model structure and the approach to determination of model parameters. These may be considered adequately determined by *a priori* knowledge and fixed accordingly. The other parameters have to be determined for the situation in question. This requires experimental design, either by creating data time series for input and corresponding output with adequate excitations or by making dedicated experiments in the laboratory or on-line at the full-scale treatment plant.

## 12.5.5 Model uncertainty

Any model prediction has an inherited uncertainty. That uncertainty can be determined by parameter estimation and by the standard deviation of a fit, but care has to be taken with extrapolation. Uncertainty of extrapolation can be estimated on the assumption that the model structure is a good approximation to reality and that the model parameters are constant, but subject to statistical variation. Any model prediction should in fact be described as a predicted mean performance and an estimated uncertainty. By today's standards, the information on uncertainty is as important as the prediction of a mean.

In estimation of model uncertainty, sensitivity analysis and uncertainty analysis are important tools.

### 12.5.5.1 *Sensitivity analysis*

In sensitivity analysis each parameter is estimated for importance regarding the ultimate result of performance. From such an analysis the importance of each parameter can be estimated for the range of operation in question. That is important for the selection of fixed parameters on the basis of *a priori* knowledge and parameters to select for dedicated experimentation.

### 12.5.5.2 *Uncertainty analysis*

The error propagation is a means for determination of the uncertainty of the prediction. It is done by choosing an uncertainty for each parameter and the input data on the basis of *a priori* knowledge and then combine the uncertainty in the model as it affects the ultimate prediction. This can be done by Monte Carlo simulations, which constitute many runs with random selection of parameters according to their *a priori* estimated statistical distribution. In this way an output statistical distribution is generated.

The difference between sensitivity analysis and uncertainty analysis is that the former investigates one model component at a time, while uncertainty analysis determines the influence of all model components and the interrelationship of their influence on model outcome.

## 12.6 TREATMENT PLANT DESIGN

### 12.6.1 Identification of problem

Design in this context is understood to be determination of the configuration and size of each component in the treatment train from inlet to outlet. It may be a virgin layout, because no treatment plant was there in advance or it may be the expansion of an existing plant, typically to include upgrading from a traditional treatment for BOD-removal to nutrient removal.

An important component in design is to choose the parameters characterizing the influent to be treated. That applies to the flow to be designed for and the typical concentrations. The existing inflow and influent concentrations can be measured accordingly. The uncertainty is the estimate of the future conditions under which the treatment plant is to be operated. The approach when little is known about future circumstances is to choose scenarios. A scenario is not a prediction of future circumstances and performance, but a plausible set of circumstances that are chosen with the expectation that it may be close to an anticipated future. Owing to uncertainty in choosing scenarios a spectrum is chosen for analysis, to assess the reaction of the system to plausible circumstances.

The level of design procedure has to be chosen:
- In very small plants, e.g. single family houses, the plant does not have to be designed, but pre-designed plants can be bought according to size
- In small plants, e.g. villages and small towns with no anticipated problems from industrial wastewater, the plants can be designed on the basis of the well-established loading rules.
- In larger plants, e.g. middle size towns or any plant with a significant industrial load, the plant can be designed on the basis of design rules, but it is wise to check the design with computer simulations of the design for expected wastewater loading.
- In plants for large cities, the plants should be designed and the design checked by computer simulations. This should include sensitivity and uncertainty analysis of the operation to various loading and parameter options, including alternative future optional development of loading. This could include pilot-plant studies to identify problems and to check the parameter values and functionality of the simulation, e.g. for nitrification (Sinkjær *et al.* 1994).

## 12.6.2 Model for biofilm system

The models for biofilms are complicated by the fact that the diffusional limitation of reaction has to be taken into account. Attempts to use activated sludge models for biofilters have failed. The reason is that the phenomena typical of biofilms simply cannot be modeled without accounting for the zonation of the biofilm. Different processes take place in each zone, because the redox conditions vary. The typical example is denitrification under aerobic conditions in the bulk water, while diffusional limitation of oxygen penetration creates anoxic conditions in a zone submerged in the biofilm.. Another example is the conditions for nitrification. The outgrowth of the nitrifiers in biofilms is a very different mechanism, compared with the wash-out and sludge age criteria for activated sludge plants (see below).

The complication of the models for biofilms is that the processes known from activated sludge have to be incorporated into a model structure, which includes the diffusion of substrates into and the products out of the biofilm, the growth of the biofilm, and the population dynamics inside the biofilm. Many of the parameters in the model are quite uncertain due to less consistent *a priori* knowledge of model structure and parameters.

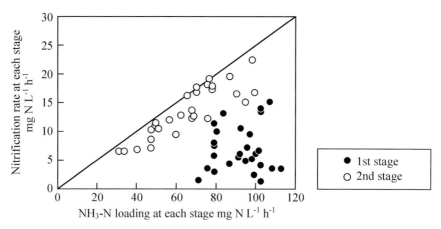

**Figure 12.3.** Shown are results from an investigation of nitrification in a biofilter. The parameters are volumetric rate of nitrification in mgN L$^{-1}$h$^{-1}$ as a function of loading of N in mgN L$^{-1}$h$^{-1}$ per unit reactor volume (modified from Yumikura *et al*. 2000). First, the reaction per unit volume is not the design parameter of choice, because the reaction is primarily a function of surface area. Second, there is no unique relationship between volumetric loading and nitrification. The rate of nitrification depends on the competition between the heterotrophs and the nitrifiers for aerobic space in the biofilm (see also Figure 12.4).

**Example 12.3**
Despite the complexity described above, an example will illustrate the need to interpret results on the basis of process understanding rather than simple loading relations.

Figure 12.3 shows results from an investigation (Yumikura *et al*. 2000) of an anoxic/aerobic bio-filter system for nitrogen removal. It shows the volumetric rate of nitrification in mg NL$^{-1}$h$^{-1}$ as a function of loading of N in mg NL$^{-1}$h$^{-1}$ per unit reactor volume. The results are shown for the first and the second stage. It is well established (Henze *et al*. 2000) that autotrophs and heterotrophs compete for space in the aerobic part of the biofilm. If the biofilm grows too fast owing to a high organic carbon content available to heterotrophs, the nitrifiers are outgrown and end up inactive in the anoxic/anaerobic depth of the biofilm. This phenomenon is essentially governed by the ratio between the concentration of organic matter and the concentration of oxygen (Hagedorn-Olsen *et al*. 1994) as illustrated in Figure 12.4 (after Henze *et al*. 2000). This relationship is quite universal. Not surprisingly, there is a significant scatter of the points based on the loading, as illustrated in Figure 12.3, because none of the essential variables have been taken into account.

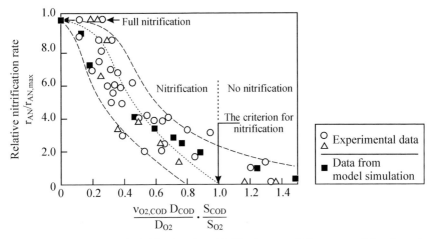

**Figure 12.4.** The rate of nitrification in a biofilm is essentially governed by the ratio between the concentration of organic matter and the concentration of oxygen, because heterotrophs and nitrifiers compete for aerobic space (from Henze *et al.* 2000). $S$ is concentration, $D$ is diffusion coefficient, $\nu$ is stoichiometric constant, and $r_{AN}$ is rate of nitrification per unit surface area of biofilm.

## 12.6.3 Pilot plant investigations

Pilot plant studies are much more expensive than model simulations. However, they do provide more substantial information about the reality in anticipation of the performance of the full-scale plant. Again, it is important to identify what the issue is. Pilot plants are mostly run for two very different reasons.

- The pilot plant study is operated entirely with to demonstrate that the configuration of the suggested design will do the job. That is the most frequent motivation. Accordingly, the pilot plant is operated for the entire period of the investigation in a fashion similar to the anticipated operation of the full-scale plant. That approach provides very little information useful for an understanding of what goes on in the pilot plant. Because of the cost of running pilot plants it is strange that there is no more incentive to generate better information than just compliance.
- The pilot plant is operated entirely to identify/verify the values of the design parameters. It has to be identified which design parameters are the essential ones. That is no easy task, because the period is short in most cases and the cost is high. One approach is to verify the standard loading figures, another to verify the parameters of the model used for simulation. There is most

perspective in the latter, because more information is gained per unit cost of pilot-plant operation. The main point is to design pilot plant operation such that the parameters looked for become identifiable. That requires loading variations, which may not resemble the loading to be experienced in full-scale operation. In the extreme, the idea is to bring the pilot plant to the brink of failure, even beyond, to identify the security against failure.

> **Example 12.4**
> Figure 12.5 shows two results of an investigation of P removal in a biofilter system for treating wastewater from a car-washing facility (Pak and Chang 2000). The two figures show P removal in $gP/m^3$ as a function of organic loading in $kgCODm^{-3}d^{-1}$ and as function of nitrogen loading in $kgT-Nm^{-3}d^{-1}$. It might just as well have been plotted as a function of hydraulic load, because the wastewater concentrations are described to be $COD = 420$ mg/l, $BOD = 118$ mg/l, $TN = 57$ mg/l and $TP = 17.5$ mg/l, with no variation at all.
>
> The result may be applicable to the design of that particular treatment plant under those particular circumstances, but there is no universal value in the data, because the input data are interrelated and it is impossible to interpret the cause–effect relationship underlying the data. The key issue has not been identified and change of just one inlet value may significantly change the design.

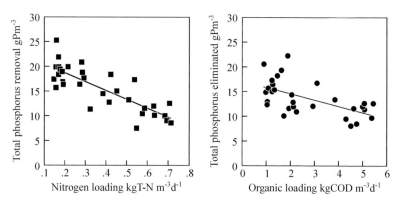

**Figure 12.5.** Results from an investigation of P removal in a biofilter. The effluent concentration from P-removal in g $P/m^3$ is shown as a function of organic loading in kg COD $m^{-3}d^{-1}$ and as function of nitrogen loading in kg T-N $m^{-3}d^{-1}$. The question is whether there is any universality shown by such relationships. Volumetric loading is not the important parameter. The concentrations were constant during the experiments and variation was based on changing hydraulic flow to the reactor. These practical results represent valuable, empirical observations for a particular set of circumstances, but it is difficult to see how the experience can be translated to other circumstances (modified from Pak and Chang (2000)).

## 12.7 ANALYSIS OF EXISTING PLANT / PILOT PLANT

Model simulation of an existing plant is a valuable tool for the analysis of performance, understanding of the reasons for abnormalities, and investigation of means for improvement.

### 12.7.1 Identification of the problem

It has to be realized that the models at disposal contain many functions and many parameters, some of them of universal applicability, some subject to local variations depending on climate, influent characteristics, bacterial population and mode of operation. It is not feasible to monitor an existing plant with respect to all functions and parameters. Some have to be chosen as universally applicable and some have to be selected for analysis by model fitting or to be determined by dedicated experimentation. There is no clear demarcation between the two, because the operation of the plant can become part of deliberate alterations with the purpose of increasing identifiability of the parameters selected for analysis: the more variation, the better.

The key is that the purpose of the analysis has to be identified and the monitoring, the operation, and the external experimentation has to be decided upon as a function of that purpose. Mere model fitting may not reveal information justifying the cost.

### 12.7.2 Design of experimental program

Many of the parameters can be determined by dedicated experimental procedures only. The report on Activated Sludge Model No. 1 (Henze *et al.* 1986) has a description of various tests to be made for identification of certain important parameters. Another approach to experimentation is to monitor the performance of the plant. That can be done by taking frequent samples for chemical analysis or by on-line sensors. Frequent sampling increases costs and efforts required, which do not make it an attractive approach. In recent years much better monitoring has become feasible by the installation of on-line sensors. There are two different approaches to on-line measurements.

1. On-line, on-stream sensors, which measure concentrations as they vary in the plant during operation. The spectrum was presented at conferences on sensor technology (Lynggaard-Jensen and Harremoës 1996).
2. On-line, off-stream sensors provide information on concentrations in the side-stream or batches taken out from the system at a certain frequency and analyzed automatically. An example is to extract a sample for on-line OUR-

analysis of the influent or to make side-stream batch-tests for measurements of rate of respiration in the activated sludge. The spectrum of options is described in the report (Spanjers *et al.* 1998).

On-line measurements give an overwhelming quantity of information, which cannot be coped with without model interpretation and automated estimation of parameters. There are new methods suitable for on-line parameter estimation, but it has to be realized that these methods can estimate a few parameters and only such parameters, which are identifiable on the basis of the data collected. Also in this case, it is important to identify the purpose of the analysis and the means by which to achieve these goals (Harremoës 1997).

## 12.8 OUTSTANDING ISSUES OF ENGINEERING SIGNIFICANCE

### 12.8.1 Key phenomena not fully understood

Many phenomena in biofilms and in biofilm reactors have been elucidated over the past 30 years, but essential concepts are still poorly understood to the extent that they pose an obstacle to proper design and operation of biofilm reactors. Such phenomena are listed below.

#### *12.8.1.1 Biomass*

A simple parameter such as biomass concentration in biofilm reactors is relatively unknown. There is no general theory explaining why the density, the concentration of bacteria, and the extracellular polymeric substances (see also Chapter 8) of the biofilm vary to the extent observed. Variability is a result of biological responses to environmental conditions, chemical and physical, which are not known in any detail. Their descriptions are purely empirical as far as application is concerned.

#### *12.8.1.2 Population dynamics*

At present, biofilm kinetics involve bacteria performing organic matter degradation, nitrification, denitrification and biological P removal, plus the much more specialized functions related to degradation of special chemicals. In engineering kinetics, the bacterial groups are in fact identified only by process, not by taxonomic identification. The experience is that the traditional taxonomic identification provides little information of engineering significance.

Modern molecular biology technology may have the potential for improving that situation as has been outlined in several chapters in this book. However, it has to be realized that, from an engineering point of view, the aim is to relate genetic characterization to engineering performance. Undoubtedly, a lot of new information is under way. The challenge is there to provide operational information that will contribute to identification, understanding and functional description of cause–effect relationships that can be applied in engineering. Understanding biological P removal is as important to biofilm performance as to activated sludge performance.

### 12.8.1.3 Particulate organic matter

Diffusional resistance to degradation of dissolved organic matter is a well-described phenomenon for biofilms and has been applied with success. However, it is a disturbing fact that wastewater, e.g. domestic, contains more organic matter in particulate form than in dissolved form. It is well known that hydrolysis of particulates is a slower process than the mineralization of dissolved organic matter. The role of particulates in biofilm reactors is poorly understood and deserves more attention (Larsen and Harremoës 1994; Confer and Logan 1997a,b; Janning *et al*. 1998; Morgenroth *et al*. 2002).

### 12.8.1.4 Unidentified reactions in biofilms

The zonation of biofilms is well described for thin biofilms. Very thick biofilms are characterized by having an outer layer of different zones with reasonably well-described processes, but what happens at greater depths of biofilms is not well described. What is the effect of anaerobic conditions in the interior of thick biofilms? Does the 'sulfur pump' provide a deeper penetration of oxygenation capacity than described by the well known processes? Does methane production have significant effects (e.g., on sloughing due to methane bubble formation)?

### 12.8.1.5 Diffusivity into biofilms

The simple zonation into well-defined diffusion layers have proved to be too simple a description. An increasing number of experimental results appear to suggest that the configuration of the biofilm is of significance to the zonation. Investigations have shown the simple zonation theory to be an adequate description for many practical purposes. However, it is not well known when and under what circumstances other phenomena influence the diffusion/transport of substrate into the biofilms.

### *12.8.1.6 Biomass control*

It is of paramount importance to the engineering of biofilm reactors that the biomass is well controlled. In too many cases, lack of control of the biofilm has led to disastrous results. The nightmare of the treatment plant operator is to face a clogged-up bioreactor, which would be as disastrous an occurrence as bulking sludge in the activated sludge system. Biomass control is a completely empirical engineering art that could be improved.

### 12.8.2 Research needs and engineering development

The engineering application of biofilm reactors was completely dominant at the beginning of the 20th century. The activated sludge system was developed in the 1920s and gained favor from the 1950s onward, but it was the demand for nutrient removal during the 1980s and 1990s that tilted the balance towards activated sludge systems, and biofilm systems fell into disfavor. Not until new technologies emerged (backwashing biofilter technology and moving bed reactors) did biofilm reactors regain popularity.

There are at least three reasons for this fact.
1. Technology with adequate facilities for removal of nutrients and adequate biomass control has not been developed until recently.
2. The technology is relatively costly and is competitive mostly under special conditions, e.g. locations with little space for expansion may favor treatment plants with a smaller 'footprint'.
3. Some of the issues mentioned above as inadequately known are of engineering significance and are detrimental to proper design until better described.

Research and development should be given priority because of the potential of biofilm reactors, a potential further emphasized by the new technology associated with membrane biofilm reactors. There is a need for both fundamental research and for engineering development. There is good reason to believe that in combination they can achieve more than they can individually. The need for bridges between science and engineering is as important as ever.

## 12.9 References

Carstensen, J., Harremoës, P. and Madsen, H. (1995) Statistical identification of monod-kinetic parameters from on-line measurements. *Wat. Sci. Tech.* **31**(2), 125–133.
Confer, D.R. and Logan, B.E. (1997a) Molecular weight distribution of hydrolysis products during biodegradation of model macromolecules in suspended and biofilm cultures .1. Bovine serum albumin. *Wat. Res.* **31**, 2127–2136.
Confer, D.R. and Logan, B.E. (1997b): Molecular weight distribution of hydrolysis products during the biodegradation of model macromolecules in suspended and biofilm cultures. 2. Dextran and dextrin. *Wat.Res.* **31**, 2137–2145.
Dupont, R. and Sinkjær, O. (1994) Optimisation of wastewater treatment plants by means of computer models. *Wat. Sci. Tech.* **30**(4), 181–190.
Hagedorn-Olsen, C., Møller, I.H., Tøttrup, H. and Harremoës, P. (1994) Oxygen reduces denitrification in biofilm reactors. *Wat. Sci. Tech.* **29**(10–11), 83–91.
Harremoës, P. (1977) Half-order reactions in biofilm and filter kinetics. *Vatten* **33**, 122–143.
Harremoës, P. (1978) Biofilm kinetics. In *Water Pollution Microbiology* (ed. R. Mitchell), vol. 2, pp. 71–109. John Wiley, New York.
Harremoës, P. and Wilderer, P.A. (1993) Fundamentals of nutrient removal in biofilters. In *9th EWPCA–ISWA Symposium, München, 11–13 May, Documentation, Liquid Waste Section*, pp. 111–126. Abwassertechnische Vereinigung e.V., St. Augustin.
Harremoës, P., Hvitved-Jacobsen, T., Lynggaard-Jensen, A. and Nielsen, B. (1994) Municipal wastewater systems, integrated approach to design, monitoring and control. *Wat. Sci. Tech.* **29**(1/2), 419–426.
Harremoës, P. (1997) Transient experimentation for process understanding and control. In *Environmental Biotechnology, International Symposium, Oostende, April 21–23, Technological Institute, Oostende*, pp. 1–8.
Harremoës, P., Haarbo, A., Winther-Nielsen, M. and Thirsing, C. (1998) Six years of pilot plant studies for design of treatment plants for nutrient removal. *Wat. Sci. Tech.* **38**(1), 219–226.
Harremoës, P. and Rauch, W. (1999) Optimal design and real-time control of the integrated urban run-off system. *Hydrobiologia* **410**, 177–184.
Harremoës, P. & Madsen, H. (1999) Fiction and reality in the modeling world – Balance between simplicity and complexity, calibration and identifiability, verification and falsification. *Wat. Sci. Tech.* **39**(9), 1–8.
Harremoës, P. (2000) Scientific incertitude in environmental analysis and decision making. 'The Heineken Lecture'. In *The Heineken Award for Environmental Sciences*, The Royal Netherlands Academy of Arts and Sciences, October 2nd, 2000.
Henze M, Grady C.P.L., Gujer W., Marais G. v. R. and Matsuo T. (1986) *Activated sludge model no. 1*, Scientific and Technical Report No. 1. IAWPRC, London.
Henze, M., Harremoës, P., Jansen, J. la Cour and Arvin, E. (eds.) (2000) *Wastewater Treatment. Biological and Chemical Processes*. 3nd edn. Springer, Berlin.
Janning, K.F., le Tallec, X. and Harremoës, P. (1998) Hydrolysis of organic wastewater particles in lab scale and pilot scale biofilm reactors under anoxic and aerobic conditions. *Wat. Sci. Tech.* **38**(8–9), 179–188.
Larsen, T.A. and Harremoës, P. (1994) Degradation mechanisms of colloidal organic matter in biofilm reactors. *Wat. Res.* **28**, 1443–1452.

Lijklema L., Tyson J.M. and Le Souef A. (eds). (1993) INTERURBA Selected proceedings of the 1st International Conference on Interactions Between Sewers, Treatment Plants and Receiving Waters in Urban Areas *Wat. Sci. Tech.* **27**(12), 1–236.

Lynggaard-Jensen A. and Harremoes P. (eds) (1996) Sensors in wastewater technology. *Wat. Sci. Tech.* **33**(1), 1–336.

Matos, J.S. (ed.) (2002) INTERURBA II Selected proceedings of the 2nd International Conference on Interactions Between Sewers, Treatment Plants and Receiving Waters in Urban Areas, held in Lisbon, Portugal, 19 – 22 February, 2001. *Wat. Sci. Tech.* **45**(3), 1–279.

Morgenroth, E, Kommedal, R. and Harremoës. P. (2002) Processes and modeling of hydrolysis of particulate organic matter in aerobic wastewater treatment – a review. *Wat. Sci. Tech.* **45**(6), 25–40.

Pak, D. and Chang, W. (2000) Factors affecting phosphorus removal in two biofilter system treating wastewater from car-washing facilities. *Wat. Sci. Tech.* **41**(4–5), 487–492.

Pöpel, F. (1964) Leistung, Berechnung und Gestaltung von Tauchtropfkörperanlagen, Stuttgarter *Beridhte zur Siedlungswasserwirtschaft*, 9., Forschungs- und Entwiklungsinstitut für Industrie- und Siedlungswasserwirtschaft, Stuttgart.

Sinkjær, O., Yndgaard, L., Harremoës, P. and Hansen, J.L. (1994) Characterisation of the nitrification process for design purposes. *Wat. Sci. Tech.* **30**(4), 47–56.

Sinkjær, O., Thirsing, C., Harremoës, P. and Jensen, K.F. (1996a) Running-in of the nitrification process with and without inoculation of adapted sludge. *Wat. Sci. Tech.* **34**(1/2), 261–268

Sinkjær, O., Bøgebjerg, P., Grüttner, H., Harremoës, P., Jensen, K.F. and Winther-Nielsen, M. (1996b) External and internal sources which inhibit the nitrification process in wastewater treatment plants. *Wat. Sci. Tech.* **33**(6), 57–66.

Spanjers H., Vanrolleghem P.A., Olsson G. and Dold P.L. (1998) *Respirometry in Control of the Activated Sludge Process: Principles.* Scientific and Technical Report No. 7. IAWQ, London.

Temple, R. (1998) *The genius of China – 3000 years of science, discovery and invention.* Prion Books Ltd., London.

Yumikura, J., Ueda, E., Mikawa K. and Emori H. (2000) Study of novel anoxic/aerobic bio-filter system for nitrogen removal. In *Proceedings of the IAWQ Conference on Biofilm Systems, October 17–20, New York, USA.*

# 13

# Effect of clay particles on biofilm composition and reactor efficiency

*Luis F. Melo and Maria J. Vieira*

## 13.1 INTRODUCTION

Most wastewaters contain suspended inorganic particles a few micrometers in size that may act as adhesion surfaces for biofilm formation or penetrate and become incorporated in the biofilm attached to a carrier in a reactor. How can such phenomena change biofilm population and architecture? And what are the consequences for the efficiency of wastewater treatment plants?

Previous reports on the role of clays in soil microbiology, in aqueous microbial suspensions and in biofilm systems have already shown that relevant effects can be obtained from the presence of such particles inside the biological matrix, through their interference with the microbial activity and the physical architecture of the biofilm. This, in turn, is expected to affect the performance of biofilm reactors, and was actually confirmed in a few cases.

© 2003 IWA Publishing. *Biofilms in wastewater treatment*. Edited by S. Wuertz, P.L. Bishop and P.A. Wilderer. ISBN: 1 84339 007 8.

There are not many studies about the effect of inorganic particles on the architecture and activity of biofilms. Most of those studies were focused on clays (such as kaolin and montmorillonite), because those particles are quite common in industrial and natural waters (Lowe 1988; Bott and Melo 1992; Srinivasan *et al.* 1995; Vieira and Melo 1995). Regarding the operation of wastewater treatment processes, work has been published on the role of inorganic particles on the efficiency of biofilm reactors, both in anaerobic digestion (Murray and van den Berg 1981; Pérez Rodriguez *et al.* 1992; Muñoz *et al.* 1994) and in nitrification (Gisvold *et al.* 2001; Vieira *et al.* 2001).

Therefore, the results published so far indicate that inorganic particles affect, usually in a positive way, the physical architecture and population structure of biofilms, as well as the bacterial activity and resistance to different stresses (hydrodynamic shear stress, temporary lack of nutrients, presence of toxics, etc.).

This chapter starts with brief reviews on the properties of clay particles and on the interactions between these particles and microbial species in suspended cultures and in soils, so as to establish the grounds for a critical assessment of the effects of clay particles in biofilms. It then leads the reader through the discussion of microhabitats created by clay particles and the possibilities of enhancing the genetic exchange within particle-containing biofilms, showing how these localized phenomena may influence the performance of biofilm reactors. It also reviews the existing information about the effects of clay particles on the toxicity of biocides, on the physical properties of microbial films, and on reactor operation, before proposing to the reader a set of questions to be addressed in future research efforts.

## 13.2 PROPERTIES OF CLAY PARTICLES

Clay minerals are crystalline hydrous aluminum silicates that are classified as type 2:1 or 1:1, according to the arrangement of the tetrahedral sheets of silicate ($Si_2O_3^{2-}$) and the octahedral sheets of aluminum hydroxide ($Al(OH)_5^{3-}$).

In 2:1 clays such as montmorillonite, vermiculite and ilite, these molecules are associated as 'unit layers' of (Si tetrahedra-Al octahedra-Si tetrahedra) loosely held together by weak van der Waals forces and the electrostatic interaction of interlayer cations. The relative weakness of these forces allows the penetration of water and other polar molecules between the unit layers. As a result, swelling of the structure occurs, giving rise to a larger surface area due to the exposure of the internal surfaces of the unit layers. It should be noted that one particle of these clays is composed of many superimposed 'unit layers'.

Conversely, in 1:1 clays, such as kaolinite and halosite, the tetrahedra of Si and the octahedra of Al are associated in a 1:1 structure (Si tetrahedra-Al octahedra), held together tightly by hydrogen bonds. Consequently, these clays do not normally expand much in contact with water and do not expose the internal surfaces of the so-called 'unit layers' as much as 2:1 type.

In the formation of some clays, structural cations such as aluminum and silicon were replaced by ions of lower valence in a process called isomorphous substitution, creating many negative charges in the structure. In aqueous media, this positive charge deficiency is compensated by the retention of foreign cations on the surface (hydrogen, aluminum, calcium, magnesium, sodium) (Michaels and Boelger 1962). In 2:1 clays, the charges to be neutralized are present both in the external and internal surface areas. If the particles are subsequently exposed to other conditions, the retained ions can be exchanged with other cations present in the environment. The cation exchange capacity (CEC) of the clay is defined as the maximum amount of cations of the medium that can be retained by a fixed mass of clay. Consequently, 1:1 clays have a smaller CEC than 2:1 clays. For example, kaolinite has a specific surface area of around 10–50 $m^2/g$ and a CEC of 2–10 milliequivalents (meq) per 100 g, whereas montmorillonite has values of 700–750 $m^2/g$ and 120–200 meq/100 g, respectively. The CEC of kaolinite is dependent on the size of the particles, in contrast to that of montmorillonite, which seems to be independent of particle size (Hattori 1973).

Clay particles are plate-like and generally smaller than 10 μm. They do not usually appear as discrete particles but have a tendency to form aggregates, although montmorillonite may be easily separated in single plate-like particles. Michaels and Boelger (1962) explained the process of floc formation in kaolinite suspensions: kaolinite particles are hexagonal plates and negatively charged (charges distributed over the basal surfaces). Under acidic conditions, alumina groups exposed at the edges of the plate apparently bind hydrogen ions and acquire positive charge, causing electrostatic attraction between edges and faces, with the formation of highly expanded 'card-house' flocs as represented in Figure 13.1 (adapted from Strenge and Sontag 1987).

328                    Biofilms in wastewater treatment

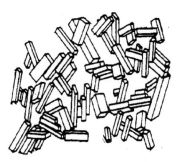

**Figure 13.1.** Schematic representation of the 'card-house' structure (adapted from Strenge and Sontag 1987).

Under alkaline conditions, the edges become either neutral or negatively charged, and the particles deflocculate if the electrolyte concentration is low; however, for high electrolyte concentrations, the particles adhere to one another along their basal surfaces, forming 'card-pack' flocs (Figure 13.2).

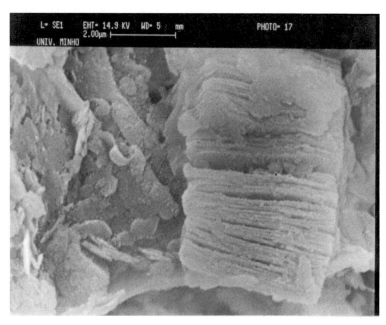

**Figure 13.2.** Card-pack type of agglomerate formed by kaolin particles at pH = 9 (a few bacteria can be seen sheltered between kaolin particles).

The properties mentioned above show that clay particles have very reactive surfaces, influencing the environment where they are immersed in several ways, by:
- modifying the pH of the medium (exchange of protons with the medium);
- supplying ions, initially present in the clay, to the medium, and thereby influencing the nutritional properties of the solution;
- retaining water (especially in 2:1 clays), thereby slowing down the process of desiccation;
- sorbing substances, namely, substrates, inhibitors, and enzymes;
- sorbing bacterial cells on the surface of the particles.

## 13.3 MICROORGANISMS IN SOILS AND IN AQUEOUS SOLUTIONS: INTERACTIONS WITH CLAYS

Many reports were published on the mutual effects of inorganic particles, substrates, and microorganisms in soils (Stotzky 1966; Burns 1975, 1986, 1989; Macura and Stotzky 1980; Huang and Yang 1995; Amin and Jayson 1996; Hommes *et al.* 1998). Other authors worked with aqueous suspensions of particles and microorganisms, both in dispersed cultures and in activated sludges (Lewandowski 1983; Filip and Hattori 1984; Bowen and Dempsey 1992; Chudoba and Pannier 1994; Magdaliniuk *et al.* 1995; Castellar *et al.* 1998). The main conclusions that can be drawn from these studies on particle–microbe interactions in liquid suspensions and in soils are as follows.

1. Depending on the pH, clay particles may present positive charges on the edges, promoting the attachment of negatively charged bacteria. Generally, as mentioned below in (2), the effect of clays on microbial activity is positive, but in one case (Magdaliniuk *et al.* 1995) the particles were much smaller (1–100 nm) than the bacteria (1–3 µm) and seem to have covered the cell surface, thereby hindering the exchange of substrates and metabolites with the medium and decreasing the substrate consumption rate;

2. Particles such as clays and zeolites (Castellar *et al.* 1998) were found to increase substrate consumption and microbial respiration, leading to higher biomass and metabolite productivity. These effects were attributed to the well known capacity of those particles to keep the pH of the medium within suitable values for microbial activity. Other authors showed that adding metal hydroxides and silica to an activated sludge increased the substrate consumption rate, possibly because of the adsorption of toxic ions ($Cu^{II}$) present in the water to the particles. However, if the substrate molecules bind too strongly to the particles (such as certain amino acids to clays), their uptake by microorganisms

may be reduced (Fletcher 1992). It is interesting that the addition of montmorillonite to cultures of *Saccharomyces cerevisiae* enhanced biomass formation, glucose consumption and ethanol production (Hattori 1973). On the contrary, the presence of bentonite in suspended cultures of some bacteria (*Pseudomonas pycocyanea*, *Bacterium megaterium*, *Staphylococcus aureus*, and others) reduced their respiratory activity, as was the case when non-growing cells of *Pseudomonas fluorescens* and *Escherichia coli* were attached to resin particles (Filip and Hattori 1984).

3. The toxicity of some pesticides towards soil microorganisms can be largely reduced when the clay content of the soil is high. These compounds become unavailable after they are adsorbed and bound to the surface of the clay. Stotzky (1992) reported that the biological activity of the organic herbicide paraquat was reduced when it was adsorbed on the interlayer surfaces of expanding clays.

We have recently searched for more detailed explanations of clay particle–bacteria interactions (Pereira *et al.* 2000; Vieira *et al.* 2001; Pereira *et al.* 2002). These studies started with bacterial suspensions and were subsequently expanded to attached biomass systems. One of the main issues in the tests with suspended bacteria was to ascertain whether the effects of clay particles on bacterial activity were due only to the well-known buffering effect of such particles on the pH of the medium, or whether other mechanisms were involved. For such purposes, buffered media were used in all tests with suspended cultures. The results of such tests were as follows.

In one set of experiments (Vieira *et al.* 2001), the activity of suspended nitrifying bacteria (ammonia and nitrite oxidizers) was clearly enhanced in the presence of kaolin particles. For example, the ammonia oxidation rate in the absence of kaolin particles was around 0.5 mg $NH^+_4$–N/kg.s, but it was much larger (15 mg $NH^+_4$–N/kg.s) in the presence of 0.5–1 g/L of kaolin particles. The ions (Ca, Mn, Si, and Al) released by the clay to the aqueous medium appear to stimulate bacterial metabolism because they can be exchanged with the $H^+$ produced during ammonium oxidation. That release is favored by microbial metabolites, which act as ion-solubilizing agents (Stotzky 1986).

In heterotrophic cells of *Pseudomonas fluorescens* at pH 5 (Pereira *et al.* 2000), it was also concluded that kaolin particles stimulated the respiratory activity of the microbial cells in suspension (Table 13.1). This appears to be a consequence of the special spatial configuration of the kaolin aggregates at pH 5, where a 'card-house' open structure is formed (Figure 13.1), conferring a larger available area for bacterial attachment inside the kaolin flocs. These flocs provide a suitable microenvironment for bacterial survival, protecting them from the adverse conditions of the medium (low pH, as compared to the optimum pH = 7).

**Table 13.1.** Enhancement of respiratory activity of suspended cultures of *Pseudomonas fluorescens* at pH 5 (buffered) in the presence of kaolin particles (adapted from Pereira *et al.* 2000).

| Time elapsed (h) | Bacterial activity increase (%) |
|---|---|
| 0.1 | 19.1 |
| 1 | 36.4 |
| 3 | 37.3 |
| 7 | 38.8 |

## 13.4 MICROHABITATS CREATED BY CLAY PARTICLES

Clay particles affect, most of the time positively, bacterial growth, activity, survival and community structure in soil environments (Burns 1989). Most of the reported effects are believed to be due to the well-established capacity of the particles to exchange cations with the surrounding medium. For example, clays may reversibly adsorb substrate molecules, which will be consumed by the microorganisms if lack of such nutrients in the environment occurs, and this will favor microbial growth and development around the clay particles. In other cases, when there is a strong adsorption to the particle surfaces, some compounds may become unavailable. If the latter are deleterious to the metabolism of microorganisms, such adsorption will reduce their toxic effects; conversely, if they are an energy source for microorganisms, it will create unfavorable conditions for microbial growth around the particles. Another important feature of clays is the ability to maintain a suitable pH for growth in their vicinity, owing to their capacity to exchange $H^+$ with the medium. The so-called 'nutritional effect' of the particles has been under discussion (Pérez Rodriguez *et al.* 1992): it appears to be due to the continuous release of inorganic ions ($K^+$, $Na^+$, etc.) which are exchanged with protons produced during microbial metabolism, or to their solubilization by organic or inorganic acids produced by microorganisms.

Although no detailed work on population dynamics in industrial biofilm systems has been published yet, it is reasonable to suppose that the microbial ecology in biofilms will be affected in a way similar to microbial aggregates in soils. In fact, when a biofilm incorporates clay particles, the protons released as a result of microbial activity in the biofilm can be exchanged with basic cations present in the clay lattice. In a multispecies biofilm this fact will affect the distribution of the microbial populations: Those microorganisms more prone to be influenced by pH will tend to grow around the particles that are able to maintain a suitable microenvironment for growth. In autotrophic metabolism (a good example because of the fact that the energy source, the cation $NH_4^+$, is adsorbed by clays, especially by those with a high cation exchange capacity), the $H^+$ produced during nitrification can constantly be exchanged with other cations retained by the clay, namely, $NH_4^+$ providing a better access of the bacteria to this ion. Therefore, in addition to the stratified microbial distribution resulting from the concentration profiles of substrates and oxygen, clay particles bring new contributions to the spatial arrangements within the biofilm matrix.

## 13.5  GENETIC EXCHANGE

Because of the metabolic state of cells and their close proximity in biofilms, the latter offer especially favorable conditions for horizontal gene transfer to the bacteria imbedded in the matrix (Wuertz 2002). In biofilm reactors this may increase their ability to survive within different microenvironments. An example is that plasmids encoding mercury resistance were transferred between bacteria in biofilms (Fletcher 1992). Several bacterial species known to be biofilm formers seem to be competent at taking up DNA and, therefore, they may undergo transformation by foreign DNA. It also seems that many bacteria coping with survival problems display an increased ability to accept extracellular DNA (Veal *et al.* 1992).

When a biofilm reactor used for wastewater treatment is subjected to toxins, such as heavy metals, genetic transfer may help in imparting an adequate resistance to the biofilm bacteria and thus prevent failure in the reactor operation. As stated by Palenik (1989), gene transfer may be an important tool in the long-term success of the biofilm. However, there are still many doubts concerning the more effective techniques to achieve this goal in biofilm reactors.

Free DNA can be present in the biofilm matrix adsorbed to the EPS. Furthermore, as indicated by Wuertz (2002), clay particles can contribute to the genetic exchange in biofilm reactors, because the survival of DNA bound to clay particles will probably also take place in biofilms. Genetic exchange has been demonstrated in a large variety of natural environments, namely soils, sediments and natural water (Veal *et al.* 1992). Khanna and Stotzky (1992) reported that DNA bound to montmorillonite was protected against degradation by DNase, supporting the assumption that 'cryptic genes' may persist in the environment when bound to particles. Moreover, they concluded that bound DNA was capable of transforming competent cells, indicating that the transforming ability of the bound DNA was not altered. Although this assertion is still considered speculative, DNA carried by clay particles can be transferred to competent bacteria which are very close or attached to particles (see Figure 13.3).

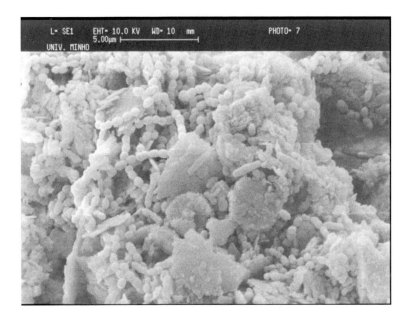

**Figure 13.3.** SEM micrograph of *Pseudomonas fluorescens* biofilm formed in the presence of kaolin particles.

## 13.6 EFFECT OF PARTICLES ON THE TOXICITY OF BIOCIDES

Srinivasan *et al.* (1995) reported one experiment where kaolin and calcium carbonate particles (50 mg/L of each) were incorporated in mixed biofilms of *Pseudomonas aeruginosa* and *Klebsiella pneumoniae*. A monochloramine biocide solution was used against this biofilm and against similar biofilms that did not contain particles. They concluded that the biofilm grown in the presence of the abiotic particles was more resistant to the biocide treatment than the biofilm that did not contain particles. Several hypotheses were advanced to explain this effect, but none was tested.

On the other hand, Gauthier *et al.* (1999) found no significant effect of goethite particles (iron hydroxide formed in corrosion processes) on the disinfection capacity of chlorine in drinking water systems where bacteria were attached to those particles.

More information on the effect of biocides on microorganisms in the presence of particles is clearly needed to explain the data. Recent work where a carbamate-based biocide was used against a suspended culture of *Pseudomonas fluorescens* (Pereira *et al.* 2000) showed that, in general, the efficiency of the biocide was remarkably reduced in the presence of kaolin particles (Table 13.2). Chemical adsorption of carbamate to the clay surfaces seemed to occur, transforming the biocide molecules to other non-toxic chemical species (Pereira *et al.* 2002).

**Table 13.2.** Biocidal effect of carbamate (300 mg/L) on suspended cultures of *Pseudomonas fluorescens* grown in the absence or presence of kaolin particles (300 mg/L) at pH 7 (adapted from Pereira *et al.* 2000).

| Biocide contact time (h) | Percent reduction in the bacterial respiratory activity without kaolin particles | Percent reduction in the bacterial respiratory activity with kaolin particles |
|---|---|---|
| 0.1 | 41.8 | 2.7 |
| 1 | 46.3 | 23.3 |
| 3 | 74.2 | 38.4 |
| 7 | 59.1 | 59.1 |

The decrease in biocidal activity shown in Table 13.2 is reinforced by the fact that the kaolin particles reduce the carbamate concentration in solution, as found in independent tests done without any bacteria in suspension (Table 13.3).

**Table 13.3.** Reduction in carbamate concentration in solution due to kaolin particles in the absence of bacterial culture (adapted from Pereira *et al.* 2002).

| Biocide contact time (h) | Residual carbamate concentration (mg/L) | |
|---|---|---|
| | Initial concentration (mg/L) | |
| | 276 | 360 |
| 0.1 | 261 | 335 |
| 1 | 258 | 339 |
| 3 | 258 | 324 |
| 7 | 236 | 320 |

At pH 5, when kaolin particles and biocide were added to the microbial suspension, the bacterial activity was even enhanced for carbamate concentrations of 100 mg/L and 200 mg/L. This surprising effect appears to be due to the spatial arrangements of these clay particles such as the 'card house' effect (Figure 13.1), which contributes to keeping the bacteria inside kaolin flocs where a protected microenvironment is formed.

It should be stressed that the results presented in Tables 13.1–13.3 were obtained in experiments where buffered media were used, which discards the hypothesis of possible changes in the pH (and therefore, in the reactivity of biocides) caused by the clay particles.

The 'protective' effect of clays against biocides was then tested in biofilms of *Pseudomonas fluorescens* formed in a flow cell (Pereira *et al.* 2002). The concentration of bacteria in the 10-day-old biofilm was similar to the bacterial concentration in the suspended cultures mentioned in Table 13.2 (about $10^8$ cells/mL). The results shown in Table 13.4 for a 10-day-old biofilm incorporating kaolin particles confirm the conclusions obtained with suspended cultures.

**Table 13.4.** Biocidal effect of carbamate (300 mg/L) on a 10-day-old biofilm of *Pseudomonas fluorescens* grown in the absence or presence of kaolin particles (300 mg/L) at pH 7 (adapted from Pereira *et al.* 2002).

| Biocide application event | Per cent reduction in the bacterial respiratory activity without kaolin particles | Per cent reduction in the bacterial respiratory activity with kaolin particles |
| --- | --- | --- |
| 1st shock | 57.3 | 15.4 |
| 2nd shock | 32.9 | 26.1 |
| 3rd shock | 31.8 | 28.6 |

The role of the biofilm polymeric matrix as a protective system against biocides is known (see also Chapter 8) and is usually explained on the basis of:
(i) The additional mass transfer resistance introduced by that matrix (Costerton *et al.* 1987); or
(ii) The effects of molecular sieve and adsorption created by the polymers (Lawrence *et al.* 1994).

The results presented above show, however, that a much greater resistance of biofilms to biocides can be obtained by incorporating clay particles in the biofilm matrix. This will enable a much more stable performance of biofilm reactors when subject to the unexpected appearance of toxic compounds in the influent.

## 13.7 EFFECTS OF PARTICLES ON BIOFILM PHYSICAL PROPERTIES

Lowe et al. (1984) and Lowe (1988) conducted experiments to investigate the scouring effects of kaolin and sand particle suspensions on fully established biofilms of *Pseudomonas fluorescens*, compared with the shearing effects of water streams alone and found that biofilm erosion depended on the shape and roughness of the particles. When aqueous kaolin suspensions were used (concentrations from 50 to 5,000 mg/L) at a shear stress of 75 $N/m^2$ corresponding to turbulent flow, the amount of biomass detached was similar to the one detached when water was used (75% of the biofilm was removed from the surface). However, when sand suspensions flowing at the same shear stress were used, higher fractions of the biofilm were removed: 90% with particle concentration of 5,000 mg/L.

Vieira and Melo (1995) found that the mechanical stability of *Pseudomonas fluorescens* biofilms under starvation conditions increased when small (10 μm) kaolin particles were incorporated in the biological matrix. The experiments were done by growing the biofilms with particles (under turbulent flow conditions) until they reached their steady-state thickness and, subsequently, by omitting the substrate from the feed. The biofilm started to lose its biomass 2.5–3 days after the omission of the substrate. Similar experiments done in the absence of kaolin particles resulted in a loss of biomass 1–2 days after substrate was removed from the feed solution. The difference between the two types of biofilm was more pronounced when they were grown under higher flow velocities (1 m/s) than under lower velocities (0.34 m/s). This may be due to the fact that more kaolin particles are transported to and incorporated in the biomass when the water flows at higher velocities (as confirmed by Lowe's measurements of the inorganic fraction of biofilms), thereby reinforcing the physical structure of the biological matrix.

There is only one report on the effect of the presence of inorganic particles on the mass transfer within biofilms (Vieira and Melo 1995), which shows that the transport of an inert tracer (LiCl) through a single-species biofilm was enhanced by more than 50% when small kaolin particles were incorporated in the biological matrix. The hydrodynamic conditions were similar in both types of experiment (with and without inorganic particles). The particles imparted a more open structure to the biological matrix, increasing the transport rate of chemical species within the biofilm. This result may also explain why biofilms containing particles faired much better under starvation conditions: their different architecture favored the access to residual nutrients and, furthermore, the adsorption of nutrients to the clays provided a reservoir that allowed bacterial metabolism to proceed longer.

## 13.8 CLAY PARTICLES IN WASTEWATER TREATMENT BIOREACTORS

The high specific surface area and the capacity to exchange ions and inorganic compounds from the medium with ions from the lattice, make clay particles attractive as 'reactive' supports in bioreactors. They contribute to immobilization of biomass, they can help in maintaining suitable local pHs, they provide inorganic ions for microbial metabolism, and they indirectly select for microorganisms in the bioreactor. Huysman *et al.* (1983), using sepiolite as support for anaerobic digestion in a biofilm reactor, found that the colonization was faster than when other supports were used, mainly because of the existence of crevices with a size similar to that of bacteria. In this case, *Methanosarcina* was the main organism forming the biofilm.

In addition to the increased number of attachment sites provided by the clay, the slow release of minerals seems to be very beneficial to the methanogens attached on the surface. Pérez Rodriguez *et al.* (1992) compared the performance of clay supports and expanded PU and PVC in anaerobic digestion and concluded that sepiolite, vermiculite, and montmorillonite increased the process efficiency, probably by providing inorganic nutrients for bacterial metabolism. Furthermore, they found that the microbial species prevailing in the biofilm and present in suspension were dependent on the type of support used, thereby affecting the efficiency of the process. Muñoz *et al.* (1994) pointed out that the highest increase in the production of methane in pilot-scale anaerobic digesters was obtained when sepiolite was used as the carrier. Sanchez *et al.* (1994) compared several support materials (sepiolite, diatomaceous earth, diabase, PVC and montmorillonite), and concluded also that methanogenesis was enhanced in the presence of sepiolite, owing to the enrichment of the culture by acetotrophic methanogens. Murray and van den Berg (1981) showed that the development of a methanogenic film was threefold faster on fired clay that on PVC plastic or etched glass, and that the film formed on the clay was thick and uniform, contrary to the other cases.

Vieira *et al.* (2001) studied the effect of kaolin particles on the performance of a nitrifying air-lift reactor. They found that the incorporation of those particles within the biofilm resulted in a significant decrease in the nitrate concentration in the reactor outlet, although the removal of ammonia and nitrite was almost complete. They tested whether this could be due to the adsorption effects of clay particles (which would act as temporary sinks), but the results showed that this did not happen. The 'disappearance' of nitrogen seemed to be due to the precipitation of ammonia salts such as struvite, which has also been reported in anaerobic digesters that use clay particles as a carrier for biomass attachment (Muñoz *et al.* 1994). This mechanism is associated with the release of ions from the clay to the aqueous medium, which form precipitating salts with the ammonia. Such a mechanism is seen as an advantage in nitrification/denitrification processes, because clay particles are therefore able to reduce the amount of nitrate to be removed in subsequent treatments (denitrification).

Gisvold *et al.* (2001) managed to keep the high efficiency of a nitrifying biofilter during peak loads of ammonium by introducing a zeolite containing expanded clay. In this study, the particles act not only as carriers for the bacteria but also as temporary sinks of nutrients which are subsequently released to the aqueous medium during low-load periods. This is particularly useful in domestic wastewater treatment where at least one peak load per day occurs.

## 13.9 SUMMARY AND FUTURE RESEARCH

The existing knowledge, although limited, shows that the modifications introduced by the clay particles in the spatial arrangements and specific microenvironments within biofilms can improve biofilm activity and biofilm reactor performance.

In recent decades, a considerable amount of work was published concerning the beneficial effects of clay particles on soil microbiology and on the growth of dispersed microorganisms in aqueous solutions. The few reports on the effects of such particles on the population structure and architecture of biofilms and on the efficiency of biofilm reactors showed that clay particles coming from the external fluid integrate into the biofilm architecture and help create a stronger and more open biological matrix, with specific microhabitats.

For example, kaolin particles were found to increase biofilm stability when the biofilm was subject to a sudden lack of substrate and also to facilitate mass transfer through the biofilm matrix, thereby influencing the efficiency of the bioreactor. The presence of such particles also resulted in an increase in the respiratory activity of the attached bacteria. Methane production in attached biomass anaerobic reactors was also enhanced when clay supports (sepiolite) were used.

Apart from the higher physical resistance of the biofilm when it is reinforced with inorganic matter, the formation of kaolin flocs of the 'card-house' type in a low pH environment also has a significant effect on biofilm performance. In fact, even if the pH of the reactor influent is controlled at a pre-fixed value, the pH inside the biofilm microenvironments is frequently quite different from the optimum value (measured and controlled in the bulk fluid), and the presence of such flocs with a more open geometry makes a decisive contribution to the activity and survival of the microbial population inside the biofilm.

Another important consequence of the presence of clay particles within biofilm matrices is that they reduce the negative impact of toxics on biofilm activity and thus provide a more stable operation of biofilm reactors. Bacteria closest to the particles are better protected, and particles thus impart greater cohesion and survival opportunities to the inner zones of the biofilm. .

The protective effect of clay particles on the DNA bound to their surfaces was experimentally established in geological sediments but these effects have yet to be tested in biofilm systems. It is also expected that the introduction of appropriate concentrations of particles of a suitable chemical nature and dimension (e.g., powdered clays) in biofilm reactors will result in higher degradation rates and higher biofilm stability against sudden changes in the wastewater composition (mainly substrate concentration and unpredictable presence of toxic compounds) and hydraulic conditions.

In summary, there is already a consistent set of results that confirms the benefits of clay particles in biofilm reactor operation through the changes they introduce in the specific microenvironments existing inside biofilms. What is now lacking? More data on the performance of different wastewater treatment biofilm processes in the presence of such particles would be useful. But much more fundamental research work is still needed. The investigation of the following aspects is therefore encouraged.

How is the spatial distribution of different microbial populations in biofilms affected by the presence of particles?

How effective can the introduction of clay particles with bound DNA in biofilm reactors be to enhance their performance (e.g., their resistance to toxins)?

Which nutrients or which toxins are physically adsorbed or chemically transformed through interactions with particles? How does this affect the population dynamics within the biofilm? In which particular situations (type, dimensions, concentrations of particles) does the interaction result in a better-performing biofilm reactor ?

The link between the fundamental information gained from biofilm studies with clay particles and the practical data obtained on bioreactor performance is still only tenuously established. In the present case, as so often happens, the information loop started with studies done in a different discipline (soil microbiology) and went on through typical engineering work, where clays were used with success in wastewater treatment bioreactors. Now the challenge is to understand more deeply the role of such particles in the properties and activity of biofilms, creating new opportunities to develop more efficient processes for the removal of pollutants through biological processes.

## 13.10 REFERENCES

Amin, S. and Jayson, G.G. (1996) Humic substances uptake by hydrotalcites and pilcs. *Wat. Res.* **2**, 299–306.

Bott, T.R. and Melo, L.F. (1992) Particle–bacteria interaction in biofilms. In *Biofilms – Science and Technology* (ed. L.F. Melo, T.R. Bott, M. Fletcher and B. Capdeville), pp. 199–206. Kluwer Academic Publishers, Dordrecht.

Bowen, R.B. and Dempsey, B.A. (1992) Improved performance of activated sludge with addition of inorganic solids. *Wat. Sci. Tech.* **26** (9–11), 2511–2514.

Burns, R.G. (1975) Factors Affecting Pesticide Loss from Soil. In *Soil Biochemistry* (ed. E.A.Paul and A.D. McLaren), pp. 103–141. Marcel Dekker, New York.

Burns, R.G. (1986) Interaction of enzymes with soil minerals and organic colloids. In *Interaction of Soil Minerals with Natural Organics and Microbes* (ed. P.M. Huang and M. Schnitzer), pp. 429–451. Soil Science Society of America, Madison, Wisconsin, USA.

Burns, R.G. (1989) Microbial and Enzyme Activities in Soil Biofilms. In *Structure and Function of Biofilms* (ed. W.G. Characklis and P.A Wilderer), pp. 333–349, Wiley-Interscience, New York.

Castellar, M R, Aires-Barros, M R, Cabral, J M S and Iborra, J L (1998) Effect of zeolite addition on ethanol production from glucose by *Saccharomyces bayanus*. *J. Chem. Tech. Biotech.* **73**, 377–384.

Chudoba, P. and Pannier, M. (1994) Use of powedered clay to upgrade activated sludge process, *Environmental Technology* **15**, 863–870.

Costerton, J.W., Cheng, K.G., Geesey, G.G., Ladd, T.I., Nicckel, J.C., Dasgupta, M. and Marrie, T.J. (1987) Bacterial biofilms in nature and disease. *Annu. Rev. Microbiol.* **41**, 453–464.

Filip, Z. and Hattori, T. (1984) Utilisation of substrates and transformation of solid substrata. In *Microbial Adhesion and Aggregation* (ed. K.C. Marshall.), pp. 251–282, Springer, Berlin.

Fletcher, M. (1992) Bacterial metabolism in biofilms. In *Biofilms-Science and Technology* (ed L.F. Melo, T.R. Bott, M. Fletcher and B. Capdeville), pp 113–124. Kluwer, London.

Gauthier, V., Redercher, S., Block, J.C. (1999) Chlorine inactivation of *Sphingomonas* cells attached to goethite particles in drinking water. *Appl. Environ. Microbiol.* **65**, 355–357.

Gisvold, B., Ödegaard, H. and Föllesdal, M. (2001) Enhanced removal of ammonium by combined nitrification/adsorption in expanded clay aggregate filters. *Wat. Sci. Tech.* **4**(4/5), 409–416.

Hattori, T. (1973) *Microbial Life in the Soil*. Institute for Agricultural Research, Marcel Dekker, New York.

Hommes, N.G., Sterling, A.R., Bottomley, P.J. and Arp, D.J. (1998) Effects of soil on ammonia, ethylene, chloroethane, and 1,1,1-Trichloroethane oxidation by *Nitrosomonas europea*. *Appl. Environ. Microbiol.* **64**, 1372–1378.

Huang, C. and Yang, Y.L. (1995) Adsorption characteristics of Cu(II) on humus–kaolin complexes. *Wat. Res.* **11**, 2455–2460.

Huysman, P., van Meenen, P., van Assche, P. and Verstraete, W. (1983) Factors affecting the colonization of non porous packing materials in model upflow methane reactors, *Biotech. Lett.* **5**, 643–648.

Khanna, M. and Stotzky, G. (1992) Transformation of *Bacillus subtilis* by DNA bound on montmorrilonite and effect of DNase on the transforming ability of bound DNA, *Appl. Environ. Microbiol.* **56**, 1930–1939.

Lawrence, J.R., Wolfaardt, G.M. and Korber, D.R. (1994) Determination of diffusion coefficients in biofilms by confocal laser microscopy, *Appl. Environ. Microbiol.* **60**, 1166–1173.

Lewandowski, Z. (1983) Nitrification process in activated sludge with suspended marble particles. *Wat. Res.* **19**, 159–164.

Lowe, M. (1988) The effect of inorganic particulate materials on the development of biological films. Ph.D. thesis, University of Birmingham, UK.

Lowe, M.J., Duddridge, J.E., Pritchard, A.M. and Bott, T.R. (1984) Biological-particulate fouling interactions: effects of suspended particles on biofilm development. *Inst. Chem. Engrs Symp. Ser.* **1**, 391–400.

Macura, J. and Stotzky, G. (1980) Effect of montmorillonite and kaolinite on nitrification in soil. *Folia Microbiol.* **25**, 90–105.

Magdaliniuk, S., Block, J.C., Leyval, C., Bottero, J.Y., Villemin, G. and Babut, M. (1995) Biodegradation of naphthalene in montmorillonite/polyacrylamide suspensions. *Wat. Sci. Tech.* **31**(1), 85–94.

Michaels, A.S., Bolger, J.C. (1962) Settling rates and sediment volumes of flocculated kaolin suspensions. *Industr. Engr Chem. Fundamentals* **1**, 24–32.

Muñoz, M.A., Sanchez, J.M., Rodriguez-Maroto, J.M., Moriñigo, M.A., Borrego, J.J. (1994) Evaluation of sepiolite to optimize the methanogenesis from anaerobic domestic sludges in laboratory conditions. *Wat. Res.* **28**, 195–200.

Murray, W.D. and Van den Berg, L. (1981) Effect of support material on the development of microbial fixed films converting acetic acid to methane. *J. Appl. Bacteriol.* **51**, 257–265.

Palenik, B. (1989) Biofilms: properties and processes, Group Report, In *Structure and Function of Biofilms*. (ed. W.G. Characklis and P.A. Wilderer), pp. 351–367. Dahlem Workshop Reports, John Wiley and Sons.

Pereira, M.O., Vieira, M.J. and Melo, L.F. (2000) The effect of clay particles on the efficacy of a biocide. *Wat. Sci. Tech.* **41**(4–5), 61–64.

Pereira, M.O., Vieira, M.J. and Melo, L.F. (2002) The role of kaolin particles in the performance of a carbamate-based biocide for water bacterial control. *Wat. Environ. Res.* **74**, 235–241.

Pérez Rodriguez, J.L.P., Maqueda, C., Lebrato, J. and Carretero, M.I. (1992) Influence of clay minerals, used as supports in anaerobic digesters, in the precipitation of struvite. *Wat. Res.* **26**, 497–506.

Sanchez, J.M., Arijo, S., Muñoz, M.A., Moriñigo, M.A. and Borrego, J.J. (1994) Microbial colonization of different support materials used to enhance the methanogenic process. *Appl. Microbiol. Biotech.* **41**, 480–486.

Srinivasan, R., Stewart, P.S., Griebe, T., Chen, C-I and Xu, X. (1995) Biofilm parameters influencing biocide efficacy. *Biotech. Bioengng* **46**, 553–560.

Stotzky, G. (1966) Influence of clay minerals on microorganisms. II. Effect of various clay species, homoionic clays and other particles on bacteria. *Can. J. Microbiol.* **12**, 831–848.

Stotzky, G (1986) Influence of soil mineral colloids on metabolic processes, growth, adhesion, and ecology of microbes and viruses. In *Interactions of Soil Minerals with Natural Organics and Microbes* (ed. P M Huang and M. Schnitzer), pp 305–428. Soil Society of America, Madison, Wisconsin.

Strenge, K. and Sontag, H. (1987) *Coagulation Kinetics and Structure Formation*. Plenum Press, New York and London.

Veal, D.A., Stokes, H.W., Daggard, G. (1992) Genetic exchange in natural microbial communities. In *Advances in Microbial Ecology* (ed. K.C. Marshall), vol 12, pp. 383–430. Plenum Press, New York and London.

Vieira, M.J. and Melo, L.F. (1995) Effect of clay particles on the behaviour of biofilms formed by *Pseudomonas fluorescens*. *Wat. Sci. Tech.* **32**, 45–52.

Vieira, M.J., Pacheco, A.P., Pinho, A. and Melo, L.F. (2001) The effect of clay particles on the activity of suspended autotrophic nitrifying bacteria and on the performance of an air-lift reactor. *Environ. Tech.* **22**, 123–135.

Wuertz, S. (2002) Gene exchange in biofilms. In *Encyclopedia of Environmental Microbiology* (ed. G. Bitton), vol. 3, pp. 1408–1420. John Wiley, New York.

# Acknowledgements

The contributions of Ph.D. student M.O. Pereira and M.Sc. student A.P. Pacheco to this work are gratefully acknowledged.

# 14

# Bioprocess engineering and microbiologists: a profit-sharing alliance

*Peter A. Wilderer and Martina Hausner*

## 14.1 POSTULATE

The technical application of metabolic processes is a cross-disciplinary exercise *par excellence*. Design and operation of bioreactors require input of various disciplines, process engineering and microbiology in the first place. Fusion of the knowledge accumulated by both disciplines leads to advanced technological solutions, to the solution of environmental problems in particular. In the past, cross-feeding between both disciplines was sub-optimal, however. Because of the lack of solid information from microbiology a trial-and-error approach had to be used by engineers in the process of developing and applying environmental biotechnology. Now, however, the situation has changed. Novel methods of relevance to process engineers have been developed in microbiology, and are being progressively used in the attempt to shed light on the 'black box' of bioreactors (Keller *et al.* 2002). An even more extensive exploitation of these methods is in the interest of engineers and users of bioreactors. But microbiologists would profit from collaborating with engineers as well. Because the environmental conditions in engineered biotopes (i.e., bioreactors) can be managed in many different ways, the interface between engineering and microbiology provides a great potential for identification of the as yet unknown, and for the expansion of the general knowledge base. In particular, factors affecting composition, structure, and function of microbial communities are to be explored at the community, species, and molecular level to gain a deeper understanding of the interactions in heterogeneously composed microbial communities, to better understand the performance of bioreactors under real environmental conditions, and to enable a straightforward further development of biotechnology. An alliance of process engineers and microbiologists is of mutual advantage for both disciplines, and needs to be cultivated (Figure 14.1).

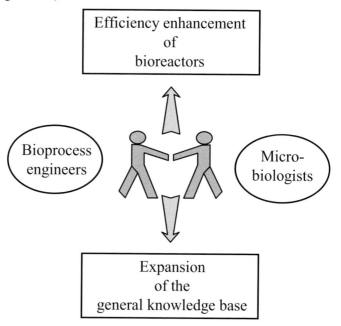

**Figure 14.1.** Postulated gains from a close collaboration between process engineers and microbiologists.

## 14.2 ANALYSIS OF THE CURRENT STATE OF BIOTECHNOLOGY

Bioprocess engineering plays an important role in our economy. In the food-processing industry, bioengineers are responsible for the design and operation of reactors in which materials of agricultural origin get converted into products of higher market value such as vinegar, wine, or beer. In the pharmaceutical industry the target products are vaccines and medicines for many different purposes. The purpose of biological wastewater treatment plants is to convert wastewater constituents into biomass that can be readily separated from the water, and into environmentally tolerable inorganic substances such as carbon dioxide and nitrogen gas. Microorganisms, and more specifically bacteria, are employed as a kind of 'labor force', capable of performing the required metabolic processes. In engineered environments (bioreactors) conditions are provided encouraging growth of bacteria in the form of either sessile aggregates (biofilms) or in a mobile form as flocs or granules.

The transformation processes required to gain the desired products could certainly be executed in chemical reactors as well because the reactions taking place in a cell are catalyzed chemical reactions by nature. In a cell, biocatalysts (enzymes) are used to enhance the reaction rates and are manufactured by the cells themselves. Many of the enzymes active in cells are known, and can be synthetically produced. In cells, catalyzed reactions are optimally done to minimize loss of energy and to make effective use of intermediate substances. The same could be done in a biochemical reactor *in vitro*, but the use of living cells has been shown to be much more effective and less costly.

The reason for employing living cells is simple. Cells self-regulate the system of enzymatic reactions, and manage impressively well their metabolism even under widely changing environmental conditions (e.g., changes in the substrate concentration in the bulk fluid, in the composition of the nutrient solution, and in the nutritive quality of the components). It would be difficult to copy that self-regulating system and implement it in a chemical reactor system designed for the treatment of a highly variable medium such as municipal wastewater. First of all, we do not have available sensors to monitor the hundreds and thousands of chemical components potentially present in wastewater. And even if sensors were available, the costs of implementing sensor-based technology, evaluating the sensor signals, and performing the required control actions would be prohibitive.

Although bacteria are capable of coping with a great variety of substrates, a single species can hardly possess the metabolic potential required to catabolize the full spectrum of pollutants that wastewater may contain. Fortunately, however, nature provides a great diversity of bacteria, distinguished by their metabolic potential. For wastewater treatment, bacteria capable of hydrolyzing particulate organic material and of metabolizing readily biodegradable soluble substances as well as poorly degradable substances – most of which are products of the chemical industry – play an especially important role. Also of interest are nitrifiers, representatives of the chemolithoautotrophic group of bacteria, which are able to convert ammonia into nitrite and nitrate. Of further interest are bacteria that can use nitrite and nitrate, instead of dissolved oxygen, as electron acceptors, and convert these substances into nitrogen gas. For anaerobic wastewater treatment, of interest are members of the domains of Bacteria and Archaea which are collectively capable of transforming organic material into fatty acids, carbon dioxide, hydrogen and finally into methane gas (see, for example, McMahon et al. 2001).

Each of the groups of microbial species mentioned above appears in natural habitats. To be able to grow, each group needs a specific niche distinguished by a set of characteristic environmental conditions such as availability of specific substrates, micronutrients, electron donors and electron acceptors, availability of sites to adhere to, protection mechanisms against grazing, and environmental parameters such as pH, temperature, hydrodynamic shear, mixing patterns, and many more.

To make use of the metabolic potential of the various groups of microbial species, the engineer needs to know the characteristics of the niche the species of interest occupy in natural habitats. Ideally, these niche parameters would have to be translated into parameters to be applied to the technical niche a bioreactor provides. As a result, a microbial community would establish itself in the reactor, and perform the various metabolic reactions, which lead to the desired result, the purified wastewater for instance.

In practice, this concept of defining niche parameters first and then translating them to process parameters was rarely applied, mainly because the required information was not available. But the wastewater produced in municipalities and in industry had to be treated, nevertheless, to avoid deterioration of the environmental quality of surface waters. Thus, engineers had to develop and operate bioreactors based on educated guesses, trials and on the experience made by failures. The main reason why empirical approaches had to be chosen was and is the uncertainty of correlation between niche parameters and the respective technical process parameters. In addition, the situation the microbial community is exposed to in a bioreactor is very often extremely complex, variable, and difficult to predict.

The conditions in bioreactors for biological wastewater treatment are further complicated by the inter-relationships between the different groups of microorganisms that will find their niche in the reactor, even if the process engineer does not provide that niche on purpose. The result is a developing microbial community that may not meet the expectations of the engineer for process efficiency and costs. Food chains may form where soluble organic material is metabolically converted into bacterial mass by bacteria, and bacterial mass is taken up and metabolized by protozoa, which themselves are grazed by higher organisms (Figure 14.2). On the other hand, food chains like this may be desirable if organic materials are to be converted into compost. In this case niche parameters in favor of higher microorganisms are to be purposely provided by the engineer.

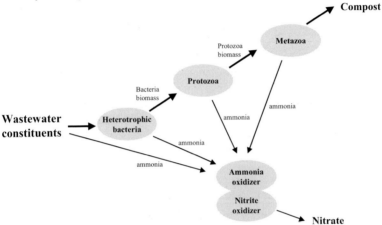

Figure 14.2. Model of a heterogeneously composed microbial community as it may appear downstream a river or in a low-loaded biofilm reactor (e.g., trickling filter). Readily biodegradable substances are taken up by heterotrophic bacteria and transformed into biomass that is subsequently taken up by protozoa. In turn, the protozoa biomass is consumed by a sequence of metazoa, and ultimately transformed into compost. Ammonia contained in the original feed (wastewater), and released by heterotrophic organisms, is taken up by nitrifying bacteria and used as the electron donor.

For wastewater, it should be noted that the substrate to be metabolized is primarily organic matter of relatively low molecular mass (readily biodegradable substances). Metabolic conversion of this category of substances is most effectively performed by heterotrophic bacteria. They may also convert particulate organic substances into soluble, readily biodegradable compounds by hydrolytic reactions. Higher organisms, however, feed on the bacteria and thus diminish the required 'labor force'. Massive growth of higher organisms is counter-productive from the engineering point of view.

Because higher organisms usually grow much slower than heterotrophic bacteria, one could speculate that by shortening the mean cell residence time in the bioreactor the bacteria would out-compete the higher organisms. Thus, the mean cell residence time plays the role of an environmental factor. As shown in Figure 14.3, any change of the factor value leads to a corresponding change of the microbial community (Hartmann 1960).

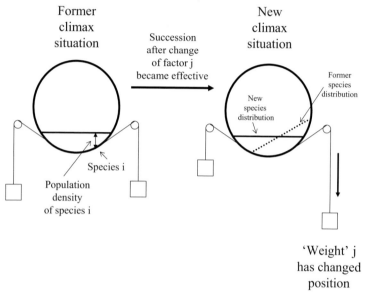

**Figure 14.3.** Hollow sphere model of a microbial community affected by the change of environmental factor j. Each point at the inner surface of the sphere represents a microbial species. The sphere is partly filled with water, and the depth of the water represents the density of the respective species populations. The sphere is kept in position by a system of strings and rolls. Each string carries weights on either side representing specific environmental factors. After changing the position of one of the weights (i.e., the value of the respective environmental factor) the sphere rotates into a new position. As a result, some of the points inside of the sphere fall dry (i.e., species disappear), others become wet (i.e., new species invade the microbial community), and the depth of the water on top of any point may change (i.e., change of the density of the respective species populations). Model adapted from Hartmann (1960).

Biological wastewater treatment plants have to reduce in concentration not only readily biodegradable substrates but also organic substances that are difficult to metabolize (recalcitrant, mostly xenobiotic substances), ammonia, nitrite, nitrate and phosphate. The capability of bacteria to eventually metabolize recalcitrant substances is restricted to species that possess the capability of synthesizing specific enzymes, not required under 'normal' conditions in nature. For example, Chakrabarty (1996) suggested that new biodegradative pathways arise as individual genes diverge or are recruited from the genomes of other microorganisms. Although some genes are chromosomal, others are plasmid-associated. The information required to synthesize such enzymes is often coded on plasmids (see, for example, Wackett *et al.* 2001). The energy gained from the degradation of recalcitrant substances is often small, and the growth rate of the bacteria relying solely on these constituents is subsequently low. Many recalcitrant substances are only metabolized by specialized bacteria, when readily biodegradable substrates already have been consumed even though co-metabolism may play a role in some cases (Bae and Rittmann 1990; Malmstead *et al.* 1995). Nitrification relies on bacteria capable of oxidizing ammonia and nitrite. Carbon dioxide serves as carbon source of the anabolic reactions. For thermodynamic reasons these bacteria also grow very slowly. Recently, improved nitrogen removal was accomplished by application of new autotrophic nitrogen-cycle bacteria, namely, those capable of anaerobic ammonium oxidation (Jetten *et al.* 2002).

To enrich a microbial community with nitrifiers and with bacteria capable of degrading recalcitrant organics a relatively high mean cell residence time has to be maintained in the bioreactor. However, that leads automatically to the enrichment of other slow growers. In particular, the composition of the biocommunity shifts as bacterial numbers are modified as a result of grazing processes (Hahn and Hofle 2001; Lebaron *et al.* 2001). Eventually, organic mass is converted into inorganic substances and into compost-like material (Figure 14.2).

In essence, the niche characterizing the situation in aerobic bioreactors for wastewater treatment must be understood as a function of a great variety of environmental factors (Figure 14.4), but primarily as a function of time. With elapsing time the niche parameters systematically shift as they do downstream a river. This fact was realized years ago (Liebmann 1962). Biological treatment plants were understood, basically from the beginning of the development of the technology, as a copy of the processes taking place in rivers. Nevertheless, the technical concept of biological wastewater treatment plants is based on empiricism rather than on principles of microbial physiology and ecology. The bioreactor is still considered to be a 'black-box' despite the fact that recent approaches in mathematical modeling (Henze *et al.* 1987; Barker *et al.* 1997; Gujer *et al.* 1999) have shed some light on this box (Figure 14.5).

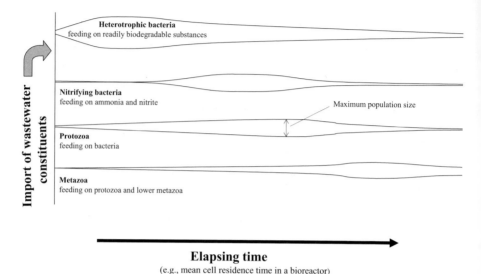

**Figure 14.4.** Shift of the microbial community composition with increasing mean cell residence time in a bioreactor fed with wastewater.

## 14.3 CHANCES AND REQUIREMENTS

To gain a better understanding of the system that develops in a bioreactor under the process conditions provided by the engineer (see Figure 14.4) a much more intense cooperation between microbiologists and engineers is needed. Both disciplines must learn to listen to each other better, respect the methodology and concerns of each side, but translate the results of their research and practical experience into the 'language' of the respective partner discipline. The complexity of the system under dispute needs to be acknowledged by both the engineers and the microbiologists and studied with the aim of intellectual control of that system. Eventually, simplifications have to be made to be able to communicate the results to practitioners. When these simplifications are based on a well-founded understanding of all the relevant scales of the system the result should lead to a novel, more effective, reliable, and robust technology.

## Bioprocess engineering and microbiologists 351

| Processes j | Components i of a microbial system | | | | | | Process equations |
|---|---|---|---|---|---|---|---|
| | 1 | 2 | 3 | 4 | 5 | 6 | |
| 1 | | | | | | | |
| 2 | | | $v_{ij}$ | | | | $P_j$ |
| 3 | | | | | | | |
| 4 | | | | | | | |
| Overall reaction rate | | | $r_i$ | | | | |

$$r_i = \Sigma \, (P_j \cdot v_{ij}) \quad \text{with} \quad v_{ij} = \text{stoichiometric factor}$$

Important components (i): readily biodegradable, problematic and recalcitrant organic substances
heterotrophic bacteria, nitrifyers, protozoa, metazoa
dissolved oxygen, ammonia, nitrite, nitrate, phosphate

Important processes (j): growth under aerobic, anoxic and anaerobic conditions
grazing, and decay as a result of starvation

**Figure 14.5.** Schematic of the kinetic matrix modern mathematical models of microbial systems are based upon.

Bioprocess engineers depend on information from microbiologists in the form of:
1. Factors controlling the composition of the microbial community (biomanipulation).
2. Factors leading to the optimization of the desired processes (biostimulation).
3. Factors allowing control of the physicochemical properties of microbial aggregates (physicochemical stimulation).
4. Parameter values describing the kinetics and stoichiometry of the metabolic processes taking place in the reactor, after manipulation and stimulation processes have been successfully completed (bioreaction control).

Optimization of the metabolic potential of heterogeneously composed microbial communities can be achieved by giving a 'competitive edge' to the microbial species required to perform the required metabolic conversion processes (see Figure 14.4). If the target species are known, the engineer needs to learn how to manipulate and stimulate them to obtain conditions that are favorable to enriching the mixed culture for the most desired species. He or she does not need to know the name of the species nor the position of that species in a phylogenetic tree. Of importance are factors such as growth rate, minimum requirements for co-substrates, micronutrients, pH, temperature, electron donors and acceptors. For example, Siegrist et al. (1999) showed that biomass decay is reduced significantly under anaerobic conditions.

The engineer should know whether the species of concern becomes readily associated with the bacterial community that comprises biofilms, granules and flocs, whether the species contributes to the accumulation of extracellular polymeric substances (EPS) or whether it forms filamentous structures which may deteriorate the sedimentation characteristics of activated sludge flocs. Of interest are whether the required species is ubiquitous or needs to be introduced into the microbial community from a specific biotope (bioaugmentation). As an example of bioaugmentation, extremely alkali-tolerant ammonia-oxidizing bacteria from Mongolian soda lakes (Sorokin et al. 2001) were inoculated into bioreactors to treat industrial wastewaters with inherent high salt concentrations.

Of interest, also, is whether the genetic information coding for specific metabolic pathways can be transferred to autochthonous members of the biocommunity by gene transfer ('*in situ* genetic manipulation'), so that the new metabolic traits become established even when the donor strains may not be able to compete in the long run (Eberl et al. 1997; Newby et al. 2000; Top et al. 1998; Wuertz 2002).

In contrast to the main interest of engineers, the emphasis of microbiologists is primarily directed towards basic research, the development of novel analytical methods, and the application of these methods to better understand microbial systems on the community, cell, and molecular level. The methods required to gain relevant information are highly sophisticated, nowadays, and thus a great deal of specialization is necessary. As a result, the field of microbiology has become a widely diversified scientific area.

Admittedly, a scientist working in a specific area is in a very delicate intellectual competition with scientists working in the same field. Discoveries and methodological developments need to be discussed within and eventually accepted by the specialized scientific community to become generally recognized. The reputation of a scientist is built upon the acceptance of his or her contributions to the specialized scientific area.

It must also be admitted, however, that a significant added value of a discovery or a new method may be gained when the results of the specialized research topic are translated and made available to a broader scientific community and to applied disciplines. This translation exercise needs to be promoted, despite the fact that communication across the borders of a specialized discipline into the field of engineering is all but trivial. It requires mutual understanding of the methods, concerns and objectives of the respective partner disciplines. It needs commitment, openness, and time, the last in particular.

The build-up of scientific enclosures was probably the main reason why in the past natural scientists did not contribute much to the development of environmental technology. Because environmental problems had to be solved, nevertheless, engineers themselves tried to gain necessary information to understand better the microbiological systems that developed in biological wastewater treatment plants, namely, in ponds, activated sludge plants, trickling filters and rotating disc contactors, membrane bioreactors, high-rate biofilters, anaerobic sludge blanket reactors, and sludge digesters. Over the years, a tremendous knowledge base has been built up. Methods have been developed that enable the engineer to design and successfully operate bioreactors. Empirical design guidelines such as the so-called guideline 'A 131' of the German Waste and Wastewater Association (ATV 1991) or from the Water Research Commission of South Africa (1984) were supplemented by dynamic mathematical models (Henze *et al.* 1987; Gujer *et al.* 1999) which take into account the various components and processes characteristic for microbial systems in reactors for biological wastewater treatment (Figure 14.4).

There is certainly no reason to underestimate the contributions made by engineers to the knowledge base of wastewater microbiology, but it must be admitted that the lack of microbiological research during the early stage of technology development and the sub-optimal communication between natural scientists and engineers has lead to severe scientific and engineering errors. Over-estimation of the role of *Acinetobacter* species in the process of enhanced biological phosphate removal (EBPR) (Wagner *et al.* 1994) and the over-emphasized role of *Nitrosomonas* and *Nitrobacter* species in the process of nitrification (Daims *et al.* 2001) are good examples of this. It is certainly not the name of the bacterial strains that matters in this context. The real problem is that in the past design parameters and operation management schematics were based on the niche parameters of species, which were believed to play an important role in practice, but in reality did not. The replacement of the niche parameters by the relevant ones now provides the opportunity to overcome obvious operational problems and to achieve higher process efficiency of bioreactors in the future. As one example, the metabolic model of Murnleitner *et al.* (1997) takes this new biological consideration for polyphosphate-accumulating organisms (PAO) into account.

The discovery that *Acinetobacter* is not the most important species in EBPR plants was a collaborative effort of bioprocess engineers and microbiologists. The examples given have been understood by our respective scientific communities. It has been realized by microbiologists that it is worthwhile to go beyond laboratory-based studies and to enter the field of technically controlled reactor systems. The door is open for a gradual replacement of empiricism by solid scientific knowledge from both engineering and natural sciences.

The questionnaire, which is presented in the Appendix of this chapter, may provide the basis for a better communication between bioprocess engineers and microbiologists interested in the analysis of multi-species biocommunities and microbial aggregates. Wilderer *et al.* (2002) published a first version of the document. The data introduced into the questionnaire describe the biography of a microbial sample examined, and the results obtained by microbial analysis. Researchers are invited to download the questionnaire as a Microsoft Word document from www.wga.bv.tum.de, to fill out the questionnaire, and to send it electronically to questionair@bv.tum.de. The information provided is introduced into a database, which can also be downloaded for further processing. For instance, correlation of process data with the results of microbial analysis may lead to information about the effects individual process parameters have on structure and function of the biocommunity and the resulting performance of the microbial system in the bioreactor. With reference to Figure 14.3, the question to be answered is which string the engineer must pull to establish the desired microbial community, and the desired reactor performance as a result.

## 14.4 CONCLUDING REMARKS

Advances in biological wastewater treatment can only be expected when engineers and scientists join forces. An alliance between disciplines needs to be formed and cultivated. Such an alliance must be based on mutual interest and understanding of the scientific culture in each of the partner disciplines, the concerns of the individual disciplines, and the role they play in the academic, social and economic theatre.

In particular, process engineers and microbiologists are called upon to intensify their collaboration. Contributions from these two disciplines are of specific importance for a sustainable development of environmental biotechnology.

It will be the task of the process engineer to address questions based on observations and operational problems encountered, and to implement answers into practical operations. The task of the microbiologist is to translate the questions raised by the engineers into a well-founded research concept, and to translate the answers found into a language readily understandable by engineers and practitioners. The attached questionnaire may assist in the build up of a database, which can be used in the process of optimization of bioreactors.

## 14.5 REFERENCES

ATV (1991) Arbeitsblatt A 131, Bemessung von einstufigen Kläranlagen ab 5000 Einwohnergleichwerten (Design guideline A 131 for single stage wastewater treatment plants serving more than 5,000 p.e.). Gesellschaft zur Förderung der Abwassertechnik, St. Augustin, Germany

Bae, W. and Rittmann B.E. (1990) Effects of electron acceptor and electron donor on biodegradation of $CCl_4$ by biofilms. In *Environmental Engineering Division*, ASCE; Environmental Engineering, Proceedings of the 1990 Specialty Conference (ed. C.R. O'Melia), pp. 390–397, Arlington, VA.

Barker, P.S. and Dold, P.L. (1997) General model for biological nutrient removal activated sludge systems: model presentation. *Wat. Environ. Res.* **69**, 985–991.

Chakrabarty, A.M. (1996) Microbial degradation of toxic chemicals: evolutionary insights and practical considerations. *ASM News* **62**(3), 130–137.

Daims, H., Purkhold, U., Bjerrum, L, Arnold, E., Wilderer, P.A. and Wagner, M. (2001) Nitrification in sequencing biofilm batch reactors: lessons from molecular approaches. *Wat. Sci. Tech.* **43**(3), 9–18.

Eberl, L., Schulze, R., Ammendola, A., Geisenberger, O., Erhart, R., Sternberg, C., Molin, S. and Amann, R. (1997) Use of green fluorescent protein as a marker for ecological studies of activated sludge communities. *FEMS Microbiol. Lett.* **149**, 77–83.

Gujer, W., Henze, M., Mino, T. and Van Loosdrecht, M. (1999) Activated sludge model no. 3. *Wat. Sci. Tech.* **39**(1), 183–193.

Hahn, M.W. and Hofle, M.G. (2001) Grazing of protozoa and its effect on populations of aquatic bacteria. *FEMS Microbiol. Ecol.* **34**(3), 255–266.

Hartmann L (1960) Die Beziehung zwischen Beschaffenheit, Leistungsfähigkeit und Lebensgemeinschaft der Belebtschlammflocke am Beispiel einer mehrstufigen Versuchsanlage. (Relationship between properties, efficiency and community of the activated sludge floc.) *Vom Wasser* **17**, 107–184.

Henze, M., Grady C.P.L. Jr, Gujer, W., Marais, G. v. R. and Matsuo, T. (1987) *Activated Sludge Model No. 1.* IAWPRC Scientific and Technical Report No. 1. IAWPRC, London.

Jetten, M.S., Schmid, M., Schmidt, I., Wubben, M., Van Dongen, U., Abma, W., Sliekers, O., Revsbech, N.P., Beaumont, H.J.E., Ottosen, L., Volcke, E., Laanbroek, H.J., Campos-Gomez, J.L., Cole, J., Van Loosdrecht, M., Mulder, J.W., Fuerts, J., Richardson, D., van de Pas, K., Mendez-Pampin, R., Third, K., Cirpus, I., van Spanning, R., Bollmann, A., Nielsen, L.P., den Camp, H.O., Schultz, C., Gundersen, J., Vanrolleghem, P., Strous, M., Wagner, M. and Kuenen, J.G. (2002) Improved nitrogen removal by application of new nitrogen-cycle bacteria. *Rev. Environ. Sci. Biotech.* **1**, 51–63.

Keller J., Zhiguo, Y. and Blackall, L.L. (2002) Integrating process engineering and microbiology tools to advance activated sludge wastewater treatment research and development. *Rev. Environ. Sci. Biotech.* **1**, 83–97.

Lebaron, P., Servais, P., Troussellier, M., Courties, C., Muyzer, G., Bernard, L., Schafer, H., Pukall, R., Stackebrandt, E., Guindulain, T, and Vives-Rego, J. (2001) Microbial community dynamics in Mediterranean nutrient-enriched seawater mesocosms: changes in abundances, activity and composition. *FEMS Microbiol. Ecol.* **34**, 255–266.

Liebmann, H. (1962) *Handbuch der Frischwasser- und Abwasserbiologie.* (Handbook of fresh and wastewater biology.). Vol. 1. Oldenbourg, Munich.

Malmstead, M.J., Brockmann, F.J., Valocchi, A.J. and Rittmann, B.E. (1995) Modeling biofilm degradation requiring co-substrates: the quinoline example. *Wat. Sci. Tech.* **31**(1), 71–84.

McMahon, K.D., Stroot, P.G., Mackie, R.I. and Raskin, L. (2001) Anaerobic codigestion of municipal solid waste and biosolids under various mixing conditions II: Microbial population dynamics. *Wat. Res.* **35**, 1817–1827.

Murnleitner, E., Kuba, T., Van Loosdrecht, M.C.M. and Heijnen, J.J. (1997) An integrated metabolic model for the aerobic and denitrifying biological phosphorus removal. *Biotech. Bioengng* **54**, 434–450.

Newby, D.T., Gentry, T.J. and Pepper, I.L. (2000) Comparison of 2,4-dichlorophenoxyacetic acid degradation and plasmid transfer in soil resulting from bioaugmentation with two different pJP4 donors. *Appl. Environ. Microbiol*. **66**, 399–3407.

Siegrist, H., Koch, G., Le-Van, C. and Phan, L.C. (1999) Reduction of biomass decay under anoxic and anaerobic conditions. *Wat. Sci. Tech.* **19**(1), 129–137.

Sorokin, D., Tourova, T., Schmid, M.C., Wagner, M., Koops, H.P., Kuenen, J.G. and Jetten, M. (2001) Isolation and properties of obligately chemolithoautotrophic and extremely alkali-tolerant ammonia-oxidizing bacteria from Mongolian soda lakes. *Arch. Microbiol.* **176**, 170–177.

Top, E.M., Van Daele, P., DeSaeyer, N. and Forney, L.J. (1998) Enhancement of 2,4-dichlorophrnoxyacetic acid (2,4-D) degradation in soil by dissemination of catabolic plasmids. *Antonie Van Leeuwenhoek* **73**, 87–94.

Wackett, L.P., Sadowsky, M.J., Martinez, B. and Shapir, N. (2001) Biodegradation of atrazine and related s-triazine compounds: from enzymes to field studies. *Appl. Microbiol. Biotech.* **58**, 39–45.

Wagner, M.., Erhart, R., Manz, W., Amann, R., Lemmer, H., Wedi, D. and Schleifer, K.-H. (1994) Development of an rRNA-targeted oligonucleotide probe specific for the genus *Acinetobacter* and its application for in situ monitoring in activated sludge. *Appl. Environ. Microbiol.* **60**, 792–800.

Water Research Commission (1984) *Theory, Design and Operation of Nutrient Removal Activated Sludge Processes*. Water Research Commission by the University of Cape Town, City Council of Johannesburg, and the National Institute for Water Research of the CSIR.

Wilderer, P.A., Bungart, H.-J., Lemmer, H., Wagner, M., Keller, J. and Wuertz, S (2002). Modern scientific methods and their potentials in wastewater science and technology. *Wat. Res.* **36**, 370–393.

Wilderer P.A. and Wuertz S. (eds) (2002) Modern scientific tools in bioprocessing. In *Developments in Water Science*, vol. 49. Elsevier Science, Amsterdam.

Wuertz, S. (2002) Gene exchange in biofilms. In *Encyclopedia of Environmental Microbiology* (ed. G. Bitton), vol. 3, pp. 1408–1420. John Wiley, New York.

# Acknowledgements

The authors thank Stefan Wuertz for critically reviewing this chapter and Elisabeth Müller for further developing the questionnaire on the biography of microbial samples. Input on modeling from Marc Wichern is greatly appreciated.

## APPENDIX: BIOGRAPHY OF MICROBIAL SAMPLES

In recent years, a great variety of novel analytical methods were developed to analyze composition, architecture and physicochemical properties of microbial aggregates such as activated sludge flocs and biofilms. The application of these novel analytical tools has led to fascinating discoveries (Wilderer and Wuertz 2002). However, correlation between results of microbial analysis and plant performance has not been well established yet. To enhance the commonly available information base, the scientific and engineering community is invited to answer the questions that are compiled below (adapted from Wilderer *et al.* 2002). The answers given may inform about the origin, history, and the nature of a biological sample with respect to milieu and reactor system from which the sample was taken, and to preparatory and analytical methods applied during examination.

To facilitate the compilation of data and the correlation of the results a form sheet can be downloaded from www.wga.bv.tum.de (see under 'database'). Authors are invited to fill in the form and send it back electronically. The data will then be introduced into the database with open access for downloading.

## PART 1. MILIEU

### TREATED WASTE STREAM
- ☐ municipal wastewater
- ☐ municipal wastewater with significant industrial proportions
- ☐ sewage sludge

synthetic wastewater (specify) ………………………
industrial wastewater (specify) ……………………..
solid wastes (specify) …………………………………
others (specify) ………………………………………..

### PARAMETERS CHARACTERIZING THE REACTOR INFLUENT

| | | |
|---|---|---|
| COD | [mg/L] | …………………………… |
| $BOD_5$ | [mg/L] | …………………………… |
| $NH_4-N$ | [mg/L] | …………………………… |
| $PO_4-P$ | [mg/L] | …………………………… |
| SS | [mg/L] | …………………………… |
| conductivity | [µS/cm] | …………………………… |
| alkalinity | [mmol/L] | …………………………… |

**MILIEU CONDITIONS IN THE REACTOR (AVERAGE)**
    temperature         [°C]……………………………..
    pH value              ……………………………..
    DO                   [mg/L]……………………………..

## PART 2. REACTOR

### REACTOR SYSTEM

<u>Scale of the reactor</u>
    ☐    lab scale
    ☐    pilot scale
    ☐    full scale

<u>Treatment plant characteristics</u>
    ☐    single stage
    ☐    multi stage

<u>Reactor configuration</u>
    ☐    single tank
    ☐    tank in series

<u>Sample location (tank)</u>
    ☐    first tank
    ☐    second tank
    ☐    third tank
    ☐    fourth tank

<u>Sample location (stage)</u>
    ☐    first stage
    ☐    second stage
    ☐    third stage

<u>Design capacity of the plant</u>
    population equivalents        ……………………………..
    average inflow rate
    (dry weather flow)            ……………………………..

<u>Actual flow conditions</u>
    actual inflow rate              ……………………………..
    recirculation flow rate        ……………………………..

Operational status
- ☐ start-up operation
- ☐ after shock loading
- ☐ after accident
- ☐ during campaign
- ☐ steady state

specific conditions: ……………………………….

Reactor examined
- ☐ activated sludge system
- ☐ biofilm reactor
- ☐ anaerobic reactor

others:…………………………….

Operation mode
- ☐ continuous flow
- ☐ intermittently fed
- ☐ batch

Intermittently fed reactors
- ☐ constant volume reactor
- ☐ variable volume reactor

Major treatment objectives
- ☐ COD removal
- ☐ nitrification
- ☐ denitrification
- ☐ phosphorus removal
- ☐ hydrolysis
- ☐ disinfection
- ☐ biogas production

others: …………………………….

Dosage of chemicals
- ☐ for pH control
- ☐ for chemical P removal
- ☐ additional C source

other reasons: ……………………………….

## BIOFILM REACTOR

Type of biofilm reactor
    ❑ trickling filter
    ❑ rotating biological contactor
    ❑ submerged fixed-bed reactor
    ❑ fluidized bed reactor
    ❑ reactor with biofilm + suspended biomass

Support media
    ❑ rock-type media
    ❑ plastic media

Structure of the media
    ❑ non-porous
    ❑ sponge-like
    ❑ closed pores
others: ……………………………….

Permeability of the media
    ❑ permeable
    ❑ non-permeable

Specific weight of the media
    ❑ heavier than water
    ❑ lighter than water

Parameter values
    size of media (diameter) ……………………………
    size of disc (diameter) ……………………………
    voids fraction [%] ……………………………
    spec. surface area ……………………………
    height of the packing ……………………………
    reactor height ……………………………
    working volume ……………………………
    cross sectional area ……………………………

Sampling of the reactor
    location (specify) ……………………………

Biofilm characteristics
    biofilm thickness [µm] ……………………………

Biofilm appearance
- ☐ patchy
- ☐ smooth surface
- ☐ rough surface
- ☐ porous

Continuous flow reactor
filter velocity
(incl. recirculation) ..................................
overall hydraulic
retention time (HRT)     [h] ..................................
HRT in
- aerobic zone     [h] ..................................
- anoxic zone     [h] ..................................
- anaerobic zone     [h] ..................................

Intermittently fed reactor
recirculation filter
velocity     ..................................
vol. exchange rate     [%] ..................................
cycle time     [h] ..................................
fill phase     [h] ..................................
anaerobic react     [h] ..................................
anoxic react     [h] ..................................
aerobic react     [h] ..................................
drain phase     [h] ..................................

Washing strategy
- ☐ continuous washing
- ☐ intermittent washing

washing frequency     $[d^{-1}]$ ..................................
operation time since
  last washing     [h] ..................................

## ACTIVATED SLUDGE SYSTEM

Operation characteristics
    sludge age                           [d] ……………………………….
    aerobic sludge age            [d] ……………………………….
    biomass concentration      [g/L] ……………………………….
    (MLSS aeration tank)
    sludge volume index (SVI)  [mL/g] ……………………………….

Aeration system
    ❐   surface aeration
    ❐   bubble aeration

Aeration strategy
    ❐   continuous aeration
    ❐   intermittent aeration

Activated sludge activity
    max. oxygen uptake rate     [mg/g,h] ……………………………
    max. nitrification rate         [mg/g,h] ……………………………
    max. denitrification rate     [mg/g,h] ……………………………
    max. P uptake rate            [mg/g,h] ……………………………
    experimental conditions
    (specify)                                 ………………………………..

Continuous flow reactor
    ❐   completely mixed reactor
    ❐   reactor cascade
    ❐   plug flow reactor
    overall hydraulic
    retention time (HRT)         [h] ……………………………….

    HRT in
    - aerobic zone                [h] ……………………………….
    - anoxic zone                [h] ……………………………….
    - anaerobic zone            [h] ……………………………….
    - clarifier                     [h] ……………………………….
    - working volume                  ……………………………….

Sequencing Batch Reactor
  working volume          ....................................
  vol. exchange rate      [%]..................................
  cycle time              [h] .................................
  fill phase              [h] .................................
  aerobic react           [h] .................................
  anoxic react            [h] .................................
  anaerobic react         [h] .................................
  settling phase          [h] .................................
  draw                    [h] .................................
  idle                    [h] .................................

## ANAEROBIC REACTOR

Reactor type
  ☐ chemostat
  ☐ reactor with sludge recycling
  ☐ biofilm reactor
  ☐ UASB

Mixing
  ☐ stirrer
  ☐ screw pump
  ☐ upflow
  ☐ gas injection

Operation characteristics
  working volume            ....................................
  overall HRT               [h] .................................
  anaerobic sludge age      [d] .................................
  biomass concentration     [g/L] ...............................
  (MLSS aeration tank)

Biomass activity
  biogas production rate    [L/h] ...............................
  methane content           [%]..................................
  acetate concentration     [mg/L] ..............................
  propionate concentration  [mg/L] ..............................

# PART 3. PREPARATORY AND ANALYTICAL METHODS

## TYPE OF BIOMASS ANALYZED
- ☐ biofilm
- ☐ activated sludge flocs
- ☐ scum
- ☐ foam
- ☐ anaerobic sludge

others (specify) ……………………………….

## MICROSCOPIC ANALYSIS
Time passed between sampling and
microscopic analysis [h] ………………………….
- ☐ Sample kept at 4–10 °C during transport and storage?
- ☐ Sample homogenized and/or centrifuged?

others (specify) ……………………………….

## CONVENTIONAL LIGHT MICROSCOPY
- ☐ structure of flocs
- ☐ filaments identified by guide (Eikelboom, Jenkins)
- ☐ protozoa identified
- ☐ metazoa identified
- ☐ fungi identified
- ☐ algae identified

simple staining procedures applied ?
specify ……………………………….
any unusual observations ?
(specify) ……………………………….

## METHODS APPLIED TO MEASURE
- ☐ total bacterial cell count (e.g. DAPI staining)
  specify ……………………………….
- ☐ viable cell count (e.g. CTC staining)
  specify ……………………………….
- ☐ others
  specify ……………………………….

## FLUORESCENT *IN SITU* HYBRIDIZATION (FISH)

Sample fixation
- ☐ 4% PFA
- ☐ Ethanol

any other treatment ……………………………….

Sample treatment
- ☐ homogenized
- ☐ not homogenized
- ☐ either of the above plus tested the same day for total bacterial cell count (ethanol-fixed cells)

Microscope used
- ☐ fluorescence microscope
- ☐ confocal laser scanning microscope
- ☐ either of above plus image analysis

Cell counts
- ☐ manual
- ☐ automatic

Fluorescent labels used
……………………………….

Has the reactor been analyzed by FISH or related methods before?
- ☐ no
- ☐ yes

Was a 16S/23S rRNA gene bank prepared?
- ☐ no
- ☐ yes

Have reactor-specific gene probes for FISH been developed?
- ☐ no
- ☐ yes

Gene probes used for FISH
specify ……………………………….

## Major Findings

Archaea ..........................................

bacteria in general................................

filamentous bacteria..............................

nitrifiers.........................................

bio-P bacteria....................................

anammox bacteria.................................

other specific bacteria............................

protozoa.........................................

metazoa..........................................

fungi............................................

algae............................................

## overall conclusions
.................................
.................................


## EPS CHARACTERIZATION

Method used
- ❒ EPS extraction
  extraction method (short description) ................................
  .................................
  analyzed substances
  .................................
- ❒ *in situ* analysis of EPS
- ❒ using specific fluorescence probes
  specify .................................

❏ exoenzyme activity
   specify ……………………………….
❏ microsensors
   specify ……………………………….

Main results

EPS extraction
   COD              [mg/g] ………………………………
   protein          [mg/g] ………………………………
   carbohydrate     [mg/g] ………………………………
   humic substances [mg/g] ………………………………
   uronic acid      [mg/g] ………………………………
   ions             [mg/g] ………………………………
   DNA              [mg/g] ………………………………
   Lipid            [mg/g] ………………………………
   other components: ………………………………………………..

*in situ* analysis of EPS
   using specific fluorescent probes
   ……………………………….
   exoenzyme activity
   ……………………………….
   microsensors
   ……………………………….


## HYDROPHOBICITY

Method used
……………………………….

Main results
……………………………….


## ENZYME ACTIVITY MEASUREMENTS

Method used
❏ colorimetric measurement
   specify ……………………………….

☐ microscopic analysis (enzyme linked fluorescent probes = ELF probes)
    specify……………………………….

Main results
……………………………….


**METABOLIC POTENTIAL**

Method used
☐ oxygen utilisation rate
    specify substrates…………………………….
☐ substrate utilization tests
☐ BIOLOG
    specify substrates…………………………….
☐ microautoradiography (MAR)
    specify substrates…………………………….

Main results
……………………………….


**POLYMERASE CHAIN REACTION (PCR)**

Method used for the detection/quantification of specific organisms
☐ conventional (reverse transcription, RT) PCR
☐ quantitative (RT) real-time PCR
☐ others (specify): …………………………….

Target organisms and genes, primers and probes, extraction and results
(if more than five organisms were tested, please use additional data sheet)

☐ if sample has been stored before extraction, specify conditions:
……………………………….
☐ direct extraction of nucleic acids

☐ direct extraction of organisms (specify): …………………………….

## Organism 1

...................................
target gene..................................

forward primer (sequence/reference) ...............................

reverse primer (sequence/reference) ................................

probe (sequence/reference) ...................................

(RT)PCR conditions...................................

☐ detection only after pre-enrichment media
   (specify): ...................................

☐ detection possible by direct extraction of nucleic acids?

☐ DNA-based PCR: specify DNA extraction
   method, or reference: ...................................

☐ PCR after reverse-transcription (RT) of
   mRNA: specify mRNA extraction method, or reference:
   ...................................

Major results

☐ qualitative findings: ...................................

☐ quantitative results: ...................................

additional remarks: ...................................


## Organism 2

...................................
target gene..................................

forward primer (sequence/reference) ...............................

reverse primer (sequence/reference) ................................

probe (sequence/reference) ...................................

(RT)PCR conditions………………………………..

☐ detection only after pre-enrichment media
   (specify): ………………………………..

☐ detection possible by direct extraction of nucleic acids?

☐ DNA-based PCR: specify DNA extraction
   method, or reference: ………………………………..

☐ PCR after reverse-transcription of mRNA:
   specify mRNA extraction method, or reference:
   ………………………………..

Major results
☐ qualitative findings: ………………………………..

☐ quantitative results: ………………………………..

additional remarks: ………………………………..

## Organism 3
………………………………..
target gene………………………………..

forward primer (sequence/reference) ………………………………..

reverse primer (sequence/reference) ………………………………..

probe (sequence/reference) ………………………………..

(RT)PCR conditions………………………………..

☐ detection only after pre-enrichment media
   (specify): ………………………………..

☐ detection possible by direct extraction of nucleic acids?

☐ DNA-based PCR: specify DNA extraction
   method, or reference: ………………………………..

❏ PCR after reverse-transcription of mRNA:
  specify mRNA extraction method, or reference:
  ................................
Major results
❏ qualitative findings: ................................

❏ quantitative results: ................................

additional remarks: ................................

## Organism 4

................................
target gene................................

forward primer (sequence/reference)
................................
reverse primer (sequence/reference)
................................
probe (sequence/reference) ................................

(RT)PCR conditions................................

❏ detection only after pre-enrichment media
  (specify): ................................

❏ detection possible by direct extraction of nucleic acids?

❏ DNA-based PCR: specify DNA extraction
  method, or reference: ................................

❏ PCR after reverse-transcription of mRNA:
  specify mRNA extraction method, or reference:
  ................................

Major results
❏ qualitative findings: ................................

❏ quantitative results: ................................

additional remarks: ................................

## Organism 5

..................................
target gene..................................

forward primer (sequence/reference)
..................................
reverse primer (sequence/reference)
..................................
probe (sequence/reference) ..................................

(RT)PCR conditions..................................

❏ detection only after pre-enrichment media
   (specify): ..................................

❏ detection possible by direct extraction of nucleic acids?

❏ DNA-based PCR: specify DNA extraction
   method, or reference: ..................................

❏ PCR after reverse-transcription of mRNA:
   specify mRNA extraction method, or reference:
..................................
Major results

❏ qualitative findings: ..................................

❏ quantitative results: ..................................
   additional remarks: ..................................

## PCR-based method used for the determination of organismic diversity

❏ PCR-DGGE (specify DNA extraction method,
   or reference): ..................................
   PCR conditions..................................
   DGGE conditions..................................

❏ RT-PCR-DGGE (specify mRNA extraction method, or reference):
   RT-PCR conditions..................................
   DGGE conditions..................................

❑ PCR-SSCP (specify DNA extraction method,
   or reference): …………………………….
   PCR conditions…………………………….
   SSCP conditions…………………………….

❑ RT-PCR-SSCP (specify mRNA extraction method,
   or reference): …………………………….
   RT-PCR conditions…………………………….
   SSCP conditions…………………………….

❑ others (specify): …………………………….
   major qualitative findings…………………………….

results for sequenced bands: dominating:
- frequent: …………………………….
- less frequent: …………………………….
- rare: …………………………….

additional remarks: …………………………….

## FINAL CONCLUSIONS

..............................................................................................................................

..............................................................................................................................

# From fundamentals to practical applications: Conclusions

*Peter A. Wilderer*

Biofilm reactors play an important role in environmental technology and may gain even greater importance as more light is shed on the 'black box' of biofilm systems. Advances in science enable us to study and eventually better understand the processes affecting architecture and function of biofilms and the resulting performance of biofilm reactors. Modern tools of microbiology, chemistry and computer science are certainly a major driving force for the further development of biofilm technology. Knowledge gain in science and the availability of novel analytical methods are not necessarily sufficient, however, to achieve better performance of biofilm reactors. A technological breakthrough can be expected only when the information acquired in science is further processed, condensed, and translated into the language of engineers and treatment-plant operators. In turn, information transfer from the practitioner's workbench to the scientific laboratory is needed to assist scientists in identifying and addressing the truly relevant research questions. Knowledge transfer between science and engineering is to be understood as a 'two-way-road'.

Scientists should realize the complexity of wastewater treatment systems including the complex composition of the wastewater to be treated, and the highly challenging discharge conditions to be met. Engineers and practitioners, on the other side, need to be open to conclusions drawn from sophisticated scientific investigations, even if these conclusions seem to be in contradiction to the experience accumulated during the past. Acceptance of a paradigm shift is often painful for practitioners. It disturbs their daily routine. One must admit, however, that without continuously shifting paradigms humanity would not have evolved technologically.

© 2003 IWA Publishing. *Biofilms in wastewater treatment*. Edited by S. Wuertz, P.L. Bishop and P.A. Wilderer. ISBN: 1 84339 007 8.

## Part 3. From fundamentals to practical applications: conclusions

In the past, researchers investigating biofilm and activated sludge systems considered mostly the soluble fraction of wastewater. It is well known, however, that wastewater does not only consist of dissolved organic substrates but also of particulate organic and inorganic substances. In fact, most of the organic wastewater pollutants are in the form of particles ranging in size from a few hundred Daltons up to several millimeters. Enzymatically catalyzed hydrolytic reactions have to be executed outside bacterial cells, before the substrates can pass through the cell membrane. Little is known about the processes governing hydrolysis and metabolism of particulate organics in biofilm systems. Even less is known about transport of particles within a biofilm. But there is some experimental evidence that particulate tracers (e.g., micro-beads) after application to a biofilm system appear within seconds within the biofilm, suggesting that advective transport occurs. How the physical and chemical properties of wastewater particles affect advective and diffusive transport rates is still unknown. Also, little is known about the role inorganic particles play. As mentioned in Chapter 13, clay particles may have an important effect on the architecture of a biofilm and on its function. The true nature of the effects is still unknown. In conclusion, scientists are well advised to direct their research closer to reality, and to consider particulate substances as the major fraction of wastewater pollutants, at least in municipal wastewaters.

COD removal, nitrification, denitrification, biological phosphorus removal – all these processes are executed by bacterial cells. It was certainly correct when in the past researchers directed their main interest to the bacterial population within biofilms and activated sludge flocs. Tremendous efforts have been made to identify the bacterial species responsible for certain metabolic reactions, investigate their position in the phylogenetic tree, and detect the kinetic and stoichiometric parameters describing uptake and metabolism of substrates. With the aid of fluorescent *in situ* hybridization (FISH) and confocal laser scanning microscopy (CLSM), valuable structural information has been obtained. For instance, we have learned that the bacterial species tend to grow in clusters rather than dispersed and intermixed. With the aid of micro-sensors patterns of concentration and substrate composition, even down to the molecular level, can be detected and made visible. Mathematical models and numerical simulation tools have been developed taking into consideration an ever-growing number of process equations and components. It is assumed that the more complex the model is, the more reliable are the predictions derivable – were it not for the uncertainty stemming from the multitude of parameters whose values are mostly unknown in most of the practical cases.

Despite the excitement that high-resolution type of information may provide, researchers should recognize the discrepancy between scientific capabilities and practical demands. The question must be asked as to what the true value of the newly acquired information may be for the engineer, whose responsibility it is to design and operate a biofilm reactor. To put it into plain words: no matter which position a bacterial species occupies in the phylogenetic tree, all that matters in the end is the treatment result the biofilm reactor delivers.

It must be realized that apart from bacteria there are also higher organisms in play, protozoa for instance, feeding on particulate material that includes bacteria and metazoa (Chapters 10 and 14). In addition, the extracellular polymeric substances (EPS) produced by the microorganisms have more functions than to glue the cells together. These functions are beginning to be understood (Chapter 8). The distribution of microbial species and the fraction of EPS in the biomass is closely related to the overall process conditions the engineer provides through design and operation of the biofilm reactor. The individual species are not just sitting next to, but interact with, each other. It would be very helpful to know more about the interactions and in particular about how to stimulate certain interactive processes or dampen others. This knowledge could enhance greatly the stability of biofilm reactors.

The opportunities to obtain additional information that is useful for engineers are there, and such knowledge may help enhance the capacity and stability of biofilm reactors. To profit from the scientific capabilities we have today, a much more intensive collaboration between microbiologists and engineers must be established. Results from the laboratories of microbiologists and from mathematical modelers must be correlated with the process parameters characterizing the biofilm reactor operated under real-world conditions. We must understand the process conditions leading to a particular structure and function of a specific biofilm system in the reactor, and learn which biofilm architecture and microbial population characteristics lead to specific reactor performance data. Only then will the light shed onto biofilm reactors be to the advantage of science and engineering. It is our hope that the questionnaire presented in the Appendix to Chapter 14 will help structure this urgently needed correlation exercise.

# Glossary

**Adhesion** Intermolecular forces that hold matter together, especially touching surfaces of neighboring media such as a liquid in contact with a solid.

**Air-lift reactor** A type of biofilm reactor with a circulating bed maintained by high air velocity. The reactor is subdivided into upflow and downflow sections.

**Alpha-mesosaprobic** A quality status of organically polluted waters characterized by an increased concentration of easily degradable organic substances, by a strong deficit in dissolved oxygen, a comparatively high turbidity caused by bacteria, and an abundance of bacteria-feeding *protozoa* (q.v.).

**Ammonia oxidizers** Obligately aerobic, *chemolithoautotrophic* (q.v.) bacteria that oxidize ammonia to nitrite to obtain energy. Phylogenetically, they belong to the *Proteobacteria* (q.v.).

**Archaea** One of the two domains of *prokaryotes* (q.v.) including methane-producing microorganisms, most extreme halophilic (salt loving) and hyperthermophilic forms (with optimal growth temperature of 80 °C or higher), and *Thermoplasma* (compare *Bacteria*).

**Bacteria** One of the two domains of *prokaryotes* (q.v.). *Bacteria* differ from the *Archaea* (formerly archaebacteria) based on the degree of similarity of *rRNA* (q.v.) sequences and other cellular characteristics such as the lack of peptidoglycan in cell walls (compare *Archaea*).

**Bacterivorous** One type of feeding. This term designates micoorganisms that preferably feed upon bacteria.

**Bioaugmentation** A bioremediation technique, involving the addition of microorganisms with specific metabolic properties to enhance biodegradation.

**Biodegradation** Breakdown of organic molecules into small inorganic molecules like carbon dioxide, water, methane, chloride, etc. caused by microorganisms or their enzymes. This process provides energy and potentially intermediates for biosynthesis of cellular material to the microorganisms.

**Bioengineering** A discipline responsible for the application of biological reactions under controlled process conditions.

**Bioenhancement** See *biostimulation*

**Biofilm** Aggregation of cells and extracellular polymeric substances at a solid–liquid, solid–gas, liquid–liquid or liquid–gas interface.

© 2003 IWA Publishing. *Biofilms in wastewater treatment.* Edited by S. Wuertz, P.L. Bishop and P.A. Wilderer. ISBN: 1 84339 007 8.

**Biofilm reactor** A reactor in which a biofilm grows on fixed, suspended or rotating carrier materials (often called 'media').

**Biofilter** Biofilm reactor in which metabolic processes and filtration are executed simultaneously.

**Bioprocesses** Metabolic reactions carried out by living cells, consortia of cells and microbial communities (e.g. substrate uptake, degradation, synthesis of cell components).

**Bioreactor** An engineered, man-made biotope designed and operated for specific purposes (e.g. for the treatment of wastewater).

**Bioremediation** A process that utilizes biological processes to enhance microbial degradation of contaminants on site (*in situ*) or after removal of the material to be treated (*ex situ*).

**Biostimulation** Also referred to as *bioenhancement* (q.v.); additions of nutrients, electron acceptors, enzymes, or biosurfactants to enhance *bioremediation* (q.v.).

**Boundary layer** L. Prandtl first introduced this concept in fluid mechanics. The idea is that for flow next to a solid boundary, a thin layer (the hydrodynamic boundary layer) develops in which friction is very important, but outside this layer the fluid behaves very much like a frictionless fluid. For concentrations a similar concept applies (concentration boundary later). Instead of friction, decline of the concentration inside the boundary layer (i.e., the concentration gradient) is caused by reactions in the biofilm. The thicknesses of both boundary layers are usually different depending on viscosity, diffusion, reaction rates, and free-flow velocity.

**Boundedness of solutions** Property of a mathematical model indicating that its *solutions* (q.v.) stay in a bounded region in a finite or infinite time horizon (i.e., they do not go to infinity, there are no explosions).

**Cellular automata** Mathematical constructs composed of individual elements (cells). Each cell can be in a state chosen from a prescribed set. The states of cells are updated in every time step using a set of rules that depend on the state of the cell and its neighbors.

**Chaotropic agents** Expression used in protein chemistry for substances which attract water and create a chaos for the protein water shell, usually resulting in precipitation of the protein. Among chaotropic agents are urea and tetra methyl urea.

**Chemolithoautotrophic** Microorganisms that obtain energy through chemical oxidation of inorganic compounds (by electron transport phosphorylation), and use carbon dioxide as the sole carbon source for the biosynthesis of cellular material.

**Chemoorganotrophic** Microorganisms that use organic compounds as carbon and energy source.
**Ciliates** Members of one subphylum of *protozoa* (q.v.) – *Ciliophora* – characterized by ciliary organelles (in at least one stage of their life cycle) for locomotion/movement and/or acquisition of food particles.
**Coarse grid operator** Part of a *multigrid* (q.v.) algorithm defining what to do on the coarse *grid* (q.v.).
**Cohesion** The attraction between molecules of a liquid that enables drops and thin films to form.
**Co-metabolism** The transformation of a compound by a microorganism when the organism is unable to grow on the compound and does not derive energy, carbon or any other nutrients from the transformation itself. The biodegradation is initiated through the action of an enzyme having little specificity and which normally attacks growth substrates.
**Computational domain** Suitable approximation (polyhedral, for example) of the actual domain for purposes of *numerical simulations* (q.v.).
**Confocal laser scanning microscopy (CLSM)** In conventional light microscopy all light passing through the specimen is detected. In CLSM only some of the light is allowed to pass through the confocal pinhole so that light above and below the plane of focus is excluded. The technique is particularly useful in conjunction with fluorescent signals such as those emitted by a fluorescently labeled nucleic acid probe bound to *rRNA* (q.v.) in a bacterial cell. If the stage on which the objective slide with the specimen is mounted is moved in the third ($z$) dimension different optical sections of the specimen are obtained. The resulting two-dimensional images from different focal planes can be stored digitally and subjected to quantitative image analysis. The method allows spatial analysis of three-dimensional aggregates.
**Co-substrate** A substrate utilized together with another substrate, usually in connection with *cometabolism* (q.v.).

**Dependent variable** In *mathematical models* (q.v.), the variables describing the *solution* (q.v.) of the model are dependent variables, because they depend on the *independent variables* (q.v.). For example, in a model describing *transport of dissolved substrate* (q.v.), the variable denoting the substrate concentration is a dependent variable, because it depends on space (and time).
**DGGE (denaturing gradient gel electrophoresis)** DNA fragments (for example of *16S rRNA* (q.v.) genes) that have been amplified by *PCR* (q.v.) can be separated according to size and sequence content by DGGE. It is one of several nucleic acid fingerprinting methods available to screen wastewater reactors. The patterns obtained are used to assess changes in the microbial

population and the amplified DNA bands are available for sequencing to verify the identity of the organism. The method is not quantitative.
**Differential equation** Equation relating functions and their derivatives.
    **-ordinary** Only one independent variable, typically time.
    **-partial** Different independent variables (typically different space coordinates or time) and, hence, partial derivatives.
**Differential operator** Mathematical operator indicating operations (including derivatives) to be applied to functions, for example the famous Laplacian

$$\Delta = \frac{\partial^2}{\partial x^2} + \frac{\partial^2}{\partial y^2} + \frac{\partial^2}{\partial z^2}.$$

**Diffusion-reaction** The interaction of diffusive mass transfer and (mostly kinetic or biochemical) reactions. In *mathematical modeling* (q.v.) this leads to *partial differential equations* (q.v.), based on the assumption that the modeled medium is a continuum.
**DLVO Theory** Named after Derjaguin, Landau, Verwey, and Overbeek and developed to explain the stability of *lyophobic* (q.v.) colloids. Accordingly, the interaction between two colloidal particles is the balance between van der Waals forces and electrostatic interactions.
**DNA (deoxyribonucleic acid)** DNA consists of deoxyribonucleotides, which are composed of a base, the sugar deoxyribose and a phosphate group. Two strands of DNA are wrapped together forming a double helix. DNA bases are adenine (A), cytosine (C), thymine (T) and guanine (G). The sugar and phosphate groups form the backbone of the molecule supporting the bases, which jut out from the chain. A and T form hydrogen bonds, as do C and G, thus holding together the double helix. DNA carries the information for the synthesis of *RNA* (q.v.) and proteins in regions called genes.

**Eden Model** One of the well-known algorithms developed for use in computer modeling of rough morphology growth on thin films, which are used typically in combination with deposition and aggregation processes.
**Electrical double-layer** Macromolecules and particles immersed in aqueous medium acquire an electrical surface charge (usually negative). They attract ions of opposite charge, which together with the surface generate the so-called electrical double layer. The electrostatic potential decreases through the electrical double-layer from its value at the surface to zero in the bulk.
**Electrostatic interactions** In aqueous medium, these interactions occur between approaching electrical double-layers. Thus, the resulting electrostatic interactions are usually repulsive.

**Eukaryote** An organism composed of one or more cells containing visibly evident nuclei and organelles (compare *prokaryote*).

**Existence and uniqueness of a solution** Two important properties of a mathematical model ensuring that there exists one and only one *analytical solution* (q.v.) of the model equations.

**Extracellular polymeric substances (EPS)** Polymers manufactured by cells and composed of polysaccharides, proteins, *DNA* (q.v.) and other substances; providing the matrix within which cells in biofilms, granules or flocs are embedded.

**Finite volume method** One of the main numerical techniques. Finite volume schemes are based on a volume integral formulation of the original partial *differential equation* (q.v.) on a set of finite (control) volumes that partition the domain.

**FISH** See *fluorescent in situ hybridization*.

**Flagellates** Members of one subphylum of *protozoa* (q.v.) – *Mastigophora*. Members of this group are motile by the action of one or more flagella.

**Floc** Aggregation of cells in an irregularly shaped form; flocs are dispersed in a liquid phase (e.g., wastewater), and have the tendency to attach to each other under reduced hydraulic shear forces (flocculation).

**Fluorescent in situ hybridization (FISH)** *Hybridization* (q.v.) procedure used to detect intact cells of microorganisms on the basis of a sequence of *RNA* (q.v.) in their natural microhabitat (*in situ*). *Oligonucleotide probes* (q.v.) used in FISH consist of 18 ($\pm$ 3) *nucleotides* (q.v.) that are labeled with a fluorescent dye. The resulting molecule is small enough to enter bacterial cells as well as cells of other microorganisms and bind to *rRNA* (q.v.) associated with ribosomes. Because rRNA is single stranded, hybridization of complementary DNA sequences (that is, sequences with the corresponding A for every U or T and G for every C, and vice versa) in intact cells can be performed. By adjusting the hybridization conditions (the stringency), only those cells which contain the complimentary rRNA sequence to the fluorescently labeled nucleic acid probe will bind the probe. Visualization can be achieved by epifluorescence microscopy or CLSM. The method is quantitative.

**Food-web relationship** An interrelationship between organisms with different modes of nutrition in which energy is transferred from one organism to the organisms by which it is consumed.

**Galerkin's approach** Basic idea of the finite element method leading to a finite system of discrete equations to be solved.

**Galerkin operator** One possibility of choosing the coarse *grid* (q.v.) operator in a *multigrid* (q.v.) scheme.

**Gaussian elimination** The standard and probably most famous *direct solver* (q.v.) for systems of linear equations.

**Gene** See *DNA*.

**Gene cloning** See *gene library*.

**Gene library** A mixture of DNA fragments (e.g. obtained after PCR) is preserved by inserting them into a *plasmid* (q.v.) and introducing the plasmid into a bacterial culture of *Escherichia coli*. This step is referred to as cloning. Normally, each plasmid will contain only one DNA fragment and each bacterial cell will take up only one plasmid molecule. Hence the different fragments are maintained in the cell culture, which serves as a repository. The individual bacterial colonies, which are formed when a portion of the culture is spread out on solid growth medium, are due to the division of one original cell. They are clones of this cell and each cell contains an identical cloned DNA fragment inside the plasmid.

**Gram-negative bacteria** Bacteria, which on the basis of the Gram-stain, appear red when viewed with a light microscope. They are differentiated from *Gram-positive bacteria* (q.v.) owing to their cell wall structure.

**Gram-positive bacteria** Bacteria, which on the basis of the Gram-stain, appear purple when viewed with a light microscope.

**Granule** Aggregation of cells in a dense spherical shape; granules are dispersed in a liquid phase (e.g., wastewater) and do not readily attach to each other under reduced hydraulic shear forces, but remain as individual objects.

**Grazing** Type of feeding using primary or secondary producers as prey; e.g. *protozoa* (q.v.), which consume bacteria or filter-feeding zooplankton.

**Grid** Structure consisting of elements (cells) such as, for example, cuboids or tetrahedral, approximating the geometry of the examined domain and the geometric allocation of computed function values.

-**unstructured** Arrangement of grid cells of a given type but of different shape without any systematic order.

-**structured** Arrangement of grid cells of a given type and shape in a systematic order.

-**Cartesian** Grid with orthogonal grid-lines.

-**regular** Only one single shape and size of grid cells.

-**adaptive** Grid with smaller cells where a higher resolution is needed.

-**cell-centered** Function values associated to the center of the grid cells.

-**collocated** Values of all stored functions associated to the same set of points within the grid.

-**staggered** Values of stored functions associated to different sets of points within the grid.

**Heterogeneous model** An early model of biofilm structure in which liquid was supposed to flow over a dense biofilm surface (compare with *pseudo-homogenous model*).

**HGC-group of Gram-positive bacteria** One of two main classes of *gram-positive bacteria* (q.v.) recognized on the basis of *rRNA* (q.v.) sequences, it comprises the actinomycetes. Members of this cluster share a high content of guanine and cytosine in the DNA.

**High pressure extrusion** Extrusion is a process that forces a substance through a shape-forming die by means of compression, in this case at high pressure.

**Homogenization** Establishing (macro) model equations describing the overall behavior of the studied system by averaging out local (micro) effects from the original model equations. The objective is to avoid computing times and storage requirements in numerical simulations exceeding available resources.

**Hybridization** *In vitro* or *in situ* formation of double-stranded nucleic acid by base pairing between single-stranded nucleic acids usually from different sources (e.g. *oligonucleotide probe* [q.v.] and *rRNA* [q.v.]), see also *fluorescent in situ hybridization*.

**Hydrophobicity** In this context, the degree of hydrophobicity is expressed as the variation of the free energy of interaction between two moieties of the same entity (i) immersed in water (w): if entity (i) is hydrophobic $\Delta G_{iwi} < 0$; if (i) is hydrophilic $\Delta G_{iwi} > 0$.

**Independent variable** A variable in a *mathematical model* (q.v.) that does not depend on anything but can take a broad range of values, e.g., the variables denoting time or a position in space (coordinates).

**Individual-based modelling** Each member of a population is tracked individually (as opposed to considering the population as a continuum). This leads to *differential equations* (q.v.), difference equations, *cellular automata* (q.v.), or *stochastic models* (q.v.).

**Initial or boundary conditions (values)** Conditions prescribing the values of a *differential equation*'s *solution* (q.v.) at the beginning of the time interval of interest or at the boundary of the problem's domain, respectively; necessary to get a unique *solution* (q.v.). Physically describe how the model system under consideration is connected to the external world. Therefore, they are an important part of the *mathematical model* (q.v.). For example, the boundary conditions to be specified for the Navier–Stokes equations describe the mechanism driving the flow.

**Interpolation, restriction** Operators defining the transfer between the coarser and the finer *grid* (q.v.) in a *multigrid* (q.v.) algorithm.

**Initial or boundary conditions (values)** Conditions prescribing the values of a *differential equation*'s *solution* (q.v.) at the beginning of the time interval of interest or at the boundary of the problem's domain, respectively; necessary to get a unique *solution* (q.v.).

**Isothermal flow** A special case of flow with heat transfer where the temperature remains constant during the process.

**Laser doppler velocimetry** A method for measuring flow velocity from the observed Doppler shift of a laser beam traversing the fluid.

**Lattice automaton** Spatially discrete model describing the behavior of the studied system as a result of rules for virtual microscopic particles and/or their residence probabilities.

**Lewis acid–base interactions (AB)** These polar interactions arise from the electron donor – electron acceptor (Lewis acid–base) interactions between polar entities.

**Lifshitz – van der Waals interactions (LW)** This designation groups the three types of electrodynamic interaction in condensed media: randomly orienting dipole–dipole interactions (Keesom); randomly orienting dipole-induced dipole interactions (Debye) and fluctuating dipole-induced dipole interactions (London). These interactions are commonly attractive.

**Lyophobic** Marked by lack of strong affinity between a dispersed phase and the liquid in which it is dispersed, such as a *lyophobic* colloid.

**Mathematical model** A description of a (natural, scientific, technical) process or system in mathematical terms, often in form of *differential equations* (q.v.), *cellular automata* (q.v.), or *probability distributions* (q.v.).

**Metazoa** Term used to name non-photosynthetic, multicellular *eukaryotes* (q.v.) (invertebrate animals).

**Microbial community** Community of microorganisms such as *bacteria* (q.v.), *protozoa* (q.v.), *metazoa* (q.v.), algae and yeasts, that develops under the environmental conditions prevailing in the natural or man-made biotope.

**Microparticle tracing** A method for investigating flow on a very small scale. It consists of following the trajectories of very small particles injected into the liquid using CLSM or video or multi-exposure still photography.

**Model parameter** A (often constant) coefficient in a *mathematical model* (q.v.), e.g., reaction rates, carrying capacity of the environment, diffusion coefficients, fluid density. To adapt the general model for a particular situation, the parameters must de adjusted. The qualitative behavior of a model can depend drastically on the actual value of a model parameter; *e.g.*, the numerical values of model parameters might determine whether the population becomes extinct or survives in a population dynamics model.

**Monodisperse** Describes a group of (spherical) particles all of which have the same diameter.

**Multigrid/multilevel methods** Fast *iterative solver*s (q.v.) for finite systems of discrete equations taking advantage of the convergence properties of common *iterative solver*s (q.v.) on grids with different *resolution*s (q.v.).

**Newtonian incompressible fluid** A fluid for which there is a linear relationship between the shear stress and the velocity gradient (rate of shear). When a fluid is said to be incompressible, its physical properties are assumed not to be affected by changes in pressure.

**Nitrification** Process in which ammonia is oxidized to nitrate via nitrite. The two steps are carried out by two different groups of obligately aerobic, *chemolithoautotrophic* (q.v.) bacteria, the *ammonia oxidizers* (q.v.) and the *nitrite oxidizers* (q.v.).

**Nitrite oxidizer** Obligately aerobic, *chemolithoautotrophic* (q.v.) bacteria capable of obtaining energy from the oxidation of nitrite to nitrate.

**Nodal and hierarchical basis** Two construction principles of defining basis functions for the finite-dimensional linear approximation spaces in finite element methods.

**Nucleotide** The smallest unit of a nucleic acid consisting of a base, a sugar and a phosphate group.

**Object-oriented programing languages** A family of programming languages putting the focus on the objects (data) and not, for example, on the instructions.

**Oligonucleotide probe (gene probe)** A single strand of (less than 100, usually about 20) *nucleotides* (q.v.) joined by phosphodiester bonds in the DNA strand. Such an oligonucleotide of known base sequence is used to detect single stranded DNA or RNA with the complementary nucleotide sequence (the target) by *hybridization* (q.v.). This is a useful way to confirm the presence or absence of certain genes or microorganisms. Gene probes may be used on extracted nucleic acids or on whole cells, for example, in *fluorescent in situ hybridization* (q.v.) of activated sludge.

**Omnivorous** The universal type of feeding; omnivorous organisms not only feed on bacteria but also may be predators (see also *predation*).

**Parallel efficiency** Measure for the quality of *parallelization* of a certain computer program or the losses by communication and additional setup expenses, respectively, given by the ratio of the sequential runtime of a program and the number of parallel processes times its parallel runtime.

**Parallelization** Process of preparing a code for a parallel *(super-)computer.*

**PCR (polymerase chain reaction)** Any *DNA* (q.v.) fragment can be generated from a template strand and amplified using the enzyme DNA polymerase from a thermophilic bacterium. Two *primers* (q.v.) are needed to make copies of a given region of DNA located between sequences complementary to the primers. A PCR cycle consists of three different steps which occur at different temperatures denaturing of the target DNA (leading to the formation of single strands), annealing of the primers (binding to the template DNA), and DNA synthesis. Typically this cycle is repeated 20–40 times during which time the target DNA is amplified exponentially.

**Peristome** See *peritrich*.

**Peritrich** The principal property of stalked *ciliates* (q.v.) evident from a corona of ciliary (thread-shaped) organelles which serve, by currents due to rotary movement, the process of acquisition of food particles (*bacteria* (q.v.)). The frontal field of the *protozoa*n (q.v.) cell surrounded by the corona with the mouth opening (=stomal) in its center is called peristome.

**Phylogenetic tree** A 'tree-diagram' illustrating the evolutionary interrelations of organisms that usually originated from a shared ancestral form; the distance of one group from the other groups indicates the degree of relationship.

**Plasmid** Autonomously replicating, usually circular molecule of accessory *DNA* (q.v.) which normally exists outside the chromosome. However, some plasmids are also able to integrate into the chromosome of their host. Plasmids often encode additional metabolic features or of special properties like antibiotic resistances.

**Poisson equation** Simple and very important *partial differential equation* (q.v.)

$$\Delta u = \frac{\partial^2 u}{\partial x^2} + \frac{\partial^2 u}{\partial y^2} + \frac{\partial^2 u}{\partial z^2} = f \; .$$

describing (among others) the shape $u$ of a membrane subject to some force $f$, or homogeneous diffusive transport under a force $f$.

**Poisson model** A *stochastic model* (q.v.) for a completely random point cloud. The different parts of the cloud are independent of each other and the cloud is spatially homogeneous.

**Polydisperse** Describes a group of (spherical) particles all of which have a different diameter.

**Potential field** Pressure field, which is calculated according to potential theory.

**Potential theory** One of the theories used in fluid mechanics (the other one is stream theory). A potential velocity is defined as satisfying the continuity equation and representing some practical flow situations.

**Predation** Acquisition of food by overwhelming smaller organisms (the prey).

**Primer** A single-stranded *RNA* (q.v.) or *DNA* (q.v.) that is complementary to a single-stranded DNA template. By binding to the latter it provides a free 3' hydroxyl end to which the enzyme DNA polymerase can add deoxynucleotides. In this way a new strand of DNA complementary to the template DNA is synthesized.

**Prokaryote** A cellular organism (as a member of the domains *Bacteria* [q.v.] or *Archaea* [q.v.]) that does not have a distinct nucleus (compare *eukaryote*).

***Proteobacteria*** A division of *Bacteria* (q.v.) as determined on the basis of comparative rRNA sequencing. This large group of phylogenetically related gram-negative bacteria comprises purple phototrophic, *chemolithoautotrophic* (q.v.) and *chemoorganotrophic* (q.v.) bacteria and is subdivided into five subgroups (alpha- [α], beta- [α], gamma- [γ], delta- [δ], and epsilon- [ε]).

**Protozoa** Unicellular *eukaryotes* (q.v.) that are non-photosynthetic and lack a rigid cell wall.

**Pseudo-homogenous model** An early model of biofilm structure in which liquid was supposed to flow through a matrix of microorganisms (compare with *heterogeneous model* (q.v.)).

**PVM Library** A software package for developing parallel programs executable on networked Unix computers. Parallel Virtual Machine (PVM) allows a heterogeneous collection of workstations and supercomputers to function as a single high-performance parallel machine.

**Quad-/octree** Hierarchical data structure for the efficient storage and handling of discretized geometries and functions in two-dimensional/three-dimensional domains based on the idea of a recursive space partitioning.

**Qualitative behavior (of a model)** Properties of the *solutions* (q.v.) of a *mathematical model* (q.v.), such as *model parameter* (q.v.) dependencies, long-term behavior (does it approach a steady state? does it show cyclic behavior?), do the *solutions* (q.v.) *exist*, are they *unique*, are they *bounded*?

**Recalcitrant substances** Organic compounds, which resist biological action and are not readily metabolized.

**Representative elementary volume** Smallest volume size in an experimental or computational domain that is statistically representative in terms of certain properties like, for example, biofilm density, porosity, heterogeneity, flow and transport properties, and so forth.

**Resolution** The density of the *grid* points in a *grid* (q.v.); typically, a high resolution means high accuracy, but also large memory requirements and long run times of the program.

**Reynolds number** Parameter in the Navier–Stokes equations proportional to the inverse viscosity and the mean velocity of the studied fluid.

**Rheological properties** General expression for the behavior of bodies towards mechanical stress; in this case mechanical stability of biofilms towards shear stress (e.g., like visco-elastic gels or highly viscous fluids)

**Rhizopodes** A synonym for the *Sarcodina* – a major taxonomic group of *Protozoa* (q.v.) characterized by their amoeboid movement, i.e., the formation of pseudopodia. This group includes naked as well as shelled sarcodines.

**Ribosome** A macromolecular assembly of *rRNA*s (q.v.) and proteins which serves as the site of protein synthesis. Bacterial ribosomes contain three types of *RNA* (q.v.), namely, 16S, 23S, and 5S.

**rDNA** The genes coding for *rRNA* (q.v.).

**RNA (ribonucleic acid)** A single-stranded type of nucleic acid, which contains the base uracil (U) instead of thymine (T) found in DNA and the sugar ribose instead of deoxyribose.

**mRNA (messenger RNA)** Genes are transcribed (or copied) from one strand to RNA. A RNA molecule carrying information for a protein is called mRNA. It has a short half-life in bacterial cells.

**rRNA (ribosomal RNA)** A class of RNA molecules serving as components of ribosomes. They are functionally conserved and present in all organisms. In *prokaryotes* (q.v.) there are 16S and 23S rRNAs. The primary structures of rRNA contain regions of higher and lower evolutionary conservation which lend themselves to the design of complementary nucleic acid probes for whole groups of bacteria in the form of phylogenetic sublineages (resulting in a crude description of the bacterial community in a specific wastewater sample) or individual genera or species. rRNAs are very common and stable.

**16S rRNA**     See rRNA
**23S rRNA**     See rRNA

**Saprobic** Pollution status derived from *sapros* (greek) = putrescence. The term describes the magnitude of organic loading ranging between a low (i.e. oligosaprobic) and a high (i.e. polysaprobic) level.

**Scale** *Resolution*-phenomena are typically restricted to a certain scale on which they can be observed and described; multiscale phenomena (like turbulence, for example) involve many different scales (tiny and large vortices, for example).

**Simulation, numerical** Solving a *mathematical model* on a computer.

**Solution (of a *mathematical model* (q.v.), model solution)**

-**analytical/exact** Function(s), exactly fulfiling the model equation(s), in general not computable explicitly.

-**closed (form)** A model solution that can be described by a formula.

**-numerical** Computed function, defined by a finite set of data, approximately fulfiling the discretized model equation(s).
**Solver** Algorithm for the determination of the *solution* (q.v.) of a finite system of discrete equations.
**-direct** Determination of the exact *solution*.
**-iterative** Determination of an approximate *solution* via a repeated updating/improvement of the current approximation.
**SOR (successive over-relaxation)** Simple *iterative solver* (q.v.) for linear systems of equations.
**Stability (in a numerical context)** Important property of numerical algorithms ensuring that the calculated approximate *solution*s (q.v.) can be interpreted as exact *solution*s (q.v.) for slightly disturbed input parameters; unstable methods may produce oscillations in the *solution*s (q.v.).
**Stability (of a model solution)** The property of a model solution where slightly different initial/boundary values lead to a solution that remains close to the original one.
**Stagnation point** When a fluid passes over an object, a flow separation occurs at the point where they come into contact. At the stagnation point the flow is splitting up and the velocity is reduced to zero.
**Steady-state / stationary – quasi-stationary – unsteady** Terms describing the time-dependence and its representation in the model and in the numerical algorithm (no dependence – influence restricted to defined points – strong dependence, taken into account completely).
**Stochastic model** A probability distribution providing random samples, which are representative for the object of interest.
**Stress** The perpendicular force per unit area applied to an object, in a way that compresses or stretches the object. Pressure stresses are caused by the pressure applied. Extensional, elongational or normal stresses are forces acting on the area perpendicular to the surface on which they act, whereas tangential stresses are the ones acting tangential to the surface.
**Sulfur pump** The circulation of sulfur back and forth between the aerobic/anoxic zone and the anaerobic zone, by which oxygenation capacity is transported deeper into the biofilm.
**Supercomputer** High-performance computer for expensive computations as for example in numerical applications or cryptology. Two main types: parallel and vector computers. Particularly suitable for *parallelized* (q.v.) or *vectorized* (q.v.) codes. Nowadays, various mixed types and further classifications.
**Surface free energy or surface tension** Is defined as half of the free energy of cohesion of the material in vacuum ($\gamma_i = -\frac{1}{2}\Delta G_{ii}$). This means that it represents the energy available for interaction.

**Time step** Time distance between two subsequent computed states of the examined system.

**Transport (of dissolved substrates)**
  -**advective** In engineering fields working with biofilms, the term advective is usually applied synonymously with *convective* (q.v.) in the context of external and internal mass transport.
  -**convective** The flux of the substrate is caused by the velocity field of the fluid carrying the substrate. Growth of the rigid biofilm matrix also results in a velocity field carrying enclosed bacteria towards the surface of the biofilm.
  -**diffusive** The flux of the substrate is caused by local differences of the substrate concentration (concentration gradients). That is, substrate is transported from a point with higher concentration to a neighboring point with lower concentration.

**Trophic level** One of the hierarchical strata of a food web characterized by organisms, which are the same number of steps removed from the primary producers. For example, *protozoa* (q.v.) feed on *bacteria* (q.v.) and present a different trophic level.

**Uronic acids** Sugars with carboxyl groups; uronic acids normally do not occur in cells but are components of extracellular polysaccharides and sometimes used as indicators for those. Alginate, for example, is a polysaccharide composed of two uronic acids, guluronic and mannuronic acid.

**Vectorization** Process of preparing a code for a vector *(super-)computer* (q.v.).

**XDLVO Theory** Extension of the *DLVO theory* (q.v.), considering in addition Lewis acid–base interactions and the repulsive effect of Brownian movements.

**Xenobiotics** Human-made compounds with a chemical structure to which microorganisms have not been exposed in the course of evolution.

# Index

AB forces *see* hydrogen bonding
abrasion 265, 266, 267
*Acineria* spp. 250
*Acinetobacter* spp. 140, 353
activated sludge flocs 193–4
activated sludge system characteristics 362–3
active biomass loss 285
actual constituent profiles 126–9
acylated homoserine lactone (acyl-HSL) 139
N-acylhomoserine lactones (AHLs) 201
adhesion forces 96, 377
   *see also* microbial adhesion
advective transport 127, 130, 273, 283, 390
agglomerates, clay particles 327–8
aggregation 308–9
   *see also* microbial aggregation
AHLs *see* N-acylhomoserine lactones
air-lift reactors 268, 377
*Alcaligenes denitrificans* 215, 216, 221, 225, 226
algorithms, computer modeling 15
alliances, professional 343–56, 374
alpha-mesosaprobicity 377
ammonia concentration 247
ammonia flux 278
ammonia-oxidizing bacteria 140, 255, 258, 347, 377
anaerobic reactor characteristics 363
analytical methods 319–20, 364–73
anionic EPS matrix properties 184
AQUASIM computer program 6, 62, 271, 273
Archaea 346, 377
architecture 91, 123–293
   biofilm reactors performance 232–63
   cellular automata model 92
   computational assumptions 64
   detachment 264–90
   determining factors 4, 117–18
   extracellular polymeric substances 178–210
   fluid flow effects 110–11
   mass transport 146, 147–77
   mathematical modeling 16–20
   microscopic examination 238–9
   models/experiments 32–48
   new models 49–50
   overview 291–3

parameters 5
physicochemical properties 211–31
sloughing off 92
substrate concentration 110–11
*Arthrobacter* spp. 245
artificial life 24–5, 28
*Aspidisca cicada* 250
attachment processes 39, 129
automatons 13, 14, 28
autotrophic bacteria 137, 276–82
average detachment rates 268
axial velocity component 106
*Azotobacter* spp. 184

backwashing
   biological filters 301
   detachment rate equation 270, 272, 284
   phosphorus-removing bacteria 285–6
   residence times 275, 277, 278, 279–80, 281, 282, 284
bacteria 346, 377
   *see also* biomass; *individual bacteria*; microorganisms
   clay particles 326, 329 37
   distributions 281
   genetic exchange 332–3
   microhabitats 326, 331–2
base thickness ($L_{\text{base thickness}}$) 274–5
BATH method 220
BET42a gene probe 236, 253, 254
bioaugmentation 377
biocide toxicity 326, 333–5
biodegradation 377
bioengineering 377
biofilm reactors 377
   characteristics 360–1
   packed-bed 88–116
   performance 232–63
biofilm thickness ($L_F$) 105–6, 269, 270–2, 281
Biofilm Workshop *see* IAWQ Biofilm Workshop in Garching
biofilters 378
biological biofilm models 66–7
biological complexity 61

© 2003 IWA Publishing. *Biofilms in wastewater treatment.* Edited by S. Wuertz, P.L. Bishop and P.A. Wilderer. ISBN: 1 84339 007 8.

biological filters 301
biological phosphorus removal *see* enhanced biological phosphorus removal
biological processes 90–1
biomass
    control 301, 320, 321
    detachment *see* detachment
    displacement velocity 40, 41
    growth
        discretization 76
        equation, biofilm dynamics 68
        factors 65, 66–7
        rule 42
    loss 285
    microbial composition 129
bioprocess engineering 343–56
bioreactors 377, 378
    aggregation 308–9
    analysis 319–20
    clay particles 325–42
    design and operation 299–324, 343–56
    deterministic design 305, 306
    efficiency 325–42
    engineering issues 302–5, 320–3
    influent characteristics 357
    knowledge base 353
    microbial community predictions 350–4
    modelling 297–8, 302–14, 319–20
    niche characteristics 345, 346, 349
    packed-bed 88–116
    performance demands 301
    population dynamics 320–1
    system characteristics 358–9
    transport of substrates 300–1
bioremediation 378
biotic structure investigation methods 234–9
boundary layers 21–2, 98–101, 378, 383, 384
boundedness of solutions 378
Brownian movement forces 217
bulk nutrient concentration 43, 44

calibration 19–20, 309–12
carbamates 334–5
carbohydrate distribution 129, 132
carbon source provision 140–1
carbon-limited medium 189
'card-house' flocs 327–8, 330, 334
'card-pack' flocs 327–8
Cartesian grids 71, 73
case studies 276–86
catechol-1,2-dioxygenase assay 259
cation bridging 223–6
cation exchange capacity (CEC) 327

cations 190–2, 224–6, 252
CEC *see* cation exchange capacity
cell-to-cell communication 139, 141, 201, 266
cells
    growth 90–1
    lethal shear stress 108–10
    self-regulation 345
cellular automaton models 118
    definition 378
    established biofilm models 54, 55
    Lattice Boltzmann Automata 73
    pseudo-homogeneous model 34, 35
    rule-based biofilm models 66–7, 68
    spatial biomass spreading models 18, 24–5
    'state of the art' 92
    terminological confusion 27
CF319a bacteria-specific probe 253
CFBR *see* Continuous Flow Biofilm Reactor
CFD *see* computational fluid dynamics
channels 43
chaos effects 9
chaotropic agents 378
chemolithoautotrophic organisms 378
chemoorganotrophic organisms 378
chloroaromatics 246, 259
ciliates 249, 250, 251, 379
classic problems 13
classical biofilm models 54
clay particles
    biocide toxicity 326, 333–5
    biofilm composition/reactor efficiency effects 325–42
    biofilm physical properties 336
    genetic exchange 332–3
    microhabitats 326, 331–2
    microorganism interactions 326, 329–31
    nutritional effect 331
    pH effects 329
    properties 326–9, 336
    reactor efficiency 325–42
    toxicity effects 326, 330, 331, 333–5
    wastewater treatment bioreactors 337–8
CLSM *see* confocal laser scanning microscope
cluster processes 56, 57
co-metabolism 379
co-substrates 379
coarse grid operator 379
COD flux 278, 282
codes, design/specification 78–80
coexistence of microorganisms 40

# Index

cohesion 187–8, 379
collaboration, professional 343–56, 374
color visualization 81, 82
communities, microorganisms 39–41, 346–54
competition 273, 274, 276–86
complex models 52, 62
composition, clay particles 325–42
compound models 53
computational biofilms 60–87
    data inputs 80–1
    embedding 63, 80–1
    implementation 63, 78–80
    mathematical models 63, 64–9
    numerical methods 63, 69–78
    validation 64, 83–4
    visualization 64, 81–3
computational domains 70, 379
computational fluid dynamics (CFD) 14
computer hardware 6, 14–15, 79
concentration gradients 147–8
conceptual models 148, 150–3, 303–4
confocal laser scanning microscopy (CLSM)
    biofilm architecture 4, 5, 239
    biofilm heterogeneity 292
    biofilm structures 138, 233
    complex models 62
    definition 379
    hypothetical bacterial biofilm 151
    mathematical models 17
    new techniques 37, 54
    statistical models 57
constant thickness assumption 277, 278
constituent profiles 126–9
contact angle 220–1
Continuous Flow Biofilm Reactor (CFBR) 241–51
continuum mechanics 66, 67
control *see* design and control
convective transport 156–9, 161, 162, 173, 390
coordinate systems 159
core region 103, 104, 105, 113
cross-disciplinary cooperation 91, 343–56, 374
cultivation methods 234
current practice 345–50
*Cytophaga–Flavobacterium* group 235, 245, 252

Damköhler number 110
DAPI *see* 4,6-diamidino-2-phenylindole
data inputs, computation 80–1
database questions 357–73
deduction, bioreactor design 305
definitions 149, 178–9, 377–90

denaturing gradient gel electrophoresis (DGGE) 379
denitrification 266, 315–16
density of biofilms 128, 130
dependent variables 379
deposition processes 39
design aspects 17, 299–324
detachment
    competition effects 276–86
    experimental determination 267–70
    heterogeneous biofilms 136
    location 271–3
    mechanisms 265–76
    process 264–90
    rate equations 269–70
    rule 42
determining factors 351
determinism 305, 306
development, *in situ* observation 297–8
dewaterability 198–9
DGGE *see* denaturing gradient gel electrophoresis
4,6-diamidino-2-phenylindole (DAPI) 245, 252, 253, 254
differential elements 166–8
differential equations 9, 380
differential operators 74–6, 380
diffusion
    *see also* effective diffusion coefficient; mass transport
    EPS effects 200
    local coefficients 16–17
    mass transport mechanisms 156–9, 162
diffusion–reaction models 9, 13, 21, 36–7, 380
diffusivities 130–5, 157, 163–4, 321
dimensionality 65–6
dimensionless variables 102
discrete layers 162–71, 174
discrete models 41–3
discretization
    biomass growth 76
    differential operators 74–6
    domains 69–71
    equations 72–6
    time 75, 76
displacement 40, 42
divalent cations 190–2, 224–6, 252
DLVO theory 214–16, 380
domains discretization 69–71
Donor-Cell scheme 74
ducks, mechanical 13, 14, 28

Dupré equation 212, 213
dynamic detachment 271, 272–3
dynamic processes 39–40
dynamic systems 53
dynamics, biomass growth equation 68

EBPR *see* enhanced biological phosphorus removal
ecology
    EPS matrix 200–2
    indicators 237–8
    microbial 274, 287
    niches 280, 346, 349
    protozoan/metazoan communities 237–8, 248–51
    structure 39
Eden Model 380
effective diffusion coefficient 36, 130
effective diffusivities 130, 133, 134, 166–8
effluent phosphate concentration 285–6
Einstein's theory of relativity 11
electrical double-layer 380
electrical surface charge 223–6
electron dispersion spectroscopy (EDS) 224
electrostatic interactions 188, 380
elephant scale example 97
elongational effects 90, 94, 95, 96
embedding 63–4, 80–1
empty relations 11
engineering
    biofilms 149
    bioreactors 302–5, 320–3
    population dynamics 320–1
    research needs 322
enhanced biological phosphorus removal (EBPR) 282–3, 284, 285
'environmental conditions' hypothesis 19
environmental factor (j) 348
environmental technology 353
enzymes
    activity measurements 367–8
    current biotechnology 345
    extracellular polymeric substances 194, 195–6, 202
    genetics 349
EPS *see* extracellular polymeric substances
equation-based biofilm models 66
equations
    biomass growth equation 68
    detachment rate equation 269–70
    differential equations 9, 380
    discretization 72–6
    Dupré equation 212, 213

Fisher equation 9
Navier–Stokes equation 14, 52, 66, 67–8, 73, 74, 101–2
Poisson equations 56, 57, 102, 386
Young's equation 213
erosion 265, 267
EUB338 bacteria-specific probe 236, 253
evolutionary algorithms 9, 10
existence of a solution 381
exopolysaccharides 180, 181, 185
expansion growth 129
experiments
    biofilm architecture 32–48
    design by modeling 312–13, 319–20
    detachment 267–70
extensional stress 90
external detachment forces 266
external mass transport 36, 43, 44, 148, 173
extracellular localization 186–7
extracellular polymeric substances (EPS) 376, 381
    cell-to-cell interaction 201
    characteristics 366–7
    composition/properties 124, 129–30, 132, 180, 182–7, 291
    definition 179–81
    dewaterability effects 198–9
    ecological aspects 200–2
    enzyme actions 194, 195–6
    foam formation 198
    functions 194–7
    packed-bed biofilm reactors 91–2
    role 178–210
    settleability effects 198
    sorption to biomass of organic/inorganic molecules 199
    technical aspects 197–200
    wastewater biofilms, chemical composition 183
extrusion 383

falsification tests 12
faster-growing organisms 136, 277–82
feeding behavior 237–8
Fick's law 16, 36
finite difference discretization 71, 72
finite element discretization 72
finite numbers of values 69
finite volume discretization 72–3, 381
finite volume method 381
FISH *see* fluorescent *in situ* hybridization
Fisher equation 9
flagellates 249

# Index

*Flavobacterium* spp. 245
flocculation
    clay particles 327–8, 330, 334
    dewaterability v. size 199
    EPS role 181, 189–90, 191, 193
    EPS yield 185
    stability 187–8
flow
    direction (z) 99, 106
    Navier–Stokes equation 67–8
    packed-bed biofilm reactors 88–116
    régimes 93, 100
    velocity 130, 156–9, 162
fluidized bed reactors 252–7, 301
fluorescence recovery after photobleaching (FRAP) 17
fluorescent *in situ* hybridization (FISH) 62, 381
    biofilm structure 233, 234–5
    characteristics 365–6
    molecular probes 140
fluxes, bacteria/substrate 278–9
foam formation 198
food chains 347
food-processing industry 345
food-web relationship 381
forcing input 306
fractal analysis 38, 39
fractionation techniques 182
FRAP *see* fluorescence recovery after photobleaching

Galerkin's approach 381–2
Galileo, Galilei 12
GAM42a bacteria-specific probe 253
Game of Life (John Conway) 24, 54
Gaussian elimination 382
gel layer formation 200
gellan 92
gene transfer 201, 332–3, 352
generalized minimal residual algorithm (GMRES) 102
genetic algorithms 9, 10
genetics 349, 352, 382
Gibbs free energy
    DLVO theory 216
    hydrophobicity 219, 221, 223
    microbial adhesion 212–14
    XDLVO theory 217–18
Gibbsian-type probability measures 56, 57
glass, adhesion tests 217–18
GMRES *see* generalized minimal residual algorithm

Gram-negative/positive bacteria 382
    *see also individual strains*
granules 88–116, 382
    *see also* particles
gravity sewer biofilms 132
gravity theory 9, 11
grazing 382
grid structures 18, 24–5, 27, 54, 70–1, 382
growth
    biomass 65, 66–7
    manipulation 351–2
    models 43, 44, 54–5
    modes 129

halosite *see* clay particles
*Hamaker* constants 215
HDPE *see* high density polyethylene
heterogeneity
    actual constituent profiles 126–9
    detachment process 136
    diffusivity 130–5
    discretizing 169–71
    metabolic processes 125–46
    models 34, 35, 55, 150–3, 383
    mosaics 43
heterotrophic bacteria 276–82, 347, 350
HGC69a bacteria-specific probe 236, 246, 253
high density polyethylene (HDPE) 215–16, 222, 225, 226
high flow velocities 171–4
history 34–5
holism 20, 23
hollow sphere model 348
homogeneous biofilm model 150–3
homogenization 383
horizontal distributions 165
humic substances 180, 181, 182–7
hybrid biofilm models 67
hybridization 383
hydrodynamics 90–6, 101–11, 156–9, 173, 174
    *see also* flow
hydrogen bonds 188, 220, 327
hydrogen peroxide 196
hydrophobicity
    definition 383
    EPS matrix 184–5, 186
    flocculation process 191
    microbial adhesion 218–23
    samples biography 367
hypotheses, falsification tests 12

IAWQ Biofilm Workshop in Garching 3, 4, 5, 45, 66
ilite *see* clay particles
immunofluorescence techniques 234
*in situ* observation 54, 297–8, 319–20
inactivation rate 22
independent variables 383
individual-based modeling 383
induction, bioreactor design 305
industrial uses, microorganisms 345
inflection point 153
inflow region 103, 104, 105, 113
initial conditions 383, 384
inner time 97, 98
inorganic particles 325–42
input computation data 80–1
interdisciplinary research 9, 25–7, 118
internal detachment forces 266
internal hydrophobic bonds 191
internal mass transport 148, 173
inverse mathematical modeling 16–17
invertebrates 91
ions 223
    *see also* cations
irregular biofilm structures 17–19
Ising model 57
isomorphous substitution 327
isothermal flow 384
iterative solutions 389

j *see* environmental factor

kaolinite *see* clay particles
Kepler's Law 11
kinetic matrix 351
*Klebsiella pneumoniae* 333
knowledge base 353

lab-scale bioreactors 89, 99, 103–7, 191
laminar wakes 93
language 9, 25–7
large intestine 55
laser Doppler velocimetry 384
lattice automata 73, 384
Lattice Boltzmann Automata 73
$L_{base\ thickness}$ *see* base thickness
lectin-like protein 192
lethal shear stress 108–10
level of aggregation 308–9
Lewis acid–base interactions 384
$L_F$ *see* biofilm thickness
Lifshitz–van der Waals (LW) interactions 384
limestone 224

limitations of simulations 85
limiting removal rate 34
Liner-Trajectory-Aggregation Model (LTA) 94
lipids 180, 181
loading approach 303–5
loading rate *see* limiting removal rate
local cell residence time distributions 280
local cell retention time 274–5, 282
local detachment rates 268
local diffusivities 164
local mass transport rates 159–62
local relative effective diffusivities 165
local residence times 275, 276, 279–80, 283–4, 287
local rules 24, 41–2, 55
local solids residence time 276
location, biofilm detachment 271–3
London (dispersion) forces 188
Lotka–Volterra systems 53
low flow velocities 173–4
LTA *see* Liner-Trajectory-Aggregation Model
Lufshitz–van der Waals (LW) interactions 213
lumped approaches 11
LW *see* Lufshitz–van der Waals
lyophobic colloids 214–16, 384

mapping domains 70
marked point processes 56
mass of bacteria, flux 278–9
mass transport 35, 36, 147–77, 390
    definition 147
    discrete layers in biofilms 162–71
    flow velocity profiles 156–9
    high flow velocity biofilms 171–4
    limitations, models/experiments 34–41
    local rates 159–62
    mathematical modeling 13, 16–18
    microbial ecology 287
    modeling for discrete layers 166–8
    nutrient uptake kinetics 153–6
    residence time distribution 274
    substrates 300–1
    surface conformation 127, 131
MATH method 220, 221
mathematical language 25–7
mathematical models 8–28, 49–58, 62–9, 118, 378
    *see also* cellular automaton models; differential equations; stochastic models

# Index

maturation growth 129
MBRs *see* membrane biofilm reactors
MCP *see* 4-monochlorophenol
mean cell residence time 350
mean pore radius 128
mean residence time distribution 284
measurement tools 16–17
mechanical models 13, 14, 28
mechanical stability 187–8
mechanical stress 89–90, 94, 107–10
membrane biofilm reactors (MBRs) 240–51
membrane systems, principles 300–1
membrane-derived vesicles 187
metabolic potential characteristics 368
metabolic processes 125–46, 345–6, 375
metazoa 292, 384
    biofilm communities 248–51, 259
    ecological indicators 237–8
    microbial communities 347, 350
*Methanosaeta* spp. 141
*Methanosarcina* spp. 141
Michaelis–Menten kinetics 34, 36
microbial adhesion 188–94
    physicochemical factors 211–31
    predictions 212–18
    surface properties relevant to 218–28
microbial aggregation 148, 188–94, 308–9
microbial samples 357–73
microbiological determining factors 351
microbiologists 350–4, 374
microcolonies
    biofilm structure models 162–3
    conceptual models 148, 150
    diffusion 156, 159, 162
    tapered when cultured in high flow velocities 171–2
microelectrodes 37, 160
microhabitats 326, 331–2
microorganisms
    *see also* biomass; *individual microorganisms*
    biofilm architecture models 4
    clay particles 326, 329–37
    communities 39–41, 346–54
    detachment rates 268
    ecology 287
    embedded in EPS 181
    EPS matrix 178–210
    genetic exchange 332–3
    growth manipulation 351–2
    heterogeneous biofilms 136–42
    inter-relationships 346
    lethal shear stress 108–10
    metabolic processes 125–46, 375
    microhabitats 326, 331–2
    nutrient concentration gradients 148
    population structure 232–63
    protozoa/metazoa as indicators 237–8
    residence time distributions 274–6
    useful metabolic attributes 345–50
microscopic examination 238–9
    *see also* confocal laser scanning microscopy; scanning electron microscopy
microsensors 153–5
modeling 8–31, 375, 384
    *see also* cellular automaton modeling; computational biofilms
    aggregation 308–9
    analytical tool 307, 319–20
    biofilm architecture 16–20, 32–48
    biofilms 315–17
    biological models 55
    biomass detachment 268–70
    bioreactors 297–8, 302–14, 319–20
    boundary conditions 21–2
    calibration 309–12
    conceptual 150–3
    control optimization 308
    definition 10–16
    discrete layers model validation 168–9
    examples 52–3
    experimental design 312–13, 319–20
    forcing input 306
    formulation/restrictions 12
    homogeneous biofilm model 151–3
    *in situ* observation 297–8, 319–20
    inverse problems/measurement tools 16–17
    kinetic matrix 351
    mathematical models 8–28, 49–58, 62–9, 118, 378
    model structure 306, 309–10
    multidimensional 15, 19–20
    parameter estimation 310, 312, 384
    pilot plants 317–18
    planning tool 307
    plant design 307, 314–15
    principles 3–7
    qualitative system behavior 17–19
    reactor design/optimal control 17
    real-time control 308
    research tool 308
    selection of model 63, 64–9
    sensitivity analysis 314

simplification appropriateness 11–12
simulation distinction 15
solution types 388–9
standard diffusion–reaction model 36–7
strategies 50–2
structural heterogeneity 41–3
types 66–7
uncertainty 313–14
verification 310, 311–12
molecular probes 139–41
molecular techniques 38
4-monochlorophenol (MCP) 241, 242, 243, 246, 248
Monod kinetics 34
Monod-like functions 42
montmorillonite *see* clay particles
morphology characterization 38–9
moving bed reactors 301
multidimensional mathematical models 15, 19–20
multigrid methods 78
mushroom biofilm architecture 4, 37, 41, 138, 151

*Nast++* computer program package 78, 79, 80
Navier–Stokes equations 14, 67–8, 101–2
    differential operators discretization 74
    Lattice Boltzmann Automata 73
    physical biofilm models 52, 66
near wall zone 103, 104
Neu23a probe 236, 255
neural nets 9, 10
Newtonian incompressible fluids 385
Newton's gravity theory 9, 11
niche parameters 346, 353
nitrate oxidizers 385
nitrification 385
    FISH studies 140
    membrane biofilm reactors 246–7
    microbial communities 346, 349, 350
    modeling 315–17
nitrifying bacteria 330
nitrifying biofilms 252–7
*Nitrobacter* spp. 255, 292, 353
nitrogen removal *see* denitrification
*Nitrosococcus* spp. 254
*Nitrosomonas* spp. 247, 254, 255, 258, 353
*Nitrospira* spp. 292
non-mechanistic thinking 23
non-uniformity *see* heterogeneity
nonlinear biological models 9, 21
normal stresses 90, 94
Nsv443 probe 236, 255
nucleic acids 380, 386, 387, 388
    *see also* polymerase chain reaction

biofilm extracts content 244
extracellular polymeric 180, 181, 184, 186, 187
flocculation process 189
gene transfer in EPS matrix 201
molecular probes 139–40
nucleotides *see* oligonucleotide probes
numerical simulations
    computational biofilms 63, 69–78
    packed-bed film reactor flow 88–116
    time aspects 98
nutrients
    clay particle effects 331
    concentration profiles 153
    consumption rate 166–8
    EPS production 185, 186
    EPS-influenced accumulation 197
    fluxes 169–71, 173
    supplies 148
    uptake kinetics 153–6

oligonucleotide probes 385
    biofilm structure determination 233–7, 239
    nitrifying films 252, 253, 258
    xenobiotic-containing wastewater treatment biofilms 243–7
one-dimensional biofilm models 11, 19–20, 54, 62
operational factors 299–324
orthogonal grids 71
outflow region 103, 104, 105, 113
over-parameterization 311–12
overall bacterial masses 277, 278
overall residence times 280
overall retention time 282
oxygen concentration 160–1
oxygen microsensors 153–5

packed-bed biofilm reactors
    fluid flow effects 88–116
    hydrodynamics 101–11
    mechanical stress 107–10
    porosity profiles 103, 104, 105
parallel efficiency 385
parameter estimation 310, 312
particle shape 95
particulate clay 325–42
particulate organics 321, 375
*Peranema* 248
performance 76–7
peritrich 386
pesticide toxicity 330

# Index

pharmaceutical industry 345
philosophical concepts 20–5
phosphorus removal 285–6, 318
   *see also* enhanced biological phosphorus removal
*Physa gyrina* 136
physical modeling approach 52, 66
physical properties 127–8, 142, 336
physicalism 20, 23
physicochemical aspects 211–31
pilot plants 89, 99, 103–7, 317–18, 319–20
pipes 55, 252, 256–7
planktonic films *see* flocculation
PMMA *see* polymethylmethacrylate
*Podophrya fixa* 250
point processes 56–8
Poisson equations 102, 386
Poisson process model 56, 57
Poisson–Boltzmann distributions 223
polarization 82–3
polymerase chain reaction (PCR) 234, 368–73, 386
polymeric materials 215–16, 222, 225, 226
polymethylmethacrylate (PMMA) 215–16, 222, 225, 226
polyphosphate 282–3, 285
polypropylene (PP) 215–16, 222, 225, 226
polypropylene tubules 252, 256–7
polysaccharides 180, 181, 184, 185, 192
polyvinyl chloride (PVC) 215–16, 222, 225, 226
population dynamics 287, 320–1
population structure 232–63, 291–3
porosity
   constituent profiles 128
   heterogeneous biofilms 129
   microbial adhesion 226–8
   packed-bed film reactors 92, 93, 103, 104, 105, 113
Potts models 56, 57
PP *see* polypropylene
predator grazing 265, 267
predictions 212–18
preparatory methods questions 364–73
primers 387
'primitive' biofilm models 51
process engineers 350–4, 374
processes, definition 149
*Prodiscophrya collini* 250
products, definition 149
professional alliances 343–56, 374
properties, biofilms 126–38
protective effects 196

proteins
   distribution 129, 132
   extracellular polymeric 180, 181, 182–7
   flocculation process 191
Proteobacteria 252–5, 387
protozoa 292, 387
   biofilm communities 248–51, 259
   ecological indicators 237–8
   microbial communities 347, 350
pseudo-homogeneous model 34, 35
*Pseudomonas* 184, 246
   *P. aeruginosa* 190, 196, 333
   *P. atlantica* 185
   *P. fluorescens* 330–1, 333–5, 336
   *P. putida* 140, 199, 224
PVC *see* polyvinyl chloride

quadratic computational domains 70
qualitative studies 17–19, 50–1
quantitative studies
   biofilm detachment 267–70
   biofilm heterogeneity 149
   flow velocity profiles 156–9
   modeling strategies 12, 50–1
   nutrient uptake kinetics 153–6
questionnaire 354

$r_{d,V}$ *see* volume detachment
$r_{d,s}$ *see* surface detachment
randomness models 53, 56
RBC *see* rotating biological contactors
*Re* *see* Reynolds number
reactor/granule diameter ratio 105, 107
reactors *see* bioreactors
'real world' behavior 10
real-time control 308
recalcitrant substances 349, 387
reductionism 20–5
relativity theory 11
representative elementary volume 387
residence time distributions 274–6, 283–4
resolution 71, 387
reverse osmosis membranes 181
Reynolds number (*Re*) 89, 93, 388
   packed-bed film reactors 99, 100, 102–6, 110, 111, 112
rheological properties 388
Rhizopodes 388
*Rhodococcus* spp. 245
*Rhodovulum* spp. 183, 189, 190
ribosomal RNA 139–40, 292, 388

ribosomal RNA-targeted oligonucleotide probes 234–7
rotating biological contactors (RBC) 186, 301, 303–4
rotating disk biofilms 35, 41
roughness *see* surface roughness
rule-based biofilm models 66–7

*Saccharomyces cerevisiae* 95, 96, 109, 330
samples, database questions 357–73
SBBR *see* Sequencing Batch Biofilm Reactor
scale of simulations 97–8
scale-up 103–6
scanning electron microscopy (SEM) 181, 233, 239, 257
scientific competition 352
second law of thermodynamics 147
self-regulation, living cells 345
semi-linear parabolic partial differential equations 9
sensitivity analysis modeling 314
separation methods 182
Sequencing Batch Biofilm Reactor (SBBR) 241–51
*Serratia marcescens* 196
settleability, EPS effects 197
shear forces 256
    detachment rate equation 270
    packed-bed biofilm reactors 90, 92, 94, 95, 108–10
Sherwood number 18
simple models 52
simplifications, mathematical 11
simulations
    *see also* computational biofilms
    code characteristics/implementation 78–80
    limitations 85
    mathematical modeling distinction 3, 15
    mechanical duck 13–14, 28
size distribution studies 193–4
skepticism 61–2
'slime' *see* extracellular polymeric substances
sloughing off
    biofilm architecture effects 92
    biofilm detachment 265–7, 277, 278, 280–2
    EPS inhibition 197
    location of detachment 269
slower-growing organisms 136, 277–82
sludge volume index (SVI) 197
SMA *see* specific methanogenic activity
snails 136
soil interactions with clays 326, 329–31
solids residence time 276

solids retention time distribution 281
soluble substrates transport 133
solving equations 76–8
sorption to biomass 199
spatial scales 97, 99
spatial variation 185
species diversity concept 13
specific methanogenic activity (SMA) 227
specific surface area 128
spherical particles 95, 96
*Sphingomonas paucimobilis* 92, 217
spore-forming microorganisms 227
SRB *see* sulfate-reducing bacteria
stacks 43, 138
staggered grids 73, 74
stagnation points 106, 113, 389
staining 192
standard diffusion–reaction model 36–7
*Staphylococcus epidermidis* 184
steady state 152
stereoscopic visualizations 82–3
stochastic models 56–8, 389
strategies for modeling 50–2
'streamers' 37, 41
structured grids 70, 71
structures
    CLSM determination 138
    discrete layers 162–71
    features definition 152
    heterogeneity, modeling 41–3
    low flow velocities 173–4
    models, microcolonies 162–3
submerged biofilm reactors 264, 266, 282
substrates
    concentration 110–11
    flux 278–9
    transport 300–1
    utilization rate 270
substratum, definition 149
sulfate-reducing bacteria (SRB) 141
sulfur pumps 389
supercomputers 79, 389
superposition principle 21
surface detachment ($r_{d,s}$) 271, 273
surface properties
    charge 223–6
    conformation 127, 131
    free energy 389
    microbial adhesion 218–28
    roughness 96, 114, 141–2, 226–8
    tension 213, 214, 222
SVI *see* sludge volume index
system performance 277–82

# Index

system theory models 53
systems, definition 149
tangential stresses 89, 90
temporal variation 185–6, 251
theoretical biology concepts 20–5
thermodynamic approach 212–14
thickness *see* biofilm thickness
three-dimensional modeling 81, 82–3, 151
three-phase systems 300–1
time aspects
    computational assumptions 65–6, 69
    discretization 75, 76
    numerical simulations 98
    physical mathematical models 52
toluene degradation 140
tools, biofilm studies 32–3
toxicity, clay particle effects 333–5
transport *see* advection; diffusivity; mass transport
treated waste stream characteristics 357
'trees' 43
trickling filters 266, 282, 297, 301
trophic level 390
two-dimensional domains 70
two-phase systems 300–1

uncertainty, modeling 313–14
unidentified reactions 321
uniqueness of a solution 381
unit layers 326–7
'universal' biofilm models 49, 51, 310
unstructured grids 70, 71
upflow bulk fluid velocity 133
uronic acids 390

validation
    computational biofilms 64, 83–4
    discrete layers structure 168–9
    modeling 51, 310, 311–12
    standard diffusion–reaction model 37
van der Waals forces 326
Vaucanson's mechanical duck 13, 14, 28
vermiculite *see* clay particles
viscoelastic polymers 171–2
visualization 64, 81–3
vitalism 23
voids 156, 158, 160, 162
volume detachment ($r_{d,V}$) 271–2, 273
volume grids 71
*Vorticella convallaria* 250

wall region 103, 104, 105

wastewater treatment 5, 240–51, 337–8
water distribution systems 265
wettability 218–23
worms 136, 256

XDLVO theory 217–18, 219, 390
xenobiotics 240–51, 258, 390

yeast cells 95, 96, 109
yield 185
Young's equation 213

$z$ *see* flow, distribution
zeolites 329–31
zero-dimensional approaches 11
zeta potential 223–6
zonation 321
*Zoogloea* spp. 189